Wireless Communications: Networks and Systems

Wireless Communications: Networks and Systems

Edited by
Ruby Long

WILLFORD PRESS

www.willfordpress.com

Published by Willford Press,
118-35 Queens Blvd., Suite 400,
Forest Hills, NY 11375, USA

ISBN: 978-1-68285-804-2

Cataloging-in-Publication Data

Wireless communications : networks and systems / edited by Ruby Long.
p. cm.
Includes bibliographical references and index.
ISBN 978-1-68285-804-2
1. Wireless communication systems. 2. Mobile communication systems. 3. Communication--Network analysis.
I. Long, Ruby.
TK5103.2 .W57 2020
621.384--dc23

For information on all Willford Press publications
visit our website at www.willfordpress.com

WILLFORD PRESS

Contents

Preface

A transfer of power or information can occur between two or more points without wires, cables or fiber optics facilitating the transmission. This is possible through wireless technology using infrared, radio-frequency, microwave or acoustic wave communication. Cell phones, remote garage-door openers, two-way radios, GPS receivers, television remote controls, etc. use wireless technology. Wireless telecommunication networks are implemented using radio communication. Wireless sensor networks, cell phone networks, wireless local area networks, etc. are examples of wireless networks. The use of wireless modems, satellites and microwave transmitters has facilitated the access to Internet. This book unfolds the innovative aspects of wireless communication. It is a valuable compilation of topics, ranging from the basic to the most complex advancements in this area of study. The coherent flow of topics, student-friendly language and extensive use of examples make it an invaluable source of knowledge.

After months of intensive research and writing, this book is the end result of all who devoted their time and efforts in the initiation and progress of this book. It will surely be a source of reference in enhancing the required knowledge of the new developments in the area. During the course of developing this book, certain measures such as accuracy, authenticity and research focused analytical studies were given preference in order to produce a comprehensive book in the area of study.

This book would not have been possible without the efforts of the authors and the publisher. I extend my sincere thanks to them. Secondly, I express my gratitude to my family and well-wishers. And most importantly, I thank my students for constantly expressing their willingness and curiosity in enhancing their knowledge in the field, which encourages me to take up further research projects for the advancement of the area.

Editor

A Secure Privacy-Preserving Data Aggregation Model in Wearable Wireless Sensor Networks

Changlun Zhang, Chao Li, and Jian Zhang

Science School, Beijing University of Civil Engineering and Architecture, Beijing 100044, China

Correspondence should be addressed to Changlun Zhang; zclun@bucea.edu.cn

Academic Editor: Jiang Zhu

With the rapid development and widespread use of wearable wireless sensors, data aggregation technique becomes one of the most important research areas. However, the sensitive data collected by sensor nodes may be leaked at the intermediate aggregator nodes. So, privacy preservation is becoming an increasingly important issue in security data aggregation. In this paper, we propose a security privacy-preserving data aggregation model, which adopts a mixed data aggregation structure. Data integrity is verified both at cluster head and at base station. Some nodes adopt slicing technology to avoid the leak of data at the cluster head in inner-cluster. Furthermore, a mechanism is given to locate the compromised nodes. The analysis shows that the model is robust to many attacks and has a lower communication overhead.

1. Introduction

Recently, the wearable wireless sensors become powerful and rapidly expanding in healthcare monitoring [1–3]. The wearable sensors can be used to collect and transmit the data to the users. Sometimes, the data collected from some near places are similar to each other. Meanwhile, the powers of sensors are limited. Therefore, the data aggregation techniques are used to reduce the communication overhead [4, 5]. In the process of data aggregation, data need to be aggregated by the aggregation nodes. Unfortunately, data aggregation is vulnerable to some attacks because the data are sensitive or privy. If the sensitive data are revealed, this may bring serious threat or economic loss. So, the security data aggregation is playing an important role in wearable sensors.

In this paper, a security privacy-preserving data aggregation model is proposed. The model adopts a mixed data aggregation structure of tree and cluster. Data integrity is verified both at cluster head and at base station. Moreover, a locating mechanism is provided, which can locate the compromised node.

The remainder of this paper is organized as follows. In Section 2, the related work is summarized. A new secure privacy-preserving data aggregation model (SPPDA) is proposed and analyzed in Section 3. In Sections 4 and 5, the security and performance of the model are analyzed. Finally, the conclusion of this paper is given.

2. Related Work

Recently, secure data aggregation is becoming an important issue for wearable sensors. Cryptographic is an efficient mechanism to secure data aggregation. Moreover, the homomorphic encryption can aggregate encrypted messages directly from sensors without decrypting so that it has a short aggregation delay.

Castelluccia et al. [6] proposed a simple and provably secure additively homomorphic stream cipher which is slightly less efficient on bandwidth than the hop-by-hop aggregation scheme described previously. Girao et al. [7] proposed an approach that conceals sensed and aggregated data end-to-end, which is feasible and frequently even more energy efficient than hop-by-hop encryption addressing a much weaker attacker model. Feng et al. [8] proposed a family of secret perturbation-based schemes, which can protect sensor data confidentiality without disrupting additive data aggregation.

All the homomorphic encryption schemes above use the symmetric key. The securities of these schemes depend on the length of the key. Meanwhile, the security of the asymmetrical secret key schemes depends on the intractability of the algorithms. So the asymmetrical secret key schemes are designed.

Boneh et al. [9] proposed a homomorphic public key encryption scheme, which improved the efficiency of election systems based on homomorphic encryption. Mykletun et al. [10] revisited and investigated the applicability of additively homomorphic public-key encryption algorithms for certain classes of wireless sensor networks and provide recommendations for selecting the most suitable public key schemes for different topologies and wireless sensor network scenarios. Girao et al. [11] provided an approach for a tiny Persistent Encrypted Data Storage (tinyPEDS) of the environmental fingerprint. Bahi et al. [12] proposed a secure end-to-end encrypted data aggregation scheme, which significantly reduces computation and communication overhead and can be practically implemented in on-the-shelf sensor platforms. Ozdemir and Xiao [13] proposed a novel integrity protecting hierarchical concealed data aggregation protocol, which is more efficient than other privacy homomorphic data aggregation schemes. Lin et al. [14] proposed a new concealed data aggregation scheme, which is robustness and efficiency. Zhou et al. [15] proposed a Secure-Enhanced Data Aggregation, which can achieve the highest security on the aggregated result compared with other asymmetric schemes.

However, the models above can only detect the compromised nodes in verifying the data integrity at most, without locating the compromised nodes. In this paper, we present a new secure privacy-preserving data aggregation model (SPPDA), which adopts a mixed data aggregation structure. The network is divided into clusters, and the data aggregation trees are used in inner-cluster and interclusters. Firstly, some of nodes adopt slicing technology to avoid the leak of data at the cluster head. Secondly, data in the cluster are aggregated and sent to the cluster head, and cluster head verifies the data integrity to restrict the range of compromised node. Lastly, the cluster heads continue to send the data to the base station, and the data integrities are verified at the base station again. Furthermore, the model gives a mechanism to locate the compromised nodes. The analysis shows that this model has lower communication overhead.

3. SPPDA Model

The model uses the cluster structure network which contains three kinds of nodes: base station, cluster heads, and cluster nodes. The network is divided into two layers: inner-cluster and intercluster. In the inner-cluster, data are sent to the cluster head, and the cluster head verifies the data integrity to restrict the range of compromised node. In the intercluster, data are sent to the base station, and the integrity is verified at the base station. Furthermore, a mechanism is proposed to locate the compromised node. SPPDA model can be divided into initialization, the key distribution, inner-data aggregation, and interdata aggregation.

3.1. Initialization. The initialization of SPPDA model includes three parts: cluster head voting, inner-cluster data aggregation tree, and intercluster data aggregation tree.

(1) Cluster Head Voting. Using the existing cluster protocols [16, 17], the network can be divided into many clusters. In the process of cluster, the trust management mechanism [18, 19] can be used to help the selection of the cluster header. Generally, it satisfied two conditions as follows:

(1) The cluster head has higher trust values.

(2) The clusters are evenly distributed in the monitoring area.

(2) Inner-Cluster Data Aggregation Tree. In each cluster, the data are sent to the cluster head along the data aggregation tree [20]. The inner-cluster data aggregation tree is structured by a certain data aggregation tree protocol. It satisfied two conditions as follows:

(1) The degree of cluster head is large enough.

(2) The number of aggregation nodes is not more than the leaf nodes.

Lastly, cluster heads set the compromising threshold h_{ch} which is used to judge whether a branch in the cluster is compromised.

(3) Intercluster Data Aggregation Tree. When the cluster heads aggregated the data of their cluster, the data in cluster heads are sent to the base station along the intercluster data aggregation tree. The intercluster data aggregation tree is similar to the structure of the inner-cluster data aggregation tree. Lastly, base station set the compromising threshold h_{ch} which is used to judge whether a branch of the BS is compromised.

3.2. The Key Distribution. In SPDSA model, there are three sets of key: BS (base station) key, CH (cluster head) key, and N (neighbor) key. The BS key is generated by the base station which is used to ensure the security of the communication between the cluster heads and the base station. The CH key is generated by each cluster head which is used to ensure the security of the communication between cluster nodes and the cluster head. The neighbors key is generated offline which is used to ensure the security of the communication between a node and its neighbors. The structure of each key is described as follows.

(1) BS Key Distribution. BS generates three primes (q_1, q_2, q_3) and $m = q_1 q_2 q_3$ order elliptic curve (E). Then, according to the degree of BS which is defined as degree_BS, degree_BS groups of points $\{X_l, Y_l, Z_l\}_{\text{degree\_BS}}$ are selected from E, and the order of those points is m.

For each group l, we get three new points according to the formula as follows:

$$P_l = q_1 q_2 X_l,$$
$$Q_l = q_2 q_3 Y_l, \quad (1)$$
$$R_l = q_3 q_1 Z_l.$$

Here, P_l is used to encrypt the aggregated data, Q_l is used to record the number of the cluster, and R_l is used to mix the encrypted result and enhance the security of the data.

Then, the BS gets a group of keys. The public key is (m, P_l, Q_l, R_l, E) and the private key is (q_1, q_2, q_3). The public key is distributed to the cluster heads in a secure way, and the private key is reserved by the BS.

(2) CH Key Distribution. When the BS generates the key, each cluster head begins to generate the CH key. For example, CH(i) generates three primes ($p_1^{(i)}$, $p_2^{(i)}$, $p_3^{(i)}$) and an elliptic curve ($E^{(i)}$) firstly. The order of $E^{(i)}$ is $m^{(i)} = p_1^{(i)} p_2^{(i)} p_3^{(i)}$. According to the degree of CH which is defined as degree_$C(i)$, degree_$C(i)$ groups of points $\{V_{j1}^{(i)}, V_{j2}^{(i)}, V_{j3}^{(i)}\}_{degree\_C(i)}$ are selected from $E^{(i)}$, and the order of those points is $m^{(i)}$.

For each group j, we get three new points according to the formula as follows:

$$F_j^{(i)} = p_1^{(i)} p_2^{(i)} V_{j1}^{(i)},$$
$$G_j^{(i)} = p_2^{(i)} p_3^{(i)} V_{j2}^{(i)}, \quad (2)$$
$$H_j^{(i)} = p_3^{(i)} p_1^{(i)} V_{j3}^{(i)}.$$

Here, $F_j^{(i)}$ is used to encrypt the aggregated data, $G_j^{(i)}$ is used to record the number of the cluster, and $H_j^{(i)}$ is used to mix the encrypted result and enhance the security of the data.

Then, CH(i) gets a group of keys. The public key is $(m^{(i)}, F_j^{(i)}, G_j^{(i)}, H_j^{(i)}, E^{(i)})_{degree\_C(i)}$ and the private key is ($p_1^{(i)}$, $p_2^{(i)}$, $p_3^{(i)}$). Lastly, the public key is distributed to the cluster nodes in a security way, and the private key is reserved by the CH(i).

(3) N Key. N key distribution consists of five steps [21]:

(1) Generation of a large pool of P keys and their key identifiers.

(2) Random drawing of k keys out of P without replacement to establish the key ring of a sensor.

(3) Loading of the key ring into the memory of each sensor.

(4) Saving of the key identifiers of a key ring and associated sensor identifier on a trusted controller node.

(5) For each node, loading the ith controller node with the key shared with that node.

Therefore, a secure link exists between two neighboring nodes only if they share a key. If two neighboring nodes

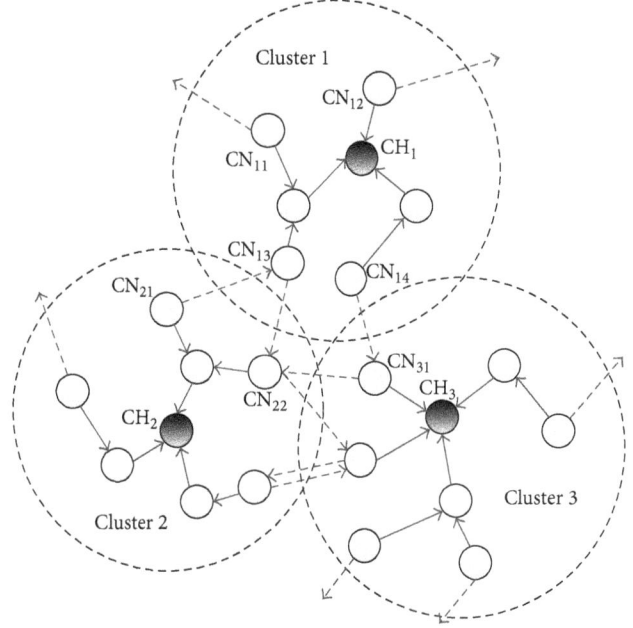

FIGURE 1: The slicing scheme.

cannot share a key but they can be connected by a link consisting of some nodes, this link can be the secure link between these two nodes.

3.3. Inner-Cluster Data Aggregation. In the inner-cluster data aggregation, the cluster heads can obtain the plaintext which is not secure enough for the data. Therefore, before the inner-cluster data aggregation, the slicing and mixing scheme [22] is used in each cluster.

(1) Slicing. In each cluster, we call one node "leaf node" if some neighbors of this node belong to other clusters. And the leaf node slice its data into two parts. One slice is sent to the other node in another cluster and the other is kept by itself. Figure 1 shows the slicing scheme. The solid line is the route in which the data is transmitted to the cluster head. The dotted line is the route in which the leaf nodes send the slices to the neighbor nodes in other clusters. In Cluster 1, there are 4 leaf nodes: CN_{11}, CN_{12}, CN_{13}, and CN_{14}. According to the rule above, these nodes divide their data into two slices. One is kept by itself; another is sent to the neighbor nodes in other clusters along the dotted line. CN_{11} and CN_{12} send the slices to the neighbor nodes in other clusters not drawn in Figure 1. CN_{13} sends the slices to the CN_{22} in Cluster 2 and receives the slices from CN_{21} in Cluster 2. CN_{14} sends the slices to CN_{31} in Cluster 3.

(2) Mixing. When all the leaf nodes send the slice, all nodes recomputed the data of it. If a node receives the slices, it adds all the slices to get a new data.

After the slicing and the mixing, the data $M_{js}^{(i)}$ is encrypted into $C_{js}^{(i)}$ according to formula (3) at each cluster node in cluster i:

$$C_{js}^{(i)} = M_{js}^{(i)} \times F_j^{(i)} + G_j^{(i)} + r_{js}^{(i)} \times H_j^{(i)}. \quad (3)$$

Here, + is the summation in elliptic curve, × is the scalar multiplication in elliptic curve, and $r_{js}^{(i)}$ is random.

Then, the encrypted data is transmitted to the cluster head. And the data are aggregated by the intermediate nodes. The aggregation of the jth branch in cluster i is

$$C_{j,\text{agg}}^{(i)} = \sum_s M_{js}^{(i)} \times F_j^{(i)} + k_j^{(i)} \times G_j^{(i)} + \sum_s r_{js}^{(i)} \times H_j^{(i)}. \quad (4)$$

$\sum_s M_{js}^{(i)}$ is the aggregation plaintext of branch j, $k_j^{(i)}$ is the number of the nodes in branch j, $\sum_s r_{js}^{(i)}$ is the aggregation of the random, and $C_{j,\text{agg}}^{(i)}$ is the ciphertext of the aggregation in branch j.

The cluster head in cluster i receives the aggregation of each branch. Then, the cluster head decrypts the $k_j^{(i)}$ of each branch using the privacy key. The plaintext $\xi_j^{(i)}$ is

$$\xi_j^{(i)} = \log_{G_j^{(i)'}} p_2^{(i)} p_3^{(i)} \times C_{j,\text{agg}}^{(i)}. \quad (5)$$

Here, $G_j^{(i)'} = p_2^{(i)} p_3^{(i)} \times G_j^{(i)}$.

The cluster head judges whether the result of each branch is compromised according to the threshold h_{ch}. If a branch is compromised, the locating mechanism is used to locate the compromised node. If not, continue to aggregation.

The cluster head gets the plaintext of the aggregation result in the cluster. That is,

$$M^{(i)} = \sum_j M_j^{(i)} = \sum_j \log_{F_j^{(i)'}} p_1^{(i)} p_2^{(i)} \times C_{j,\text{agg}}^{(i)}. \quad (6)$$

Here, $F_j^{(i)'} = p_1^{(i)} p_2^{(i)} \times F_j^{(i)}$.

At last, the data $M^{(i)}$ is encrypted into $C_{\text{agg}}^{(i)}$ by the cluster node according to formula (7) in cluster i:

$$C_{\text{agg}}^{(i)} = M^{(i)} \times P^{(i)} + k^{(i)} \times Q^{(i)} + r \times R^{(i)}. \quad (7)$$

Here, $k^{(i)}$ is the number of the cluster nodes in cluster i. r is random.

3.4. Intercluster Data Aggregation.
After the inner-cluster data aggregation, the encrypted data is transmitted to the base station. And the data are aggregated by the intermediate nodes. The aggregation of the gth branch of base station is

$$C_{g,\text{agg}} = \sum_s M_{gs} \times F_g + k_g \times G_g + \sum_s r_{gs} \times H_g. \quad (8)$$

$\sum_s M_{gs}$ is the aggregation plaintext of branch j, k_g is the number of the nodes in branch j, $\sum_s r_{gs}$ is the aggregation of the random, and $C_{g,\text{agg}}^{(i)}$ is the ciphertext of the aggregation in branch g.

The base station receives the aggregation of each branch. Then, the base station decrypts k_g of each branch using the privacy key. The plaintext ξ_g is

$$\xi_g = \log_{Q_i'} q_2 q_3 \times C_{g,\text{agg}}. \quad (9)$$

Here, $Q_i' = q_2 q_3 \times Q_i$.

The base station judges whether the result of each branch is compromised according to the threshold h_{ch}. If a branch is compromised, the locating mechanism is used to locate the compromised node. If not, continue to aggregation.

The cluster head gets the plaintext of the aggregation result in the cluster. That is,

$$M_g = \sum_s M = \sum_s \log_{P_i'} q_1 q_2 \times C_{g,\text{agg}}. \quad (10)$$

Here, $P_i' = q_1 q_2 \times P_i$.

3.5. Locating Mechanism.
Locating mechanism is used to locate the compromised nodes in the intermediate nodes. The locating mechanism works as follows.

We assume that the numbers of leaf nodes and intermediate nodes are m and n. Then we have $m \geq n$. The branch which does not pass the integrity verification is reconstructed into n branches, where there is only one intermediate node in each branch. The new intermediate nodes are the same as in old branch. And the data integrity is verified in the root node. If one branch does not pass the verification, the intermediate node in this branch is a compromised node and the locating mechanism ends.

Figure 2 shows the locating mechanism in a cluster. In the left part of Figure 2, CH finds a branch which consists of the red compromised nodes. So, this branch needs to be reconstructed. Obviously, CH_1 and CH_4 are two intermediate nodes. Therefore, this branch is divided into two new branches. CH_1 and CH_4 are also the intermediate nodes, and they are in the different branches. Then, these two branches transmit the data to the CH according to the rule described in inner-cluster data aggregation. And the CH checks their integrities. If a branch is still compromised, the only intermediate node in this branch is the compromised node.

3.6. A Case Study.
In this section, we give a detailed example of SPPDA model with initialization, the key distribution, inner-cluster data aggregation, and intercluster data aggregation.

(1) Initialization. In Figure 3, there are 25 sensor nodes distributed in the monitor area, and the base station is located in the left of the monitor area. These nodes are divided into 5 clusters. Then, the inner-cluster data aggregation tree and the intercluster data aggregation tree are constructed. In the intercluster data aggregation tree, there are 2 branches which are BSB_1 and BSB_2 from BS. BSB_1 consisted of BS, CH_1, CH_2, and CH_3. BSB_2 consisted of BS, CH_4, and CH_5. In each cluster, there are 4 CNs and 1 CH. Then, the cluster nodes are divided into 2 branches. Using the ith cluster as an example, the branches are CB_{i1} and CB_{i2}. The CB_{i1} consisted of CH_i, CN_{i1}, and CN_{i2}. The CB_{i2} consisted of CH_i, CN_{i3}, and CN_{i4}. When the data aggregation trees are completed, CH records the amount of the CNs in its cluster, and the BS records the amount of the CHs in the network.

(2) The Keys Distribution. According to the structure of the network in Figure 3, the BS generates 2 pairs of keys.

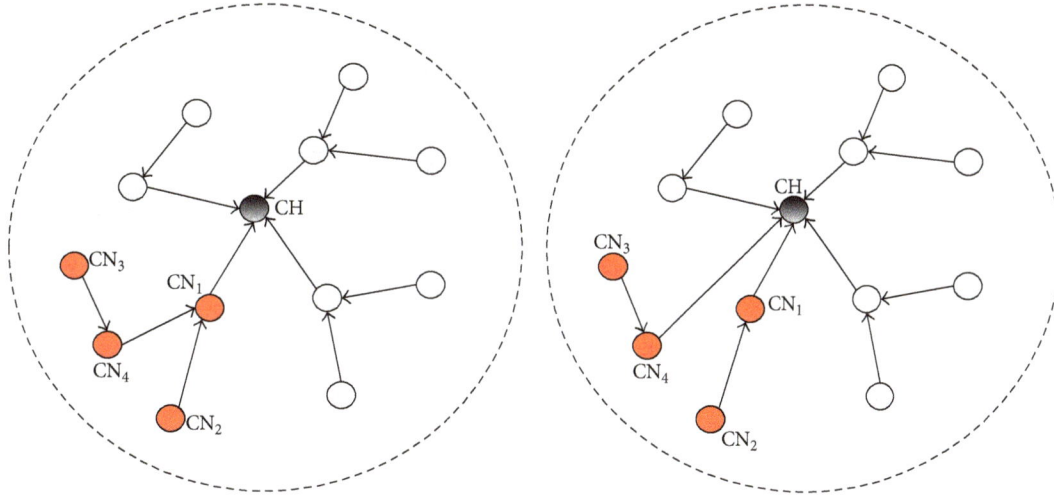

FIGURE 2: Locating mechanism.

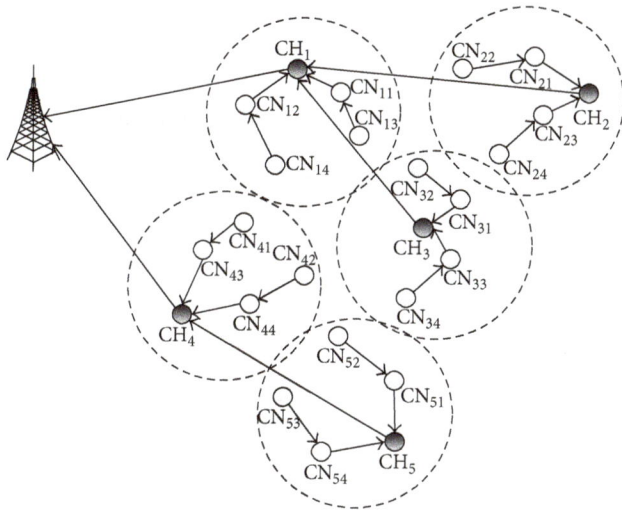

FIGURE 3: Initialization.

TABLE 1: The values of the major parameters.

Parameters	Values
$E = E'$	$m = q_1 q_2 q_3 = 4199$
$(P_l, Q_l, R_l)_2$	$q_1 = 13, q_2 = 19, q_3 = 17$
$(F_j^{(i)}, G_j^{(i)}, H_j^{(i)})_2$	$p_1^{(i)} = 13, p_2^{(i)} = 19, p_3^{(i)} = 17$

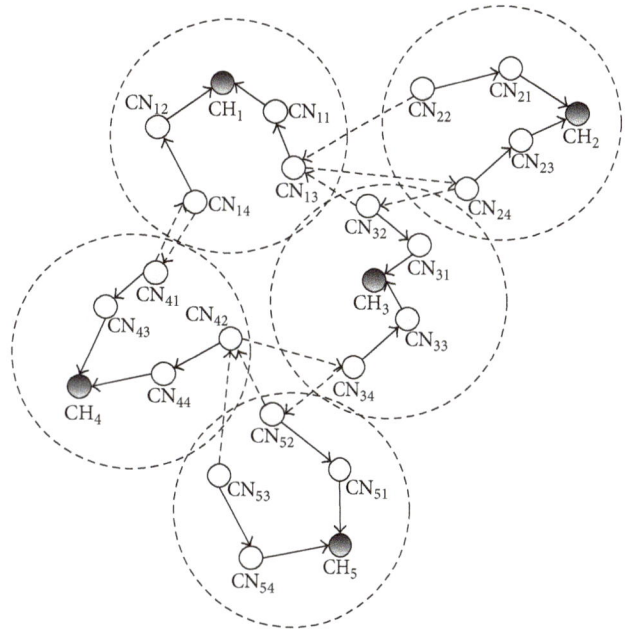

FIGURE 4: The slices of the edge nodes.

The public keys are (m, P_1, Q_1, R_1, E) and (m, P_2, Q_2, R_2, E), and the privacy keys are always (q_1, q_2, q_3). Meanwhile, according to the amount of the branches, the ith cluster head CH_i generates 2 pairs of keys. The public keys are $(m^{(i)}, F_1^{(i)}, G_1^{(i)}, H_1^{(i)}, E^{(i)})$ and $(m^{(i)}, F_2^{(i)}, G_2^{(i)}, H_2^{(i)}, E^{(i)})$. The privacy keys are always $(p_1^{(i)}, p_2^{(i)}, p_3^{(i)})$.

In order to reduce the computing overhead, the points P_l, Q_l, and R_l, $(l = 1, 2)$ use some small prime numbers. And $p_1^{(i)} = q_1, p_2^{(i)} = q_2, p_3^{(i)} = q_3$, and the elliptical curve E is same as E'. Table 1 shows the values of those major parameters. The orders of P_1, P_2, $F_1^{(i)}$, and $F_2^{(i)}$ are 17. The orders of Q_1, Q_2, $G_1^{(i)}$, and $G_2^{(i)}$ are 13. The orders of R_1, R_2, $H_1^{(i)}$, and $H_2^{(i)}$ are 19. The orders of two elliptical curves are 4199.

(3) Inner-Cluster Data Aggregation. Firstly, the edge nodes are confirmed in each cluster by its CH. In this case, the edge nodes are CN_{13}, CN_{14}, CN_{22}, CN_{24}, CN_{32}, CN_{33}, CN_{41},

CN_{42}, CN_{52}, and CN_{53}. Secondly, each edge node generates a slice from its data. Then, each edge node sends its slice to its neighbor randomly which belongs to a different cluster. Figure 4 shows the process of the slicing. The full lines express the inner-cluster data aggregation tree, and the dash lines express the flow of the slices. After slicing, the nodes which receive the slices add them into their data. In Table 2,

TABLE 2: Data processing of edge nodes.

Cluster nodes	Original data	Slices	Mixing data
CN_{13}	$MP_3^{(1)} = 6$	$S_{13} = 5$	$M_3^{(1)} = 5$
CN_{14}	$MP_4^{(1)} = 7$	$S_{14} = 5$	$M_4^{(1)} = 4$
CN_{22}	$MP_2^{(1)} = 8$	$S_{22} = 3$	$M_4^{(2)} = 5$
CN_{24}	$MP_1^{(2)} = 6$	$S_{21} = 4$	$M_1^{(2)} = 7$
CN_{32}	$MP_2^{(3)} = 7$	$S_{32} = 1$	$M_2^{(3)} = 10$
CN_{34}	$MP_4^{(3)} = 7$	$S_{34} = 4$	$M_4^{(3)} = 6$
CN_{41}	$MP_1^{(4)} = 7$	$S_{41} = 2$	$M_1^{(4)} = 10$
CN_{42}	$MP_2^{(4)} = 5$	$S_{42} = 3$	$M_2^{(4)} = 4$
CN_{52}	$MP_2^{(5)} = 2$	$S_{52} = 1$	$M_2^{(5)} = 5$
CN_{53}	$MP_2^{(5)} = 2$	$S_{52} = 1$	$M_2^{(5)} = 1$

TABLE 3: The encryption in inner-cluster data aggregation.

Cluster nodes	Plaintext $M^{(1)}$	Ciphertext $C^{(1)}$
CN_{11}	$M_1^{(1)} = 4$	$C_1^{(1)} = 4F_1^{(1)} + G_1^{(1)} + 9H_1^{(1)}$
CN_{12}	$M_2^{(1)} = 6$	$C_2^{(1)} = 6F_1^{(1)} + G_1^{(1)} + 4H_1^{(1)}$
CN_{13}	$M_3^{(1)} = 5$	$C_3^{(1)} = 5F_2^{(1)} + G_2^{(1)} + 5H_2^{(1)}$
CN_{14}	$M_4^{(1)} = 4$	$C_4^{(1)} = 4F_2^{(1)} + G_2^{(1)} + 2H_2^{(1)}$

TABLE 4: The intercluster data aggregation.

Cluster branch	Aggregation results
CB_{11}	$C_{1,\text{agg}}^{(1)} = 10F_1^{(1)} + 2G_1^{(1)} + 13H_1^{(1)}$
CB_{12}	$C_{2,\text{agg}}^{(1)} = 9F_2^{(1)} + 2G_2^{(1)} + 7H_2^{(1)}$

TABLE 5: The encryption in intercluster data aggregation.

Cluster heads	Plaintext M	Ciphertext C
CH_1	$M^{(1)} = 19$	$C^{(1)} = 19P_1 + 4Q_1 + 23R_1$
CH_2	$M^{(2)} = 19$	$C^{(2)} = 19P_1 + 4Q_1 + 17R_1$
CH_3	$M^{(3)} = 23$	$C^{(3)} = 23P_1 + 4Q_1 + 21R_1$
CH_4	$M^{(4)} = 17$	$C^{(4)} = 17P_2 + 4Q_2 + 19R_2$
CH_5	$M^{(5)} = 20$	$C^{(5)} = 20P_2 + 4Q_2 + 12R_2$

TABLE 6: The intercluster data aggregation.

Base station branch	Aggregation results
BSB_{11}	$C_{1,\text{agg}} = 61P_1 + 12Q_1 + 61R_1$
BSB_{12}	$C_{2,\text{agg}} = 37P_2 + 8Q_2 + 31R_2$

the operations of slicing and mixing are shown with specific numbers.

Using the first cluster as an example, the inner-cluster data aggregation is shown as follows. There are 4 CNs in the first cluster, and these CNs collect the data around them. Then, the data is encrypted according to formula (3). The plaintext and ciphertext of data are shown in Table 3.

After the encryption, all of the CNs send their encrypted data to the CH along the inner-cluster aggregation tree. Then, the CH receives two aggregation data items from its two branches. Table 4 shows the aggregation results in each branch.

When the CH receives the aggregation data, it decrypts t aggregation data according to formula (5). We get the amount of CNs in two branches as follows:

$$323C_{1,\text{agg}}^{(1)} = 3230F_1^{(1)} + 646G_1^{(1)} + 4199H_1^{(1)},$$
$$323C_{2,\text{agg}}^{(1)} = 2907F_2^{(1)} + 646G_2^{(1)} + 2261H_2^{(1)}. \tag{11}$$

According to the orders of these nodes, we have $17F_j^{(1)} = 0$, $13G_j^{(1)} = 0$, and $19H_j^{(1)} = 0$. So,

$$323C_{1,\text{agg}}^{(1)} = 646G_1^{(1)},$$
$$323C_{2,\text{agg}}^{(1)} = 646G_2^{(1)}. \tag{12}$$

Then, CH decrypts the aggregation again according to formula (6). We get the aggregation data of two branches which is 10 and 9. CH aggregates these two data items and encrypts them with the public key from BS:

$$C^{(1)} = 19P_1 + 4Q_1 + 23R_1. \tag{13}$$

The inner-cluster data aggregation in other four clusters is done in the same way. Table 5 shows the plaintext and ciphertext of aggregation data in those five clusters.

(4) Intercluster Aggregation Data. After the encryption, all of the CHs send their encrypted data to the BS along the intercluster aggregation tree. Then, the BS receives two aggregation data items from its two branches. Table 6 shows the aggregation results in each branch.

When the BS receives the aggregation data, it decrypts these two aggregation data items according to formula (9). We get the amount of CHs in two branches as follows:

$$323C_{1,\text{agg}} = 19703P_1 + 3876Q_1 + 19703R_1,$$
$$323C_{2,\text{agg}} = 11951P_2 + 2584Q_2 + 10013R_2. \tag{14}$$

According to the orders of these nodes, we have

$$17P_j = 0,$$
$$13Q_j = 0, \tag{15}$$
$$19R_j = 0.$$

So,

$$323C_{1,\text{agg}} = 3876Q_1,$$
$$323C_{2,\text{agg}} = 2584Q_2. \tag{16}$$

CH decrypts the aggregation again according to formula (10). We get the aggregation data of two branches which is 61 and 37. BS aggregates these two data items and gets the aggregation data of the whole network which is 87.

TABLE 7: The computation overhead of IPHCDA and SPPDA.

Operation	IPHCDA	SPPDA
Encryption	$N \cdot (E_{\text{Add}} + 2E_{\text{Mul}})$	$N \cdot (2E_{\text{Add}} + 3E_{\text{Mul}})$
Aggregation	$(N-1) \cdot E_{\text{Add}} + G \cdot E_{\text{MAC}} + k \cdot E_{\oplus}$	$(N-1) \cdot E_{\text{Add}}$
Decryption	$G \cdot E_{\log}$	$2G \cdot E_{\log}$

4. The Security Analysis

4.1. Ciphertext Only Attack. Ciphertext only attack is a basic attack in wearable sensors. When attackers use this attack, they only can try to get the plaintext by analyzing the ciphertext.

SPPDA model uses the elliptic curve cryptography, which is an asymmetric encryption model. Its security is based on the intractability in decomposition of large prime numbers. So SPPDA model can resist this attack well as long as the suitable prime numbers are used.

4.2. Chosen-Plaintext Attack. In chosen-plaintext attack, attackers can get some plaintexts and the ciphertexts. Attackers want to get the secret key by analyzing these texts so that the other ciphertexts can be cracked rapidly by using this secret key.

SPPDA model uses the elliptic curve encryption with three parameters, and one of them is used to add the random disturbance. In this way, even the same plaintexts can be encrypted to the different ciphertexts. So, no matter how many plaintext-ciphertexts the attackers get, they cannot get the secret key by analyzing the plaintext-ciphertexts.

4.3. Data Injection Attack. In data injection attack, the attackers send the unauthorized data to the aggregation node. If the aggregation aggregates this data, the result will be different from the real result. So the base station gets a fault result.

SPPDA model uses the elliptic curve encryption. So the ciphertext is satisfied with the structure of the elliptic curve encryption. If the attackers send the data which lacks standardization, the aggregation can recognize it easily and remove it by the aggregation node.

4.4. Aggregation Node Compromised Attack. In the node compromised attack model, attackers can compromise some aggregation nodes in the wearable sensors. Then, attackers get the key of these nodes and perform unauthorized aggregation. So, the base station gets the fault result.

SPPDA model verifies the data integrities both in cluster heads and in base station. If the aggregation node in cluster is compromised, cluster head can recognize the fault of branch at which the compromised node stays. If the cluster head is compromised, base station can recognize the fault of branch at which the compromised cluster head stays. Then, the cluster head or base station uses the locating mechanism to locate the compromised node and remove it.

5. Performance Analysis

In this section, the computation overhead and the communication overhead of SPPDA model are analyzed and compared with the IPHCDA model.

5.1. The Computation Overhead. The computation overhead includes encryption, aggregation, and decryption. We assume that the overhead of addition, scalar multiplication, MAC, XOR, and the decryption are expressed as E_{Add}, E_{Mul}, E_{MAC}, $E_{\oplus}$, and $E_{\log}$, G is the amount of clusters, and N is the amount of the nodes in wearable sensors. Table 7 shows the computation overhead in IPHCDA model and SPDA model.

In encryption operation, IPHCDA model needs twice E_{Mul} and once E_{Mul} in each node, while SPPDA model needs three times E_{Mul} and twice E_{Mul}. In aggregation operation, IPHCDA model needs $(N-1)$ times E_{Add}, G times E_{MAC}, and k times $E_{\oplus}$, while SPPDA model only needs $(N-1)$ times E_{Add}. The number of XOR operations is decided by the structure of the aggregation tree. The constant k is no less than 1 and no more than $G-1$. In decryption operation, IPHCDA model needs G times $E_{\log}$, while SPPDA model needs $2G$ times $E_{\log}$.

In general, the computation overhead of IPHCDA model is lower than SPPDA model in encryption and decryption. The computation overhead of SPPDA model is lower than IPHCDA model in aggregation. But, there are two aspects not described in Table 7.

(1) The orders of the elliptic curve are not the same in both models. The order in IPHCDA is larger than in SPPDA. So the E_{Add}, E_{Mul}, and $E_{\log}$ in IPHCDA model are larger.

(2) The computation overhead which is extra in SPPDA model is undertaken by the whole network, so the average overhead to each node is lower.

So, the computation overheads in both models are almost the same.

5.2. The Communication Overhead. In this section, the communication overhead between SPPDA model and IPHCDA model is compared. It is assumed that these two models are used in the same network structure. Therefore, the comparison of the communication is the same as the comparison of length of ciphertext.

It is assumed λ is the length of each prime in both models, and the number of the clusters in the network is G. So the length of ciphertext in IPHCDA model is $(G+1)\lambda$, and the length of ciphertext in SPPDA model is 3λ. In general case,

TABLE 8: The length of ciphertext in two models ($\lambda = 256$, unit is bit).

Models	$G = 1$	$G = 2$	$G = 3$	$G = 4$
SPPDA	768	768	768	768
IPHCDA	512	768	1024	1280

$\lambda = 256$ is safe enough to a ciphertext, and Table 1 shows the comparison of the length of ciphertext in two models when $\lambda = 256$.

In Table 8, the length of ciphertext increases with G in IPHCDA model, and the length of ciphertext is constant 768 when G increases. So, when $G > 2$, the length of ciphertext in IPHCDA model is larger than that in SPPDA model; that means the communication overhead of IPHCDA model is larger. Actually, a cluster-based network usually consists of plenty of clusters. Therefore, the SPPDA model has lower communication overhead.

6. Conclusion

In this paper, we present a new secure privacy-preserving data aggregation model, which adopts a mixed data aggregation structure of tree and cluster. The proposed model verifies the data integrity both at the cluster nodes and at the base station. Meanwhile, the model gives a mechanism to locate the compromised nodes. Lastly, the detail analysis shows that this model is robust to many attacks and has lower communication overhead.

Conflict of Interests

The authors declare that there is no conflict of interests regarding the publication of this paper.

Acknowledgments

This work is supported by Beijing Natural Science Foundation under Grant 4132057, National Natural Science Foundation of China under Grant 61201159, Beijing Municipal Education Commission on Projects (SQKM201510016013), and Foundation of MOHURD (2015-K8-029).

References

[1] J. Yick, B. Mukherjee, and D. Ghosal, "Wireless sensor network survey," *Computer Networks*, vol. 52, no. 12, pp. 2292–2330, 2008.

[2] C. J. Deepu and Y. Lian, "A Joint QRS detection and data compression scheme for wearable sensors," *IEEE Transactions on Biomedical Engineering*, vol. 62, no. 1, pp. 165–175, 2015.

[3] P. Picazo-Sanchez, J. E. Tapiador, P. Peris-Lopez, and G. Suarez-Tangi, "Secure publish-subscribe protocols for heterogeneous medical wireless body area networks," *Sensors (Switzerland)*, vol. 14, no. 12, pp. 22619–22642, 2014.

[4] R. Di Pietro, P. Michiardi, and R. Molva, "Confidentiality and integrity for data aggregation in WSN using peer monitoring," *Security & Communication Networks*, vol. 2, no. 2, pp. 181–194, 2009.

[5] A. Zambrano, F. Derogarian, R. Dias et al., "A wearable sensor network for human locomotion data capture," in *pHealth*, vol. 177 of *Studies in Health Technology and Informatics*, pp. 216–223, IOS Press, Amsterdam, The Netherlands, 2012.

[6] C. Castelluccia, E. Mykletun, and G. Tsudik, "Efficient aggregation of encrypted data in wireless sensor networks," in *Proceedings of the 2nd Annual International Conference on Mobile and Ubiquitous Systems—Networking and Services (MobiQuitous '05)*, pp. 109–117, New York, NY, USA, July 2005.

[7] J. Girao, D. Westhoff, and M. Schneider, "CDA: concealed data aggregation for reverse multicast traffic in wireless sensor networks," in *Proceedings of the IEEE International Conference on Communications (ICC '05)*, pp. 3044–3049, May 2005.

[8] T. Feng, C. Wang, W. Zhang, and L. Ruan, "Confidentiality protection for distributed sensor data aggregation," in *Proceedings of the 27th IEEE Conference on Computer Communications (INFOCOM '08)*, pp. 56–60, IEEE, Phoenix, Ariz, USA, April 2008.

[9] D. Boneh, E.-J. Goh, and K. Nissim, "Evaluating 2-dnf formulas on ciphertexts," in *Theory of Cryptography: Second Theory of Cryptography Conference, TCC 2005, Cambridge, MA, USA, February 10-12, 2005. Proceedings*, Lecture Notes in Computer Science, pp. 325–341, 2005.

[10] E. Mykletun, J. Girao, and D. Westhoff, "Public key based cryptoschemes for data concealment in wireless sensor networks," in *Proceedings of the IEEE International Conference on Communications (ICC '06)*, vol. 5, pp. 2288–2295, Istanbul, Turkey, July 2006.

[11] J. Girao, D. Westhoff, E. Mykletun, and T. Araki, "TinyPEDS: tiny persistent encrypted data storage in asynchronous wireless sensor networks," *Ad Hoc Networks*, vol. 5, no. 7, pp. 1073–1089, 2007.

[12] J. M. Bahi, C. Guyeux, and A. Makhoul, "Efficient and robust secure aggregation of encrypted data in sensor networks," in *Proceedings of 4th International Conference on Sensor Technologies and Applications (SENSORCOMM '10)*, pp. 472–477, Venice, Italy, July 2010.

[13] S. Ozdemir and Y. Xiao, "Integrity protecting hierarchical concealed data aggregation for wireless sensor networks," *Computer Networks*, vol. 55, no. 8, pp. 1735–1746, 2011.

[14] Y.-H. Lin, S.-Y. Chang, and H.-M. Sun, "CDAMA: concealed data aggregation scheme for multiple applications in wireless sensor networks," *IEEE Transactions on Knowledge and Data Engineering*, vol. 25, no. 7, pp. 1471–1483, 2013.

[15] Q. Zhou, G. Yang, and L. He, "A secure-enhanced data aggregation based on ECC in wireless sensor networks," *Sensors*, vol. 14, no. 4, pp. 6701–6721, 2014.

[16] O. Younis and S. Fahmy, "HEED: a hybrid, energy-efficient, distributed clustering approach for ad hoc sensor networks," *IEEE Transactions on Mobile Computing*, vol. 3, no. 4, pp. 366–379, 2004.

[17] S. Lindsey and C. S. Raghavendra, "PEGASIS: power-efficient gathering in sensor information systems," in *Proceedings of the EEE Aerospace Conference*, vol. 3, pp. 1125–1130, IEEE, Big Sky, Mont, USA, March 2002.

[18] Y. Rebahi, V. E. Mujica-V, and D. Sisalem, "A reputation-based trust mechanism for ad hoc networks," in *Proceedings of the 10th IEEE Symposium on Computers and Communications (ISCC '05)*, pp. 37–42, Cartagena, Spain, June 2005.

[19] L. I. Yong-Jun, "Research on trust mechanism for peer-to-peer network," *Chinese Journal of Computers*, vol. 40, no. 5, pp. 805–809, 2010.

[20] S. Madden, M. J. Franklin, and J. M. Hellerstein, "TAG: a tiny aggregation service for ad-hoc sensor networks," in *Proceedings of the 5th Usenix Symposium on Operating Sysems Design & Implementation (OSDI '02)*, vol. 3, pp. 131–146, Boston, Mass, USA, December 2002.

[21] L. Eschenauer and V. D. Gligo, "A key-management scheme for distributed sensor networks," in *Proceedings of the 9th ACM Conference on Computer and Communications Security (CCS '02)*, pp. 41–47, Washington, DC, USA, November 2002.

[22] W. He, X. Liu, H. Nguyen, K. Nahrstedt, and T. Abdelzaher, "PDA: privacy-preserving data aggregation in wireless sensor networks," in *Proceedings of the 26th IEEE Conference on Computer Communications*, pp. 2045–2053, Anchorage, Alaska, USA, May 2007.

2

An Analysis of QoS in ZigBee Network based on Deviated Node Priority

Md. Jaminul Haque Biddut, Nazrul Islam,
Md. Maksudul Karim, and Mohammad Badrul Alam Miah

Department of Information and Communication Technology (ICT), Mawlana Bhashani Science and Technology University, Santosh, Tangail 1902, Bangladesh

Correspondence should be addressed to Md. Jaminul Haque Biddut; biddutict@gmail.com

Academic Editor: Tho Le-Ngoc

ZigBee is an IEEE 802.15.4 standardized communication protocol. It forms a flawless Wireless Sensor Network (WSN) standard for interoperability at all levels of the network, particularly the application level which most closely touches the user. A large number of devices from different vendors can work seamlessly. These devices act as a network and send huge data traffic to the Coordinator. End devices at different zones have different roles in communication with each other. There has been a lack in executing their requests in a synchronized way based on task priority. This lack leads to massive data traffic loss and degrades the Quality of Service (QoS). One of the challenges is to analyze the QoS parameters in ZigBee network that help to detect the overall network performance. The contribution of this paper is twofold; first, a ZigBee Network is implemented based on node priority. It demonstrates a method to generate a new priority of devices with respect to their existing priority and zones' priority as well. Second, the QoS is analyzed based on the new priority status for tasks preference purposes. The outcome of this paper shows that the QoS of the network is more conspicuous than non-priority based network.

1. Introduction

ZigBee is a sensor based special network which pulls the trigger to establish it in wireless network standard. ZigBee was designed to provide high data throughput in applications where the lower duty cycle and power consumption are an important consideration. ZigBee specifications are maintained and updated by the ZigBee Alliance based on the Institute of Electrical and Electronics Engineers (IEEE) Standards Association's 802.15 specifications. It operates under the IEEE 802.15.4 physical radio frequency specification and in unlicensed radio frequency bands, including 2.4 GHz, 900 MHz, and 868 MHz. ZigBee Alliance states that it is often applied in industrial automation and physical plant operation, commercial building automation, home automation, personal, home and hospital care, smart energy, telecom applications, and so on [1–3]. It is frequently associated with machine-to-machine (M2M) communication and the Internet of Things. Internet of Things is the network in which objects are embedded with electronics, software, sensors, and connectivity to enable greater service by exchanging data with other connected devices. Each object is uniquely identifiable and able to interoperate within the existing Internet infrastructure via embedded computing system [1, 4].

However, the sequence of creating and forming a ZigBee network is simple as follows. The personal area network (PAN) Coordinator turns on a ZigBee network; it first looks for a suitable radio channel (usually the one which has the least activity) and then creates the network of self-assigning a PAN ID. When the network is formed, a router or terminal device can bring together the network through a nearby parent node (i.e., the Coordinator or another router) by sending a join request, and the parent can accept or decline the request based on its address space availability and other criteria [1, 5]. Due to ZigBee's network formation simplicity, a gigantic sum of devices can be connected together. The ZigBee Coordinator can conceal 65,535 devices in a single network [4]. The end devices are placed at diverse spots. The

Coordinator may be broken down to operate the network maintaining convenient QoS. The QoS maintenance and measurement are a challenging issue for ZigBee network. Mainly, it refers to the improvement of overall network performance.

Moreover, to increase the network performance a large network was partitioned into several zones or areas. Each zone has priority and containing devices have also local priorities in their corresponding zones. The priority indicates their task's importance. These devices send huge application traffic to the routers and Coordinator for processing. For QoS maintenance, this huge amount of traffic processing and end devices requests synchronization is needed. Priority scheduling is the promising mechanism in this regard. Beforehand, several strategic approaches had been adopted which worked on ZigBee network performance analysis in terms of routing performance and effective data flow and scheduling has been discussed [2, 6–9]. Yet, there is a lack in forming a ZigBee network efficiently. In that respect, further study is required for QoS analysis on ZigBee network as well as improvement of network performance.

This paper evolves new priority for each device considering local device and corresponding zone priority. To the best of our knowledge, this work has not been practiced yet. The goal of this paper is to construct a reliable ZigBee network for device synchronization using the node priority mechanism and study its QoS.

The paper is organized as follows. Section 2 describes the background and related works. In Section 3, the research methodology is described. In Section 4, proposed network simulation and implementation in Riverbed are illustrated. Section 5 evaluates the performance to show that results meet up with the designed objectives. Finally, Section 6 is a conclusion that presents the outcome of this work and suggests future work.

2. Background and Related Work

ZigBee is a sensible network. The ZigBee network gets worldwide popularity for its variety of usages. Since ZigBee is used in various automation system [10–12], therefore, an effective network formation is on demand. Moreover, reliability and effectiveness of a ZigBee network service are assessed based on QoS. QoS is the set of observations of ZigBee network characteristics. According to Chen et al., QoS analysis is the vital measurement of network performance. QoS in wireless sensor and actor networks in the context of critical infrastructure protection (CPI) was presented in the survey paper [13]. In this paper, the characteristics are classified into both user level and low-level specification in the context of the wireless network. The service ZigBee network mainly depends on its formation methods. Among the methods, priority is the most reliable for request scheduling purposes, event handling, synchronization, and so on. Priority based network construction is handy on the performance issue. Priority technique was applied to evaluate performance in terms of balancing energy in the study [1].

According to Islam et al., priority is an effective technique to synchronize requests from different regions. And the network performance is better than without-priority based ZigBee network with tree routing method [14]. The priority based scheme is used for event monitoring. Priority shows efficiency in handling related problematic delay and ensuring delivery of high priority packets analyzed in the article [15]. For home and M2M communication, the total home area is divided into subareas based on potential applications and radio services. And QoS improved in this strategy [10]. Again, QoS was enhanced for the smart grid. In the ZigBee-based smart grid, priority based differentiated traffic service was analyzed in the research [16]. In addition, different routing methods also play a vital role in the QoS issue. Tree, Mesh, and Ad Hoc On-Demand Distance Vector (AODV) had been used in ZigBee network. Besides, the research paper [17] describes an agent assisted routing to ensure better QoS in WSN. Localized routing for QoS of WSNs based on traffic differentiation is illustrated in the study [18]. Furthermore, QoS improvement for IEEE 802.15.4 based wireless body sensor network for health care is analyzed in the research article [19]. Moreover, room storage problem of tree based ZigBee network and neighbor table based shortcut tree routing is analyzed, respectively, in [20, 21].

In accordance with related works, priority is used for efficient network formation. And QoS is highly endorsed to analyze any network. From this context, a motivation comes to build a new priority calculation strategy for ZigBee devices where each device and zone have priority. This new priority is useful for synchronizing ZigBee devices from different zones. Since the new priority technique has not been practiced previously, QoS has been scrutinized to validate the performance of the prospective ZigBee network.

3. Methodology

A literature review is performed to find out the wide range use of wireless sensor network. Based on the literature review, ZigBee is selected to develop its versatile use of forming a smart network for home automation and office automation and performing Internet of Things concept. A careful study is done on how ZigBee performs Internet of Things and how to enhance its QoS. Afterward, an enhanced way has been found in the ZigBee network. In this study, zones and device request synchronization is performed based on the proposed priority method to construct an efficient ZigBee network.

A supported simulation tool was needed to design and implement a complete network model. A deep study has been performed on many simulation tools such as Network Simulator 2 (NS2), Network Simulator 3 (NS3), OMNET++, Riverbed [22], and MATLAB, which support ZigBee network development features. The Riverbed is chosen to design and implement ZigBee network as a simulator for its user-friendly interface and its wide range of acceptance. Moreover, Riverbed simulator provides the fastest discrete event simulation used to evaluate various parameters of network performance [23]. The selected tool is installed in an Intel Core i5 processor-based workstation.

Simulated data are collected properly and carefully from the different ZigBee protocol layer. After analyzing the simulated data, appropriate graphs are plotted. Careful study of the plots is expected to offer a quantitative measure of the effect of different network parameters. Finally, QoS has been measured of the designed ZigBee network analyzing various QoS parameters like end-to-end delays, MAC delay, MAC load, and MAC throughput.

4. Proposed Network Simulation

The section provides an overview of the proposed ZigBee network simulation configuration. Furthermore, it illustrates the analysis of QoS in this ZigBee network for the reliable and robust network.

4.1. Problem Formulation. In this subsection, formulation of the zone synchronization problem is conducted for competing for end devices. At the beginning, the ZigBee network has been marked in different zones. These zones have the same number of devices. In the ZigBee network, various types of end devices are set up in different ranges. All nodes are connected to several routers having a common central processor named ZigBee Coordinator. These nodes try to communicate with other nodes via Coordinator or routers. ZigBee end devices send a large number of requests to the routers or Coordinator and vice versa.

Each zone has a priority value. The higher priority value makes the devices of the zone executed first in the Coordinator. The zones' priority value and execution of these nodes according to the zone value are considered as global processing. On the other hand, every zone consists of numerous ZigBee nodes and each node has to adjust priority values for their local processing. The local processing concerns the end devices application traffic processing based on their respective priority values.

When two or more devices have same local priority how should the application traffic be treated? Moreover, Zone's priority values are different. How the application traffic of two different prioritized devices from two different zones will be processed? This is the primary concern. In this case, which requests are processed first before the others? The central goal of this research is to implement a method for application traffic synchronization of ZigBee nodes based on zone and individual priority and study its QoS.

4.2. Priority Implementation in ZigBee Network. A ZigBee Coordinator is responsible for forming a ZigBee network. End devices and routers have always a distance from ZigBee Coordinator. This distance around ZigBee Coordinator is defined as different area factors. The area factor is called a zone and each zone has many end devices under a router. Each zone has a set of priority attributes for defining its priority. In addition, the end nodes have also priority values in their zone. When nodes of different zones are trying to communicate via intermediate nodes, then routers or Coordinator always checks their priority values. In this mechanism, a child node (end node or router) connects to a parent node according to the child nodes willingness to pay for the connection. The paying preference is considered as a priority. We assume that each end device in the network has a

TABLE 1: Proposed zone payment scheme.

Zone name	Unit price	Available nodes
Zone_1	10	node_1, node_18, node_19, node_20, node_21
Zone_2	11	node_2, node_15, node_16, node_17, node_23
Zone_3	12	node_3, node_5, node_6, node_7, node_24
Zone_4	13	node_4, node_8, node_9, node_10, node_25
Zone_5	14	node_26, node_27, node_28, node_29, node_30
Zone_6	15	node_0, node_11, node_12, node_13, node_14

certain willingness to pay and its zone has also a priority. End devices with higher priority zones have a higher willingness to pay and vice versa [1].

At first, the ZigBee network has been divided into six zones and set the priority values to the corresponding zones. The zones and participating nodes in these zones are listed in Table 1.

In Table 1, Zone_6 has the highest priority value 15 and lowest priority value 10 for Zone_1. The devices in the highest priority zones always get preference at the Coordinator for processing. The corresponding nodes in each zone have set priority value D_P (1, 2, 3, 4, and 5) according to their appearance on the table. This priority value is true for every zone and all participating nodes. Furthermore, a priority resolution technique is described in the following.

4.2.1. Priority Calculation. Locally, the higher prioritized nodes get a chance to send application traffic via a router to the other end devices. There are two sets of priority values. One is zone's priority and the other is an individual node's priority. When end nodes try to communicate beyond their respective zone at the same time, then priority resolution is needed in Coordinator for global access. As the two sets of priorities, a generalization has been done between the individual node and zone priorities. Standard deviation calculation is the process to find out general deviation from a set of values. In this context, standard deviation relates to this two-set priority with the same deviation value [24]. That paves the way for generating new priority for global device-to-device communication. Simply, in the global priority calculation is as follows.

The sum of all device priorities in each zone is

$$S_1 = \sum_1^5 D_P = 15,$$

$$\text{average, } x_1 = \frac{S_1}{N},$$

$$\text{so, } x_1 = 3,$$

(1)

where N is number of elements.

The standard deviation of the device priority values is

$$\sigma^2 = \left(\frac{1}{N} \sum_{i=1}^N x_i^2 \right) - x_1^2.$$

(2)

Here, x_i represents individual node priority values.

The value of standard deviation of individual devices' priority is $\sigma = 1.41$.

The difference (D_i) between individual zone priorities and the sum of all device priorities (S_1) in each zone is as follows:

$$D_1 = S_1 - \text{Zone\_1 priority value},$$

$$D_2 = S_1 - \text{Zone\_2 priority value},$$

$$\vdots \qquad (3)$$

$$D_6 = S_1 - \text{Zone\_6 priority value},$$

$$\text{thus, } D_i = S_1 - \text{Zone}_i \text{ priority value}.$$

From the above, the difference is in each zone, respectively, $D_1 = 15 - 10 = 5$. Similarly, $D_2 = 4$, $D_3 = 3$, $D_4 = 2$, $D_5 = 1$, and $D_6 = 0$.

It is clear that Zone_6 has the always highest priority with no difference in total priority of the devices in this zone. As a result, all the devices in Zone_6 have the highest priority the same as the global zone priority value of Zone_6.

Now, consider the other five zones as follows: these zones have different zone values from their total sum-up of the values of corresponding devices. Now, the sum of priority values of Zone_1 to Zone_5 is given below:

$$S_2 = \sum_1^5 \text{Zone}_i \text{ Priority Value}. \qquad (4)$$

The sum of the zones priorities is $S_2 = 60$. Thus, the average is $x_2 = S_2/N = 12$.

Again, the standard deviation has been computed according to (2) for the corresponding zone values. Hence, the standard deviation of corresponding zones' priority is $\sigma = 1.41$.

The deviation is the same for both the devices' priority and zones' priority. The standard deviation method has been applied for the two sets of priorities in order to make it generalized for further calculation.

When a node tries to connect globally, then its possibility to be executed in the Coordinator depends on the node's respective priority value and its zone value. As its zone priority value is in a generalized form which is the standard deviation 1.41, it is quite easy to compute the new priority for different nodes in their global participation. The following equation generates new priority for all the participating devices, computing all the individual node priorities along with its corresponding zone priority:

$$N_p = D_P \times \sigma, \qquad (5)$$

where N_p is new priority (per device), D_P is corresponding device priority, and σ is standard deviation.

This equation simplifies the early described problems.

4.2.2. New Priority Status. Equation (5) can change the existing priority status. In addition, it helps Coordinator to make the decision for request processing in terms of equal priority devices. Different observations are described below.

Observation 1 (different priority devices). During this observation, a lower priority (node_1) and a higher priority (node_10) device have been considered from Zone_1 and Zone_4, respectively. Initial priorities of node_1 and node_10 are 1 and 4 accordingly. After computing with (5), the new priority (N_p) values are 1.41 and 5.64 for node_1 and node_10, respectively. Therefore, node_10 gets the procession first in this case.

Again, the updated priority scenario has been observed for node_23 and node_29 from Zone_2 and Zone_6. The initial priorities are 5 and 4 for node_23 and node_29 sequentially. After computation, the new priority (N_p) figures have been found at 7.05 and 5.64 for node_23 and node_29, respectively. Though node_23 is from the lower priority zone, the new priority allocates it to get privilege for the first processing compared to node_29 to the Coordinator.

Observation 2 (the same priority devices). This observation mainly concerns the same prioritized devices which are from different zones. Here, both node_5 and node_8 have the same priority in their corresponding zone. The initial priority value is 2 for both the devices. The new priority (N_p) has been computed according to (5) and it has been found to be 2.82 for both node_5 and node_8.

Thus, the new priorities are also equal for both the devices. The Coordinator checks their priority and respective zone priority. Then, the Coordinator gives access to node_8 first because of its higher zone priority. All the zones and their nodes synchronize in this manner.

Equation (5) computes the new priority for every device. Based on these new priorities, ZigBee Coordinator makes a decision when certain conditions arise in synchronizing various device application traffic from diverse zone regarding the above two observations.

4.3. Simulation Configuration. According to the proposed implementation method, a simulation scenario has been created. The scenario consists of a ZigBee Coordinator, six routers, and several ZigBee end devices. After that, all the devices are organized into different zones. The proposed ZigBee network model scenario is organized on the basis of Table 2. The zones and their devices have set of predefined priorities according to Table 1. The new priority calculation has been performed using Riverbed's "Utility" service. The service sets new priority after analyzing communicating devices.

The synchronization of different proposed zones is done on the basis of priority values for processing their nodes request. According to Table 1, the priority values have been fixed for simulation in different zones and their corresponding devices. Almost all the devices were active in communication when Figure 1 had been captured. The node_0 is a central node which is ZigBee Coordinator whereas five routers (node_1, node_2, node_3, node_4, and node_5) are placed at a certain distance from the Coordinator. The routers are intermediary nodes and they hand over the application

TABLE 2: ZigBee network scenario overview.

Parameters	Descriptions
Number of nodes (overall)	30
Number of Coordinators	1
Number of routers	5
Number of end devices	24
Network dimension	300 m × 300 m
Simulation time	600 sec

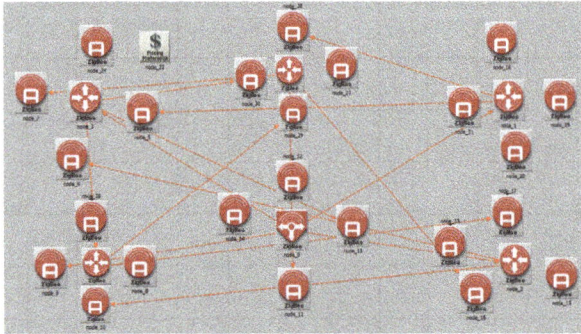

FIGURE 1: ZigBee network simulation scenario.

traffic to a particular destination. On the contrary, ZigBee Coordinator is a full-function device which is responsible for all functionalities as well as performance issues. After the 600 sec simulation, ZigBee nodes start communication as depicted in Figure 1.

5. Simulation Results

This section gives a detailed description of the analysis of the obtained simulation result. The result is prepared based on the ZigBee's (application and MAC) layer attainment analysis.

The graphs are generated by analyzing different QoS parameters (i.e., end-to-end delay in the application layer, MAC load, and MAC delay in MAC layer).

5.1. With versus without Priority. ZigBee network shows more efficiency on priority mechanism for setting up nodes in various zones of a home or office. If any zone has higher priority, then its activity will increase compared to zones with lower priority or without priority. The investigation of various QoS parameters is in progress for the proposed ZigBee network.

5.1.1. End-to-End Delay. End-to-end delay or one-way delay refers to the time taken for a packet to be transmitted across a network from source to destination.

Delay components are as follows:

$$D_{\text{end-end}} = N \left[D_{\text{trans}} + D_{\text{prop}} + D_{\text{proc}} \right], \quad (6)$$

where $D_{\text{end-end}}$ is end-to-end delay; D_{trans} is transmission delay; D_{prop} is propagation delay; D_{proc} is processing delay; and N is number of links (number of routers + 1).

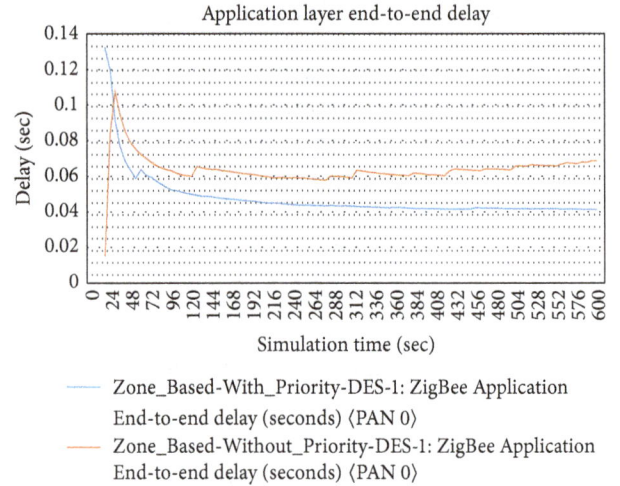

— Zone_Based-With_Priority-DES-1: ZigBee Application End-to-end delay (seconds) ⟨PAN 0⟩
— Zone_Based-Without_Priority-DES-1: ZigBee Application End-to-end delay (seconds) ⟨PAN 0⟩

FIGURE 2: End-to-end delay (sec) for the global model.

Each router has its own D_{trans}, D_{prop}, and D_{proc}. Hence, this formula gives a rough estimate [25, 26].

It describes that when nodes try to communicate with other zones' node, then the request is processed on the priority based mechanism. In this sense, when one node communicates with a node in lower priority zone compared to its priority zone, then the total delay between creation and reception of application packets generated by this node increases.

Figure 2 represents that a global end-to-end delay is higher for network without priority compared to the priority based network. The priority based network faces on average 0.05 sec delay approximately. On the other hand, non-priority based network faces 0.07 sec delay approximately. The average end-to-end delay is slightly higher in the non-priority based network.

5.1.2. MAC Delay. Most of the requests are dropped on the basis of their priority and a limited delay occurs in accessing medium. It represents the total end-to-end delays (sec) of all the packets received by the 802.15.4 MACs of all WPAN nodes in the network and forwarded to the higher layer.

In Figure 3 (both for global network and for Coordinator), it is evident that without-priority based network has slightly higher MAC delay than the priority based network. The average MAC delay in ZigBee Coordinator (node_0) is about 0.12 sec and 0.14 sec for priority based network and without-priority based network, respectively. As a consequence, the average delay in the global network is 0.11 sec and 0.12 sec approximately for the prioritized and nonprioritized network, respectively.

5.1.3. MAC Load. It represents the total load (in bits/sec) submitted to the 802.15.4 MAC by all higher layers in all WPAN nodes in the network. Figure 4 points that without-priority based networks have higher global MAC load than the priority based network. In the priority based network, Coordinator (node_0) processes all the requests. For that

MAC layer delay

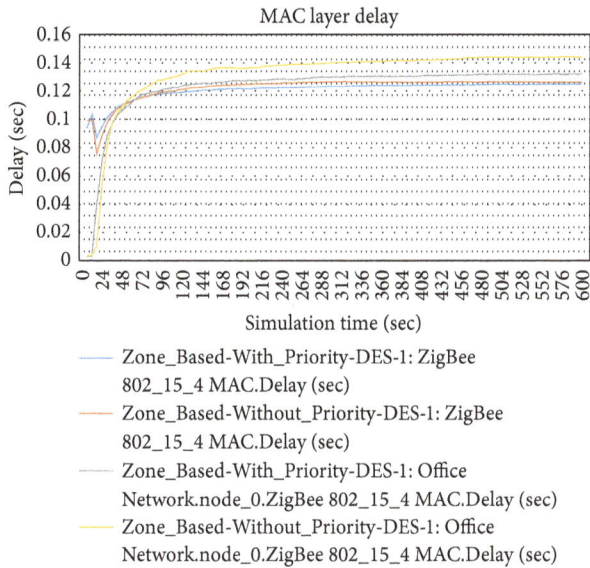

FIGURE 3: MAC delay (sec) for global model and Coordinator (node_0).

MAC layer load

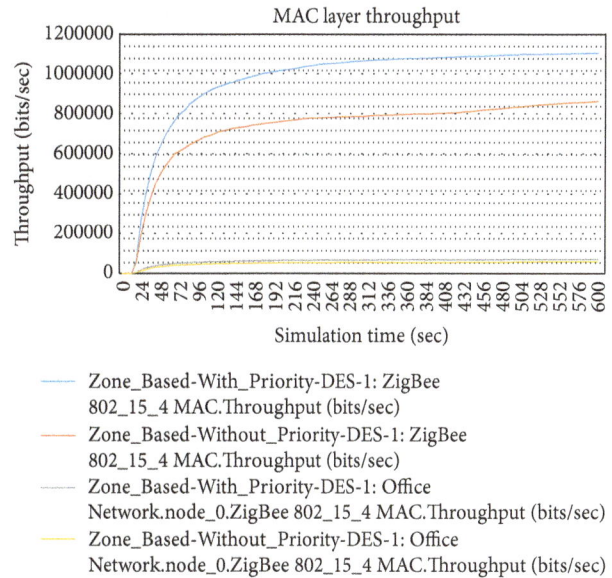

FIGURE 4: MAC load (bits/sec) for global model and Coordinator (node_0).

reason, priority based networks have more MAC load for Coordinator than without-priority based network as shown in Figure 4.

The MAC load in global network is roughly 9,000 bits/sec and 5,000 bits/sec for without-priority and priority based network. The without-priority networks carry a huge load for lack of node's request synchronization. But Coordinator faces almost similar load in both networks.

5.1.4. Throughput. In the context of a communication network, throughput is the rate of successful message delivery over the communication channel. The priority based network

MAC layer throughput

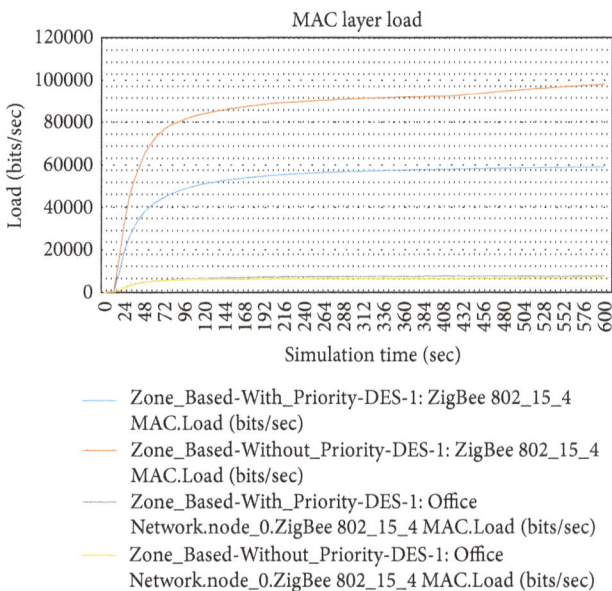

FIGURE 5: MAC throughput (bits/sec) for global model and Coordinator (node_0).

has high throughputs compared to without-priority based network.

Figure 5 represents throughput for both the global network and the ZigBee Coordinator (node_0) much higher for priority based network compared to without-priority based network.

The average throughput in a global network for the priority based network is almost 1,100,000 bits/sec. On the other hand, for the without-priority based network it is only about 700,000 bits/sec. Throughput for the Coordinator (node_0) is relatively higher in the priority based network compared to the without-priority based network.

Various QoS characters of ZigBee network are observed for performance analysis of with-priority versus without-priority mode. Almost in all the statistics shown here, priority based network performance is going prominent compared to without-priority based network.

6. Conclusion and Outlook

This paper presented a set of observations with the emphasis on the priority of different zones and their nodes. The results were obtained by performing a simulation study. This research comes to several conclusions from ZigBee network along with its QoS parameters.

First, application layer's parameter (end-to-end delay) is considered for user level service performance measurement. About 14% less end-to-end delay has been found for the priority based network in the application layer. Second, MAC layer's parameters (MAC load, MAC delay, and throughput) have been considered in MAC performance measurement. In the global network perspective, the priority based network has almost 14% and 44% less MAC delay and load, respectively. Moreover, there was 36% higher throughput compared to the without-priority based network. All the QoS

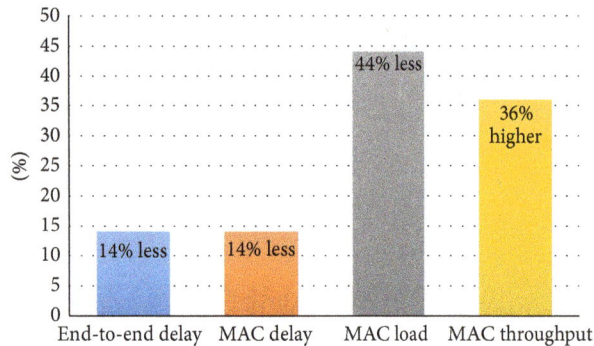

FIGURE 6: Overall QoS performance summary under proposed scheme.

parameters are shown in a percentage bar graph regarding their QoS improvement as shown in Figure 6.

The analysis of ZigBee layer's parameters contrasts significant response for QoS. After this study, the proposed priority based ZigBee network's performance is identified to be more effective than without-priority based ZigBee network with respect to these improved QoS parameters. The proposed approach is handy in differentiating or synchronizing different zones and their devices with different priorities in the ZigBee network.

The future work will address the ZigBee network performance analysis based on different queuing along with the mathematical model. This is ongoing work, including network latency and an efficient load balancing technique for ZigBee network. Moreover, this work intended to extend to security issue to model a reliable large scale ZigBee network.

Competing Interests

The authors declare that they have no competing interests.

References

[1] J. Wang, M. Chen, and V. C. M. Leung, "Forming priority based and energy balanced ZigBee networks—a pricing approach," *Telecommunication Systems*, vol. 52, no. 2, pp. 1281–1292, 2013.

[2] C. Zhang and W. Luo, "Topology performance analysis of zigbee network in the smart home environment," in *Proceedings of the 5th International Conference on Intelligent Human-Machine Systems and Cybernetics (IHMSC '13)*, vol. 2, pp. 437–440, Hangzhou, China, August 2013.

[3] J. Baviskar, A. Mulla, M. Upadhye, J. Desai, and A. Bhovad, "Performance analysis of ZigBee based real time Home Automation system," in *Proceedings of the International Conference on Communication, Information and Computing Technology (ICCICT '15)*, pp. 1–6, IEEE, Mumbai, India, January 2015.

[4] ZigBee Specification, ZigBee Alliance, http://www.ZigBee.org.

[5] W. Wang, G. He, and J. Wan, "Research on Zigbee wireless communication technology," in *Proceedings of the 2nd Annual Conference on Electrical and Control Engineering (ICECE '11)*, pp. 1245–1249, Yichang, China, September 2011.

[6] Deepika and M. Sharma, "Effective data flow in ZigBee network using OPNET," in *Proceedings of the 3rd International Conference on Communication and Signal Processing (ICCSP '14)*, pp. 1155–1158, Melmaruvathur, India, April 2014.

[7] Z. Wu, H. Chu, Y. Pan, and X. Yang, "Bus priority control system based on Wireless Sensor Network (WSN) and Zigbee," in *Proceedings of the IEEE International Conference on Vehicular Electronics and Safety (ICVES '06)*, pp. 148–151, IEEE, Beijing, China, December 2006.

[8] S. Rao, S. Keshri, D. Gangwar, P. Sundar, and V. Geetha, "A survey and comparison of GTS allocation and scheduling algorithms in IEEE 802.15.4 wireless sensor networks," in *Proceedings of the IEEE Conference on Information and Communication Technologies (ICT '13)*, pp. 98–103, IEEE, Jeju Island, South Korea, April 2013.

[9] K. Dong, *Performance and fairness enhancement in ZigBee networks [Ph.D. dissertation]*, TU Delft, Delft University of Technology, Delft, The Netherlands, 2011.

[10] Y. Zhang, R. Yu, S. Xie, W. Yao, Y. Xiao, and M. Guizani, "Home M2M networks: architectures, standards, and QoS improvement," *IEEE Communications Magazine*, vol. 49, no. 4, pp. 44–52, 2011.

[11] Y. T. Park, P. Sthapit, and J.-Y. Pyun, "Smart digital door lock for the home automation," in *Proceedings of the IEEE Region 10 Conference (TENCON '09)*, pp. 1–6, IEEE, Singapore, 2009.

[12] J. Byun, B. Jeon, J. Noh, Y. Kim, and S. Park, "An intelligent self-adjusting sensor for smart home services based on ZigBee communications," *IEEE Transactions on Consumer Electronics*, vol. 58, no. 3, pp. 794–802, 2012.

[13] J. Chen, M. Díaz, L. Llopis, B. Rubio, and J. M. Troya, "A survey on quality of service support in wireless sensor and actor networks: requirements and challenges in the context of critical infrastructure protection," *Journal of Network and Computer Applications*, vol. 34, no. 4, pp. 1225–1239, 2011.

[14] N. Islam, M. J. H. Biddut, A. I. Swapna, and M. H. R. Jany, "A study on priority based ZigBee network performance analysis with tree routing method," *Journal of Computer and Communications*, vol. 3, no. 8, pp. 1–10, 2015.

[15] T. H. Kim and S. Choi, "Priority-based delay mitigation for event-monitoring IEEE 802.15.4 LR-WPANs," *IEEE Communications Letters*, vol. 10, no. 3, pp. 213–215, 2006.

[16] W. Sun, X. Yuan, J. Wang, D. Han, and C. Zhang, "Quality of service networking for smart grid distribution monitoring," in *Proceedings of the 1st IEEE International Conference on Smart Grid Communications (SmartGridComm '10)*, pp. 373–378, IEEE, Gaithersburg, Md, USA, October 2010.

[17] M. Liu, S. Xu, and S. Sun, "An agent-assisted QoS-based routing algorithm for wireless sensor networks," *Journal of Network and Computer Applications*, vol. 35, no. 1, pp. 29–36, 2012.

[18] D. Djenouri and I. Balasingham, "Traffic-differentiation-based modular QoS localized routing for wireless sensor networks," *IEEE Transactions on Mobile Computing*, vol. 10, no. 6, pp. 797–809, 2011.

[19] J. J. Garcia and T. Falck, "Quality of service for IEEE 802.15.4-based wireless body sensor networks," in *Proceedings of the IEEE 3rd International Conference on Pervasive Computing Technologies for Healthcare*, pp. 1–6, London, UK, April 2009.

[20] L.-H. Yen and W.-T. Tsai, "The room shortage problem of tree-based ZigBee/IEEE 802.15.4 wireless networks," *Computer Communications*, vol. 33, no. 4, pp. 454–462, 2010.

[21] T. Kim, S. H. Kim, J. Yang, S.-E. Yoo, and D. Kim, "Neighbor table based shortcut tree routing in ZigBee wireless networks,"

IEEE Transactions on Parallel and Distributed Systems, vol. 25, no. 3, pp. 706–716, 2014.

[22] Riverbed, http://www.riverbed.com/.

[23] I. S. Hammoodi, B. G. Stewart, A. Kocian, and S. G. McMeekin, "A comprehensive performance study of OPNET modeler for ZigBee wireless sensor networks," in *Proceedings of the 3rd International Conference on Next Generation Mobile Applications, Services and Technologies (NGMAST '09)*, pp. 357–362, IEEE, Cardiff, UK, September 2009.

[24] J. Han, M. Kamber, and J. Pei, *Data Mining: Concepts and Techniques*, Elsevier, New York, NY, USA, 2011.

[25] A. Kaur, J. Kaur, and G. Singh, "Simulation and investigation of Zigbee sensor network with mobility support," in *Proceedings of the IEEE International Advance Computing Conference (IACC '14)*, pp. 176–181, IEEE, Gurgaon, India, February 2014.

[26] J. H. Biddut, N. Islam, M. H. Jany, and A. I. Swapna, "Performance analysis of large scale ZigBee network design through geometric structure," in *Proceedings of the 2nd International Conference on Electrical Information and Communication Technologies (EICT '15)*, pp. 263–268, Khulna, Bangladesh, December 2015.

3

Performance Analysis of SNR-Based HDAF M2M Cooperative Networks

Lingwei Xu,[1] Hao Zhang,[1,2] and T. Aaron Gulliver[2]

[1]College of Information Science and Engineering, Ocean University of China, Qingdao 266100, China
[2]Department of Electrical and Computer Engineering, University of Victoria, Victoria, BC, Canada V8W 2Y2

Correspondence should be addressed to Lingwei Xu; gaomilaojia2009@163.com

Academic Editor: Adam Panagos

The lower bound on outage probability (OP) of mobile-to-mobile (M2M) cooperative networks over N-Nakagami fading channels is derived for SNR-based hybrid decode-amplify-forward (HDAF) relaying. The OP performance under different conditions is evaluated through numerical simulation to verify the accuracy of the analysis. These results show that the fading coefficient, number of cascaded components, relative geometric gain, and power-allocation are important parameters that influence this performance.

1. Introduction

Mobile-to-mobile (M2M) communications have attracted significant research interest in recent years because they are widely employed in many wireless communication systems, such as mobile ad-hoc networks and vehicle-to-vehicle networks [1]. When both the transmitter and receiver are in motion, the double-Rayleigh fading model has been found to be suitable [2]. Extending this model by characterizing the fading between each pair of transmit and receive antennas as Nakagami, the double-Nakagami fading model has also been considered [3]. The N-Nakagami distribution was introduced in [4] as the product of N statistically independent, but not necessarily identically distributed, Nakagami random variables.

Cooperative diversity has been proposed for the high data-rate coverage required in M2M communication networks. Using amplify-and-forward (AF) relaying, the pairwise error probability (PEP) was investigated in [5] for cooperative intervehicular communication (IVC) systems over double-Nakagami fading channels. In [6], the exact symbol error rate (SER) and asymptotic SER expressions were derived for a M2M system with decode-and-forward (DF) relaying using the well-known moment generating function (MGF) approach over double-Nakagami fading channels. Symbol error probability (SEP) expressions were

obtained in [7] using this approach for multiple-mobile-relay M2M systems employing adaptive DF (ADF) relaying and fixed-gain AF (FAF) relaying over double-Nakagami fading channels.

In [8], a novel cooperative diversity protocol called hybrid decode-amplify-forward (HDAF) was proposed. This protocol combines AF and ADF relaying. When the quality of the received signal is sufficient, the relay performs ADF relaying; otherwise AF relaying is employed instead of remaining silent. However, only the SEP performance was considered, and the analysis is based on the assumption that the relay can determine whether each received symbol is correctly detected or not, which is not practical in real systems. To provide a practical HDAF protocol, in [9] the forwarding decisions at the relay were based on the signal-to-noise ratio (SNR) of the received signal. An SNR-based HDAF relaying scheme was also proposed. Further, closed-form expressions for the bit error probability of SNR-based HDAF relaying over independent nonidentical flat Rayleigh fading channels with maximum ratio combining (MRC) were derived.

To the best of our knowledge, the outage probability (OP) performance of SNR-based HDAF relaying M2M cooperative networks over N-Nakagami fading channels has not been considered in the literature. Thus in this paper, we present the analysis for the N-Nakagami case which subsumes the double-Nakagami results in [5–7] as special cases. Exact

OP expressions are derived for SNR-based HDAF relaying over N-Nakagami fading channels. The influence of the fading coefficient, number of cascaded components, relative geometric gain, and power-allocation on the OP performance is investigated.

The remainder of this paper is organized as follows. The SNR-based HDAF relaying model is presented in Section 2. Section 3 provides exact OP expressions for SNR-based HDAF relaying. Monte Carlo simulation results are presented in Section 4. Finally, some concluding remarks are given in Section 5.

2. System Model

We consider a three node cooperation model with a mobile source (MS), a mobile relay (MR), and a mobile destination (MD). These nodes operate in half-duplex mode and are equipped with a single pair of transmit and receive antennas.

According to [5], let d_{SD}, d_{SR}, and d_{RD} represent the MS to MD, MS to MR, and MR to MD links, respectively. Assuming the path loss between the MS and MD to be unity, the relative gain of the MS to MR and MR to MD links is defined as $G_{SR} = (d_{SD}/d_{SR})^{\nu}$ and $G_{RD} = (d_{SD}/d_{RD})^{\nu}$, respectively, where ν is the path loss coefficient [10]. Further, define the relative geometric gain $\mu = G_{SR}/G_{RD}$ (in dB), which is determined by the location of the relay with respect to the source and destination [5]. When the relay is close to the destination, the value of μ is negative. When the relay is close to the source, the value of μ is positive. When the relay has the same distance to the source and destination nodes, μ is 0 dB.

Let $h = h_k, k \in \{SD, SR, RD\}$, represent the complex channel coefficients of the MS to MD, MS to MR, and MR to MD links, respectively, which follow an N-Nakagami distribution. Therefore h is the product of N statistically independent, but not necessarily identically distributed, independent random variables:

$$h = \prod_{i=1}^{N} a_i, \qquad (1)$$

where N is the number of cascaded components and a_i is a Nakagami distributed random variable with probability density function (PDF):

$$f(a) = \frac{2m^m}{\Omega^m \Gamma(m)} a^{2m-1} \exp\left(-\frac{m}{\Omega} a^2\right), \qquad (2)$$

where $\Gamma(\cdot)$ is the Gamma function, m is the fading coefficient, and Ω is the scaling factor.

The PDF of h is given by [4]:

$$f_h(h) = \frac{2}{h \prod_{i=1}^{N} \Gamma(m_i)} G_{0,N}^{N,0}\left[h^2 \prod_{i=1}^{N} \frac{m_i}{\Omega_i} \Bigg|_{m_1,\ldots,m_N}^{-} \right], \qquad (3)$$

where $G[\cdot]$ is Meijer's G-function.

Let $y = |h_k|^2$, $k \in \{SD, SR, RD\}$, so that $y_{SD} = |h_{SD}|^2$, $y_{SR} = |h_{SR}|^2$, and $y_{RD} = |h_{RD}|^2$. The corresponding cumulative density function (CDF) of y can be derived as [4]

$$F_y(y) = \frac{1}{\prod_{i=1}^{N} \Gamma(m_i)} G_{1,N+1}^{N,1}\left[y \prod_{i=1}^{N} \frac{m_i}{\Omega_i} \Bigg|_{m_1,\ldots,m_N,0}^{1} \right]. \qquad (4)$$

By taking the first derivative of (4) with respect to y, the corresponding PDF can be obtained as [4]:

$$f_y(y) = \frac{1}{y \prod_{i=1}^{N} \Gamma(m_i)} G_{0,N}^{N,0}\left[y \prod_{i=1}^{N} \frac{m_i}{\Omega_i} \Bigg|_{m_1,\ldots,m_N}^{-} \right]. \qquad (5)$$

Communication in an SNR-based hybrid decode-amplify-forward (HDAF) relaying system can be described as follows. During the first time slot, the MS broadcasts to the MD and relay. The received signals r_{SD} and r_{SR} at the MD and MR can then be written as

$$r_{SD} = \sqrt{KE} h_{SD} x + n_D,$$
$$r_{SR} = \sqrt{G_{SR} KE} h_{SR} x + n_{SR}, \qquad (6)$$

where x denotes the transmitted signal and n_D and n_{SR} are zero-mean complex Gaussian random variables with variance $N_0/2$ per dimension. Here, E is the total energy used by both the source and the relay during the two time slots. K is the power-allocation parameter that controls the fraction of power reserved for the broadcast phase. If $K = 0.5$, equal power allocation (EPA) is used.

During the second time slot, by comparing γ_{SR} with a threshold γ_T, the MR decides whether DF or AF cooperation is utilized to forward the received signal. γ_{SR} denotes the instantaneous SNR of the MS to MR link. If $\gamma_{SR} > \gamma_T$, the MR decodes the received signal and generates a signal x_1 which is forwarded to the MD. With DF cooperation, the received signal at the MD is given by

$$r_{RD} = \sqrt{(1-K) G_{RD} E} h_{RD} x_1 + n_{RD}, \qquad (7)$$

where n_{RD} is a conditionally zero-mean complex Gaussian random variable with variance $N_0/2$ per dimension.

If selection combining (SC) is used at the MD, the output SNR is

$$\gamma_{SC} = \max(\gamma_{SD}, \gamma_{RD}), \qquad (8)$$

where

$$\gamma_{SD} = \frac{K |h_{SD}|^2 E}{N_0} = K |h_{SD}|^2 \bar{\gamma},$$
$$\gamma_{RD} = \frac{(1-K) G_{RD} |h_{RD}|^2 E}{N_0} = (1-K) G_{RD} |h_{RD}|^2 \bar{\gamma}. \qquad (9)$$

If $\gamma_{SR} < \gamma_T$, the MR amplifies and forwards the signal to the MD. Based on the AF cooperation protocol, the received signal at the MD is then given by

$$r_{RD} = \sqrt{cE} h_{SR} h_{RD} x + n_{RD}, \qquad (10)$$

where

$$c = \frac{K(1-K)G_{SR}G_{RD}E/N_0}{1 + KG_{SR}|h_{SR}|^2 E/N_0 + (1-K)G_{RD}|h_{RD}|^2 E/N_0}. \quad (11)$$

If selection combining (SC) is employed at the MD, the output SNR at the MD is

$$\gamma_{SCC} = \max(\gamma_{SD}, \gamma_{SRD}), \quad (12)$$

where

$$\gamma_{SRD} = \frac{\gamma_{SR}\gamma_{RD}}{1 + \gamma_{SR} + \gamma_{RD}}, \quad (13)$$

$$\gamma_{SR} = \frac{KG_{SR}|h_{SR}|^2 E}{N_0} = KG_{SR}|h_{SR}|^2 \bar{\gamma}.$$

3. OP of M2M Cooperative Networks

In this section, the OP for M2M cooperative networks is evaluated. The output SNR at the MD is

$$P_{out} = \Pr(\gamma_{SR} > \gamma_T, \gamma_{SC} < \gamma_{th}) + \Pr(\gamma_{SR} < \gamma_T, \gamma_{SCC} < \gamma_{th})$$

$$= I_1 + I_2, \quad (14)$$

where γ_{th} is the threshold.

Next, I_1 and I_2 are evaluated. First, consider I_1. As γ_{SD}, γ_{SR}, and γ_{RD} are mutually independent random variables, I_1 can be simplified as follows:

$$I_1 = \Pr(\gamma_{SR} > \gamma_T, \gamma_{SC} < \gamma_{th})$$

$$= \Pr(\gamma_{SR} > \gamma_T)\Pr(\gamma_{SD} < \gamma_{th})\Pr(\gamma_{RD} < \gamma_{th})$$

$$= (1 - \Pr(\gamma_{SR} \le \gamma_T))\Pr(\gamma_{SD} < \gamma_{th})\Pr(\gamma_{RD} < \gamma_{th}) \quad (15)$$

$$= (1 - F_{\gamma_{SR}}(\gamma_T))F_{\gamma_{SD}}(\gamma_{th})F_{\gamma_{RD}}(\gamma_{th}).$$

The CDF of γ_{SD} can be expressed as

$$F_{\gamma_{SD}}(r) = \frac{1}{\prod_{i=1}^N \Gamma(m_i)} G_{1,N+1}^{N,1}\left[\frac{r}{\overline{\gamma_{SD}}}\prod_{i=1}^N \frac{m_i}{\Omega_i}\Big|_{m_1,\dots,m_N,0}^1\right], \quad (16)$$

where

$$\overline{\gamma_{SD}} = K\bar{\gamma}. \quad (17)$$

The CDF of γ_{SR} is then

$$F_{\gamma_{SR}}(r) = \frac{1}{\prod_{t=1}^N \Gamma(m_t)} G_{1,N+1}^{N,1}\left[\frac{r}{\overline{\gamma_{SR}}}\prod_{t=1}^N \frac{m_t}{\Omega_t}\Big|_{m_1,\dots,m_N,0}^1\right], \quad (18)$$

where

$$\overline{\gamma_{SR}} = KG_{SR}\bar{\gamma}. \quad (19)$$

The CDF of γ_{RD} is given by

$$F_{\gamma_{RD}}(r) = \frac{1}{\prod_{tt=1}^N \Gamma(m_{tt})} G_{1,N+1}^{N,1}\left[\frac{r}{\overline{\gamma_{RD}}}\prod_{tt=1}^N \frac{m_{tt}}{\Omega_{tt}}\Big|_{m_1,\dots,m_N,0}^1\right], \quad (20)$$

where

$$\overline{\gamma_{RD}} = (1-K)G_{RD}\bar{\gamma}. \quad (21)$$

As γ_{SR} and γ_{SRD} are not mutually independent random variables, I_2 can be expressed as

$$I_2 = \Pr(\gamma_{SD} < \gamma_{th})\Pr(\gamma_{SR} < \gamma_T, \gamma_{SRD} < \gamma_{th}). \quad (22)$$

It is difficult to obtain the OP using γ_{SRD}, but a lower bound can be obtained. This provides a lower bound on the OP of a M2M cooperative network.

Using the well-known inequality in [11], γ_{SRD} can be approximated as

$$\gamma_{SRD} < \gamma_{up} = \min(\gamma_{SR}, \gamma_{RD}). \quad (23)$$

This approximation is used in the OP derivation instead of γ_{SRD} since it is more tractable analytically. I_2 can be approximated as

$$II_2 = \Pr(\gamma_{SD} < \gamma_{th})\Pr(\gamma_{SR} < \gamma_T, \gamma_{up} < \gamma_{th})$$

$$= \Pr(\gamma_{SD} < \gamma_{th})\Pr(\gamma_{SR} < \gamma_T, \min(\gamma_{SR}, \gamma_{RD}) < \gamma_{th}). \quad (24)$$

Case 1. It is obvious that $\min(\gamma_{SR}, \gamma_{RD}) \le \gamma_{SR}$, so if $\gamma_{SR} < \gamma_T$, then $\min(\gamma_{SR}, \gamma_{RD}) < \gamma_T$. If $\gamma_{th} > \gamma_T$, then $\min(\gamma_{SR}, \gamma_{RD}) < \gamma_{th}$. Therefore, (24) can be expressed as

$$II_2 = \Pr(\gamma_{SD} < \gamma_{th})\Pr(\gamma_{SR} < \gamma_T) = F_{\gamma_{SD}}(\gamma_{th})F_{\gamma_{SR}}(\gamma_T). \quad (25)$$

Substituting (15) and (25) into (14), the lower bound on the OP of the M2M cooperative network is

$$P_{lower} = (1 - F_{\gamma_{SR}}(\gamma_T))F_{\gamma_{SD}}(\gamma_{th})F_{\gamma_{RD}}(\gamma_{th})$$

$$+ F_{\gamma_{SD}}(\gamma_{th})F_{\gamma_{SR}}(\gamma_T). \quad (26)$$

Case 2. If $\gamma_{\text{th}} < \gamma_T$, using the Total Probability Theorem [12, Equation (2.36)], we have

$$Q = \Pr\left(\gamma_{\text{SR}} < \gamma_T, \min\left(\gamma_{\text{SR}}, \gamma_{\text{RD}}\right) < \gamma_{\text{th}}\right)$$

$$= \Pr\left(\gamma_{\text{SR}} < \gamma_T, \gamma_{\text{SR}} > \gamma_{\text{RD}}, \gamma_{\text{RD}} < \gamma_{\text{th}}\right)$$

$$\quad + \Pr\left(\gamma_{\text{SR}} < \gamma_T, \gamma_{\text{SR}} < \gamma_{\text{RD}}, \gamma_{\text{SR}} < \gamma_{\text{th}}\right)$$

$$= \Pr\left(\gamma_{\text{RD}} < \gamma_{\text{SR}} < \gamma_T, \gamma_{\text{RD}} < \gamma_{\text{th}}\right)$$

$$\quad + \Pr\left(\gamma_{\text{SR}} < \gamma_{\text{RD}}, \gamma_{\text{SR}} < \gamma_{\text{th}}\right)$$

$$= \int_0^{\gamma_{\text{th}}} \int_{\gamma_{\text{RD}}}^{\gamma_T} f_{\gamma_{\text{SR}}}(y)\,dy f_{\gamma_{\text{RD}}}(z)\,dz$$

$$\quad + \int_0^{\gamma_{\text{th}}} \int_{\gamma_{\text{SR}}}^{\infty} f_{\gamma_{\text{RD}}}(y)\,dy f_{\gamma_{\text{SR}}}(z)\,dz$$

$$= \int_0^{\gamma_{\text{th}}} \left(\int_0^{\gamma_T} f_{\gamma_{\text{SR}}}(y)\,dy - \int_0^{\gamma_{\text{RD}}} f_{\gamma_{\text{SR}}}(y)\,dy \right) f_{\gamma_{\text{RD}}}(z)\,dz$$

$$\quad + \int_0^{\gamma_{\text{th}}} \left(\int_0^{\infty} f_{\gamma_{\text{RD}}}(y)\,dy - \int_0^{\gamma_{\text{SR}}} f_{\gamma_{\text{RD}}}(y)\,dy \right) f_{\gamma_{\text{SR}}}(z)\,dz. \tag{27}$$

From the Appendix, (24) can be simplified as

$$II_2 = F_{\gamma_{\text{SD}}}(\gamma_{\text{th}})\, Q. \tag{28}$$

Substituting (15) and (28) into (14), the lower bound on the OP of the M2M cooperative network is

$$P_{\text{lower}} = \left(1 - F_{\gamma_{\text{SR}}}(\gamma_T)\right) F_{\gamma_{\text{SD}}}(\gamma_{\text{th}}) F_{\gamma_{\text{RD}}}(\gamma_{\text{th}}) + F_{\gamma_{\text{SD}}}(\gamma_{\text{th}})\, Q. \tag{29}$$

4. Numerical Results

In this section, numerical results are presented to illustrate and verify the OP analysis given in the previous sections.

Figure 1 presents the OP performance of a M2M cooperative network when $\gamma_{\text{th}} > \gamma_T$. The relative geometric gain is $\mu = 0$ dB, the power-allocation parameter is $K = 0.5$, and the thresholds are $\gamma_{\text{th}} = 4$ dB and $\gamma_T = 2$ dB. The following cases are considered based on the number of cascaded components N and the fading coefficient m.

Scenario 1. $m_{\text{SD}} = 1, m_{\text{SR}} = 1, m_{\text{RD}} = 1$ and $N_{\text{SD}} = 2, N_{\text{SR}} = N_{\text{RD}} = 2$.

Scenario 2. $m_{\text{SD}} = 2, m_{\text{SR}} = 2, m_{\text{RD}} = 2$ and $N_{\text{SD}} = 2, N_{\text{SR}} = N_{\text{RD}} = 2$.

Figure 1 shows that the numerical simulation results coincide with the theoretical results, which verifies the accuracy of the analysis. As the SNR increases, the OP performance is improved. For example, in Case 2, when SNR = 16 dB, the OP is 3×10^{-3}, and when SNR = 20 dB, the OP is decreased to 2×10^{-4}. The OP performance is also improved with a larger fading coefficient m. When SNR = 12 dB and $m = 1$, the OP is 1.8×10^{-1}, and when $m = 2$, the OP is 3×10^{-2}.

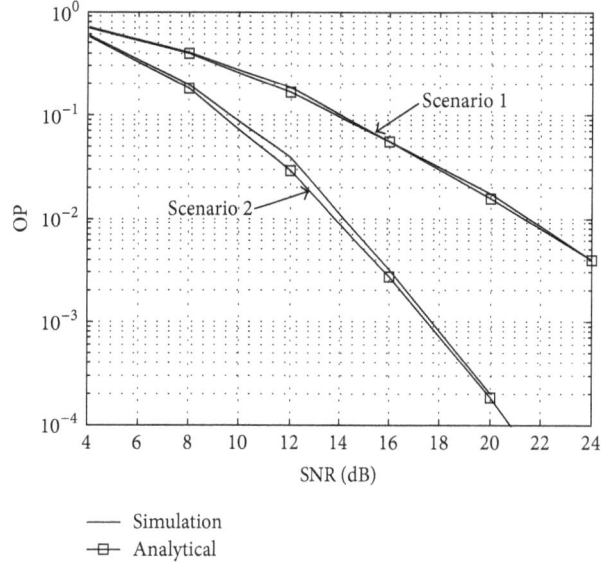

FIGURE 1: The OP performance over N-Nakagami fading channels when $\gamma_{\text{th}} > \gamma_T$.

Figure 2 presents the OP performance of the M2M cooperative network when $\gamma_{\text{th}} < \gamma_T$. The relative geometric gain is $\mu = 0$ dB, the power-allocation parameter is $K = 0.5$, and the thresholds are $\gamma_{\text{th}} = 2$ dB and $\gamma_T = 4$ dB. The following cases are considered based on the number of cascaded components N and the fading coefficient m.

Scenario 1. $m_{\text{SD}} = 1, m_{\text{SR}} = 1, m_{\text{RD}} = 1$ and $N_{\text{SD}} = 2, N_{\text{SR}} = N_{\text{RD}} = 2$.

Scenario 2. $m_{\text{SD}} = 2, m_{\text{SR}} = 2, m_{\text{RD}} = 2$ and $N_{\text{SD}} = 2, N_{\text{SR}} = N_{\text{RD}} = 2$.

Scenario 3. $m_{\text{SD}} = 3, m_{\text{SR}} = 3, m_{\text{RD}} = 3$ and $N_{\text{SD}} = 2, N_{\text{SR}} = N_{\text{RD}} = 2$.

Figure 2 shows that the numerical simulation results coincide with the theoretical results, which verifies the analysis. As the SNR increases, the OP performance improves, as expected. For example, in Case 2, when SNR = 12 dB, the OP is 1.5×10^{-2}, but when SNR = 16 dB, the OP is 1×10^{-3}. The OP performance also improves if the fading coefficient m is increased. For example, when SNR = 12 dB and $m = 1$, the OP is 1×10^{-1}, but when $m = 2$, the OP is 1.5×10^{-2}, and when $m = 3$, the OP is 2×10^{-3}.

Figure 3 presents the effect of the power-allocation parameter K on the OP performance of the M2M cooperative network over N-Nakagami fading channels versus the SNR. The number of cascaded components is $N = 2$, and the fading coefficient is $m = 2$. The relative geometric gain is $\mu = 0$ dB, and the thresholds are $\gamma_{\text{th}} = 4$ dB and $\gamma_T = 2$ dB. These results show that the OP performance is improved when the SNR is increased. For example, when $K = 0.6$ and SNR = 10 dB, the OP is 1.8×10^{-2}, when SNR = 15 dB, the OP is 9×10^{-4}, and when SNR = 20 dB, the OP is 2.5×10^{-5}. For SNR = 10 dB, the optimal value of K is approximately 0.5, for SNR = 15 dB, the

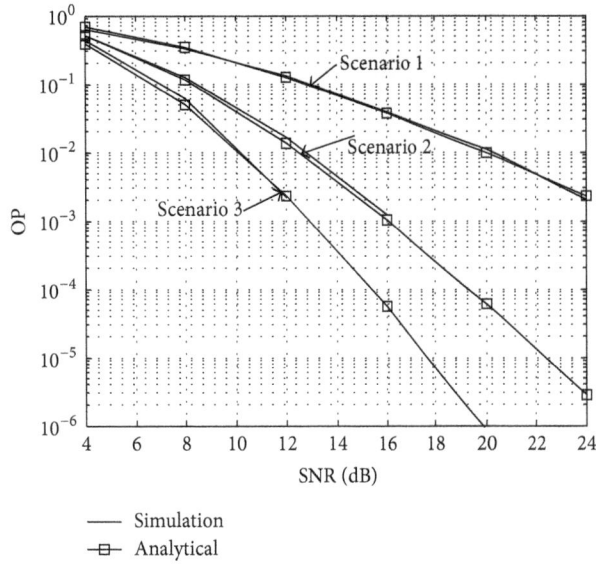

FIGURE 2: The OP performance over N-Nakagami fading channels when $\gamma_{\text{th}} < \gamma_T$.

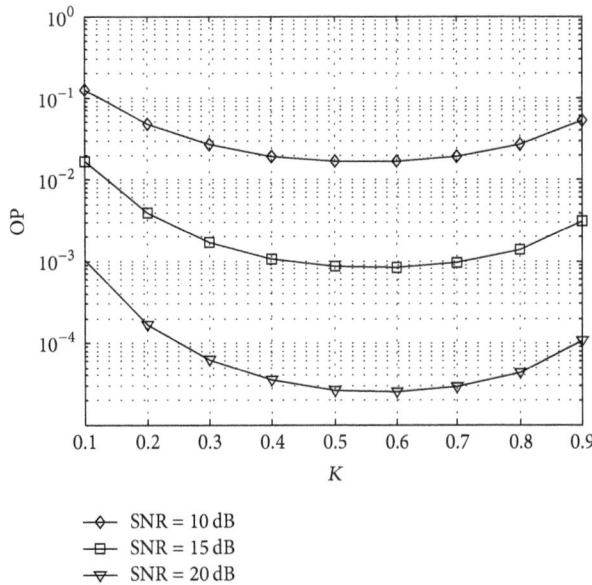

FIGURE 4: The effect of the relative geometric gain μ on the OP performance.

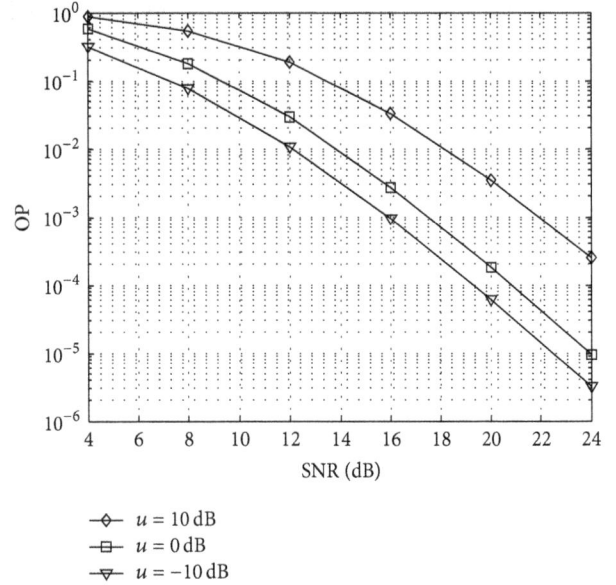

FIGURE 3: The effect of the power-allocation parameter K on the OP performance.

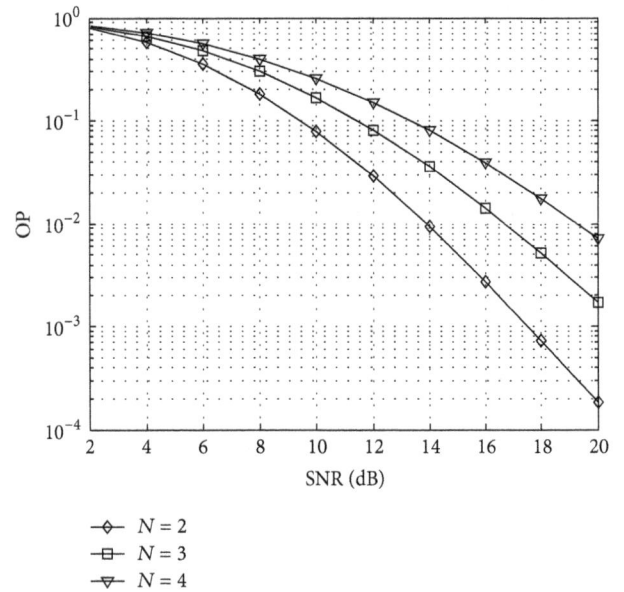

FIGURE 5: The effect of the number of cascaded components N on the OP performance.

optimal value of K is approximately 0.6, and for SNR = 20 dB, the optimal value of K is also approximately 0.6.

Figure 4 presents the effect of the relative geometric gain μ on the OP performance of the M2M cooperative network over N-Nakagami fading channels. The number of cascaded components is $N = 2$, and the fading coefficient is $m = 2$. The relative geometric gains considered are $\mu = 10$ dB, 0 dB, and -10 dB. The thresholds are $\gamma_{\text{th}} = 4$ dB and $\gamma_T = 2$ dB, and the power-allocation parameter is $K = 0.5$. These results show that the OP performance is improved as μ is reduced. For example, when SNR = 12 dB and $\mu = 10$ dB, the OP is 2×10^{-1}, when $\mu = 0$ dB the OP is 3×10^{-2}, and when $\mu = -10$ dB,

the OP is 1×10^{-2}. As the SNR increases, the OP gradually reduces.

Figure 5 presents the effect of the number of cascaded components N on the OP performance of the M2M cooperative network over N-Nakagami fading channels. The number of cascaded components is $N = 2, 3, 4$, which denote 2-Nakagami, 3-Nakagami, and 4-Nakagami fading channels, respectively. The fading coefficient is $m = 2$, the relative geometric gain is $\mu = 0$ dB, and the thresholds are $\gamma_{\text{th}} = 4$ dB and $\gamma_T = 2$ dB. The power-allocation parameter is $K = 0.5$. These results show that the OP performance is degraded as N is increased. For example, when SNR = 12 dB and $N = 2$, the OP is 3×10^{-2}, when $N = 3$ the OP is 8×10^{-2}, and when $N = 4$ the OP is 1.5×10^{-1}. This is because the fading severity for the cascaded channels increases as N is increased. For fixed N, an increase in the SNR reduces the OP gradually.

5. Conclusion

A lower bound on the outage probability (OP) of SNR-based HDAF M2M cooperative network over N-Nakagami fading channels was derived. Performance results were presented which show that the fading coefficient m, number of cascaded components N, relative geometric gain μ, and power-allocation parameter K have a significant influence on the OP. The expressions derived in this paper are simple to compute and thus complete and accurate performance results can easily be obtained with minimal computational effort. In the future, the impact of correlated channels on the OP performance of M2M cooperative networks can be considered.

Appendix

Equation (27) can be simplified as follows:

$$
\begin{aligned}
Q &= \int_0^{\gamma_{th}} \left(\int_0^{\gamma_T} f_{\gamma_{SR}}(y)\,dy - \int_0^{\gamma_{RD}} f_{\gamma_{SR}}(y)\,dy \right) f_{\gamma_{RD}}(z)\,dz \\
&\quad + \int_0^{\gamma_{th}} \left(\int_0^{\infty} f_{\gamma_{RD}}(y)\,dy - \int_0^{\gamma_{SR}} f_{\gamma_{RD}}(y)\,dy \right) f_{\gamma_{SR}}(z)\,dz \\
&= \int_0^{\gamma_{th}} \int_0^{\gamma_T} f_{\gamma_{SR}}(y)\,dy f_{\gamma_{RD}}(z)\,dz \\
&\quad - \int_0^{\gamma_{th}} \int_0^{\gamma_{RD}} f_{\gamma_{SR}}(y)\,dy f_{\gamma_{RD}}(z)\,dz \\
&\quad + \int_0^{\gamma_{th}} \int_0^{\infty} f_{\gamma_{RD}}(y)\,dy f_{\gamma_{SR}}(z)\,dz \\
&\quad - \int_0^{\gamma_{th}} \int_0^{\gamma_{SR}} f_{\gamma_{RD}}(y)\,dy f_{\gamma_{SR}}(z)\,dz \\
&= A - B + C - D.
\end{aligned}
\tag{A.1}
$$

First consider part A, which is given by

$$
\begin{aligned}
&\int_0^{\gamma_T} f_{\gamma_{SR}}(y)\,dy \\
&= \int_0^{\gamma_T} \frac{1}{y \prod_{t=1}^N \Gamma(m_t)} G_{0,N}^{N,0} \left[\frac{y}{\gamma_{SR}} \prod_{t=1}^N \frac{m_t}{\Omega_t} \Big|_{m_1,\dots,m_N}^{-} \right] dy.
\end{aligned}
\tag{A.2}
$$

To evaluate the integral in (A.2), the following integral function can be employed [13]:

$$
\begin{aligned}
&\int_0^y x^{a-1} G_{p,q}^{m,n} \left[wx \Big|_{b_1,\dots,b_q}^{a_1,\dots,a_p} \right] dx \\
&= y^a G_{p+1,q+1}^{m,n+1} \left[wy \Big|_{b_1,\dots,b_m,-a,b_{m+1},\dots,b_q}^{a_1,\dots,a_n,1-a,a_n,\dots,a_p} \right].
\end{aligned}
\tag{A.3}
$$

Equation (A.2) can then be expressed as

$$
\begin{aligned}
&\int_0^{\gamma_T} f_{\gamma_{SR}}(y)\,dy = \frac{1}{\prod_{t=1}^N \Gamma(m_t)} \\
&\times G_{1,N+1}^{N,1} \left[\frac{\gamma_T}{\gamma_{SR}} \prod_{t=1}^N \frac{m_t}{\Omega_t} \Big|_{m_1,\dots,m_N,0}^{1} \right],
\end{aligned}
\tag{A.4}
$$

so that A is given by

$$
\begin{aligned}
A &= \frac{1}{\prod_{t=1}^N \Gamma(m_t) \prod_{tt=1}^N \Gamma(m_{tt})} \\
&\times G_{1,N+1}^{N,1} \left[\frac{\gamma_T}{\gamma_{SR}} \prod_{t=1}^N \frac{m_t}{\Omega_t} \Big|_{m_1,\dots,m_N,0}^{1} \right] \\
&\times G_{1,N+1}^{N,1} \left[\frac{\gamma_{th}}{\gamma_{RD}} \prod_{tt=1}^N \frac{m_{tt}}{\Omega_{tt}} \Big|_{m_1,\dots,m_N,0}^{1} \right].
\end{aligned}
\tag{A.5}
$$

Next, consider part B. Following a procedure similar to that for (A.2) yields

$$
\begin{aligned}
&\int_0^{\gamma_{RD}} f_{\gamma_{SR}}(y)\,dy = \frac{1}{\prod_{t=1}^N \Gamma(m_t)} \\
&\times G_{1,N+1}^{N,1} \left[\frac{\gamma_{RD}}{\gamma_{SR}} \prod_{t=1}^N \frac{m_t}{\Omega_t} \Big|_{m_1,\dots,m_N,0}^{1} \right],
\end{aligned}
\tag{A.6}
$$

so that B is given by

$$
\begin{aligned}
B &= \int_0^{\gamma_{th}} \int_0^{\gamma_{RD}} f_{\gamma_{SR}}(y)\,dy f_{\gamma_{RD}}(z)\,dz \\
&= \frac{1}{\prod_{t=1}^N \Gamma(m_t) \prod_{tt=1}^N \Gamma(m_{tt})} \\
&\times \int_0^{\gamma_{th}} \frac{1}{z} G_{1,N+1}^{N,1} \left[\frac{z}{\gamma_{SR}} \prod_{t=1}^N \frac{m_t}{\Omega_t} \Big|_{m_1,\dots,m_N,0}^{1} \right] \\
&\times G_{0,N}^{N,0} \left[\frac{z}{\gamma_{RD}} \prod_{tt=1}^N \frac{m_{tt}}{\Omega_{tt}} \Big|_{m_1,\dots,m_N}^{-} \right] dz.
\end{aligned}
\tag{A.7}
$$

Next, consider part C. Since

$$
\int_0^{\infty} f_{\gamma_{RD}}(y)\,dy = F_{\gamma_{RD}}(\infty) = 1
\tag{A.8}
$$

C can be expressed as

$$C = \frac{1}{\prod_{t=1}^{N} \Gamma(m_t)} G_{1,N+1}^{N,1} \left[\frac{\gamma_{\text{th}}}{\gamma_{\text{SR}}} \prod_{t=1}^{N} \frac{m_t}{\Omega_t} \Big|_{m_1,\ldots,m_N,0}^{1} \right]. \quad \text{(A.9)}$$

Finally, consider part D. Following a procedure similar to that for (A.2) yields

$$\int_0^{\gamma_{\text{SR}}} f_{\gamma_{\text{RD}}}(y)\,dy = \frac{1}{\prod_{tt=1}^{N} \Gamma(m_{tt})}$$
$$\times G_{1,N+1}^{N,1} \left[\frac{\gamma_{\text{SR}}}{\gamma_{\text{RD}}} \prod_{tt=1}^{N} \frac{m_{tt}}{\Omega_{tt}} \Big|_{m_1,\ldots,m_N,0}^{1} \right],$$
$$\text{(A.10)}$$

so that D is given by

$$D = \int_0^{\gamma_{\text{th}}} \int_0^{\gamma_{\text{SR}}} f_{\gamma_{\text{RD}}}(y)\,dy f_{\gamma_{\text{SR}}}(z)\,dz$$
$$= \frac{1}{\prod_{t=1}^{N} \Gamma(m_t) \prod_{tt=1}^{N} \Gamma(m_{tt})}$$
$$\times \int_0^{\gamma_{\text{th}}} \frac{1}{z} G_{1,N+1}^{N,1} \left[\frac{z}{\gamma_{\text{RD}}} \prod_{tt=1}^{N} \frac{m_{tt}}{\Omega_{tt}} \Big|_{m_1,\ldots,m_N,0}^{1} \right] \quad \text{(A.11)}$$
$$\times G_{0,N}^{N,0} \left[\frac{z}{\gamma_{\text{SR}}} \prod_{t=1}^{N} \frac{m_t}{\Omega_t} \Big|_{m_1,\ldots,m_N}^{-} \right] dz.$$

Conflict of Interests

The authors declare that there is no conflict of interests regarding the publication of this paper.

Acknowledgments

The authors would like to thank the referees and editor for providing helpful comments and suggestions. This project was supported by the National Natural Science Foundation of China (no. 61304222 and no. 60902005), the Natural Science Foundation of Shandong Province (no. ZR2012FQ021), the Shandong Province Higher Educational Science and Technology Program (no. J12LN88), and the International Science and Technology Cooperation Program of Qingdao (no. 12-1-4-137-hz).

References

[1] S. Mumtaz, K. M. Saidul Huq, and J. Rodriguez, "Direct mobile-to-mobile communication: paradigm for 5G," *IEEE Wireless Communications*, vol. 21, no. 5, pp. 14–23, 2014.

[2] L. Z. Xun, H. H. Ying, R. X. Kun, and C. W. Kui, "Influences of double-Rayleigh transmission system performance," *Journal of XiDian University*, vol. 38, no. 5, pp. 172–177, 2011.

[3] H. Ilhan, I. Altunbaş, and M. Uysal, "Moment generating function-based performance evaluation of amplify-and-forward relaying in N* Nakagami fading channels," *IET Communications*, vol. 5, no. 3, pp. 253–263, 2011.

[4] G. K. Karagiannidis, N. C. Sagias, and P. T. Mathiopoulos, "N* Nakagami: a novel stochastic model for cascaded fading channels," *IEEE Transactions on Communications*, vol. 55, no. 8, pp. 1453–1458, 2007.

[5] H. Ilhan, M. Uysal, and I. Altunbaş, "Cooperative diversity for intervehicular communication: performance analysis and optimization," *IEEE Transactions on Vehicular Technology*, vol. 58, no. 7, pp. 3301–3310, 2009.

[6] F. K. Gong, J. Ge, and N. Zhang, "SER analysis of the mobile-relay-based M2M communication over double Nakagami-m fading channels," *IEEE Communications Letters*, vol. 15, no. 1, pp. 34–36, 2011.

[7] F. K. Gong, P. Ye, Y. Wang, and N. Zhang, "Cooperative mobile-to-mobile communications over double Nakagami-m fading channels," *IET Communications*, vol. 6, no. 18, pp. 3165–3175, 2012.

[8] T. Q. Duong and H.-J. Zepernick, "On the performance gain of hybrid decode-amplify-forward cooperative communications," *Eurasip Journal on Wireless Communications and Networking*, vol. 2009, Article ID 479463, 10 pages, 2009.

[9] H. Chen, J. Liu, C. Zhai, and L. Zheng, "Performance analysis of SNR-based hybrid decode-amplify-forward cooperative diversity networks over rayleigh fading channels," in *Proceedings of the IEEE Wireless Communications and Networking Conference*, pp. 1–6, Sydney, Australia, April 2010.

[10] W. F. Su, A. K. Sadek, and K. J. R. Liu, "Cooperative communication protocols in wireless networks: performance analysis and optimum power allocation," *Wireless Personal Communications*, vol. 44, no. 2, pp. 181–217, 2008.

[11] P. A. Anghel and M. Kaveh, "Exact symbol error probability of a cooperative network in a rayleigh-fading environment," *IEEE Transactions on Wireless Communications*, vol. 3, no. 5, pp. 1416–1421, 2004.

[12] A. Papoulis and S. U. Pillai, *Random Variables and Stochastic Processes*, McGraw-Hill, New York, NY, USA, 4th edition, 2002.

[13] A. M. Mathai and R. K. Saxena, *Generalized Hypergeometric Functions with Applications in Statistics and Physical Sciences*, Springer, New York, NY, USA, 1973.

An Efficient Node Localization Approach with RSSI for Randomly Deployed Wireless Sensor Networks

Xihai Zhang, Junlong Fang, and Fanfeng Meng

School of Electrical and Information Engineering, Northeast Agricultural University, Harbin 150030, China

Correspondence should be addressed to Junlong Fang; jlfang@neau.edu.cn

Academic Editor: Iickho Song

An efficient path planning approach in mobile beacon localization for the randomly deployed wireless sensor nodes is proposed in this paper. Firstly, in order to improve localization accuracy, the weighting function based on the distance between nodes is constructed. Moreover, an iterative multilateration algorithm is also presented to avoid decreasing the localization accuracy. Furthermore, a path planning algorithm based on grid scan which can traverse entirely in sensor field is described. At the same time, the start conditions of localization algorithm are also proposed to improve localization accuracy. To evaluate the proposed path planning algorithm, the localization results of beacon nodes randomly deployed in sensor field are also provided. The proposed approach can provide the deployment uniformly of virtual beacon nodes among the sensor fields and the lower computational complexity of path planning compared with method which utilizes only mobile beacons on the basis of a random movement. The performance evaluation shows that the proposed approach can reduce the beacon movement distance and the number of virtual mobile beacon nodes by comparison with other methods.

1. Introduction

Wireless sensor networks (WSNs) are closely associated with the physical phenomena in their surroundings. The gathered information needs to be associated with the position of sensor nodes to provide an accurate view of the observed sensor field. The localization is important in most applications, such as environment sensing, search and rescue, and geographical routing and tracking; the position of each node should be known [1]. These requirements motivate the development of efficient localization algorithms for WSNs.

Over the course of the past decade, there have been a large number of researches on localization for WSNs [2–8]. They share the same main idea that nodes with unknown coordinates are utilized by one or more GPS-equipped nodes with known coordinates in order to estimate their positions. Most of these works consider the static beacon. While GPS provides highly accurate location information, it may not be feasible for most randomly deployed WSNs. Firstly, GPS available for WSNs are very costly, exceeding the cost of a sensor node. Moreover, GPS operation has a high energy consumption profile, which may impose additional constraints

on the lifetime of WSNs. Furthermore, WSNs are usually static and localization algorithms may be required to run only during initialization of the network. Consequently, GPS operation may not be cost-effective for many WSNs realizations.

Therefore, to obtain location information, we need a technique which incurs lesser cost and provides more accurate location. A promising method to localize randomly deployed WSNs is to use one mobile beacon [9–16]. The localization approach using mobile beacon utilizes a beacon node equipped with GPS to traverse the region of interest (ROI). This beacon node broadcasts periodically the packets including its position, and unknown nodes estimate their positions using the received packets. The problem of localization by mobile beacon in WSNs has attracted extensive interest in the literatures [17–24]. Sichitiu and Ramadurai [12] propose a range-based localization method in which the sensor nodes estimate their positions by applying an RSSI technique. Xia and Chen [19] propose a TDOA-based localization scheme with mobile anchors in which the sensor nodes perform trilateration to estimate their positions. Galstyan et al. [11] propose a range-free mobile anchor-based localization scheme based on radio connectivity constraints to reduce the uncertainty of the estimated sensor location. Dong and Severance [23] develop an iterative localization approach based on the mobile anchor scheme in which the localization accuracy is progressively improved each time a new beacon message is obtained from an anchor node. Kim and Lee [24] propose a novel range-based localization scheme which involves a movement strategy with a low computational complexity of mobile beacon, called mobile beacon-assisted localization (MBAL). MBAL also applies RSSI for ranging to get the distance between nodes or between each node and the mobile beacon to assist localization of all nodes.

However, all approaches of mobile-assisted localization face the same problem, which is the optimum beacon path selection problem. Notice that the problem is quite difficult since the position of unknown nodes is not known. For example, some approaches [12] do not propose any specific movement strategy for the path of mobile beacon, and some approaches [25] just suggest that the beacon moves in the straight lines and then a loop. The random way point and Monte Carlo localization mobile model are utilized in most of localization algorithms based on a mobile beacon [26]. However, for those mobile models, the uniform deployment of virtual beacon in the sensor field is hard to realize, which results in emergence of localization blind area and then reduces localization accuracy. Although much effort is being spent on improving these weaknesses, the effective method has yet to be developed.

Therefore, a path planning algorithm based on grid scan which is to traverse entirely in sensor field is proposed in this paper. In order to make the beacon nodes closer to the unknown node have greater weight and to improve the localization accuracy, the weighting function is constructed based on the distance between the nodes. Furthermore, an iterative multilateration algorithm is also proposed to avoid decrease in the localization accuracy when the unknown node position is estimated. At the same time, the start conditions of localization algorithm are also proposed to improve localization accuracy. To evaluate the path planning algorithm based grid scan, the results of static beacon randomly deployed and RWP mobile path in sensor field are also provided in this paper.

2. Localization in Sensor Network

2.1. RSSI. Generally, the localization algorithms have been proposed, which can be mainly classified into two categories: range-based and range-free. Several range-based techniques estimate an unknown node distance by three or more beacon nodes. Based on the range information, the location of a node is determined. Some of the range-based localization algorithms include received signal strength indicator (RSSI), angle of arrival (AOA), time of arrival (TOA), and time difference of arrival (TDOA) [27]. On the other hand, range-free algorithms, such as Centroid Localization Algorithm, Distance Vector-Hop (DV-Hop), Approximate Point-in-Triangulation Test (APIT), and Rendered Path, only use the connectivity or proximity information to localize the unknown nodes. The most common range-based technique

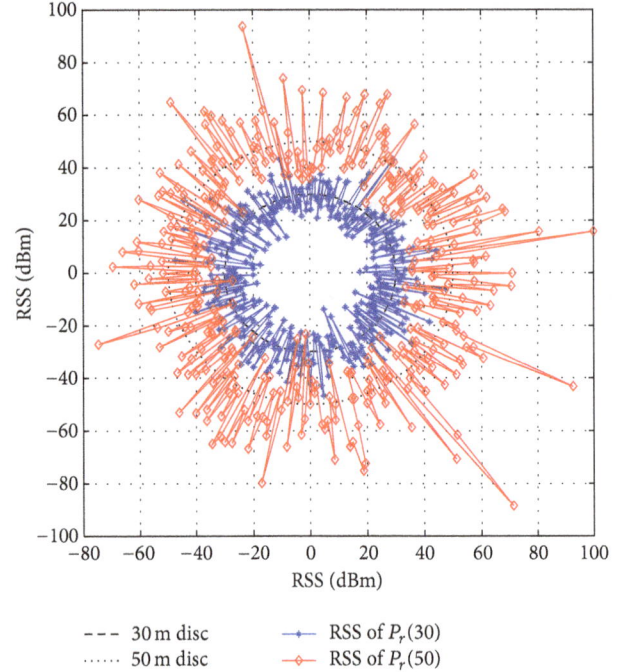

FIGURE 1: Relationship between RSS and distance.

is based on RSSI measurements. Since each sensor node is equipped with a radio and in most cases is able to send the received signal strength of an incoming packet, the main idea is to estimate the distance of a transmitter to a receiver using the power of received signal, knowledge of the transmitted power, and the path loss model.

In this scheme, the beacon node broadcasts periodically the messages including its own position, which are used to estimate the distance from the beacon node to unknown node. The power of the received signal is communicated by the transceiver circuitry through the RSSI. The received signal strength from sensor node i at node j at time t is represented by $P_r(d)$, which is formulated as

$$P_r(d) = P_T - \text{PL}(d_0) - 10\eta\log_{10}\left(\frac{d}{d_0}\right) + X_\delta, \qquad (1)$$

where $P_r(d)$ is the received signal power, P_T is the transmit power, $\text{PL}(d_0)$ is the path loss for reference distance of d_0, η is the attenuation constant, and $X_\delta = N(0, \delta^2)$ is the uncertainty factor due to multipath and shadowing. Generally, the typical value of parameters is as follows: $P_T = 0 \sim 4$ dBm, $\text{PL}(d_0) = 55$ dB ($d_0 = 1$ m), $\eta = (2, 4)$, and $\delta = (4, 10)$.

The accuracy of the RSSI-based ranging technique is limited. Firstly, the effects of shadowing and multipath as modeled by the term X_δ in (1) may be severe and require multiple ranging measurements. For example, the parameters of (1) are as follows: $P_T = 0$, $\text{PL}(d_0) = 55$ dB, $d_0 = 1$ m, $\eta = 4$, $\delta = 4$, and $d = 30$ m, 50 m, which is applied in Section 4; then the relationship between the RSS and the node distance is illustrated in Figure 1. It shows that the receiving node which is far away from the sending node has relatively large impact on X_δ. Therefore, the estimated distance based on RSSI has

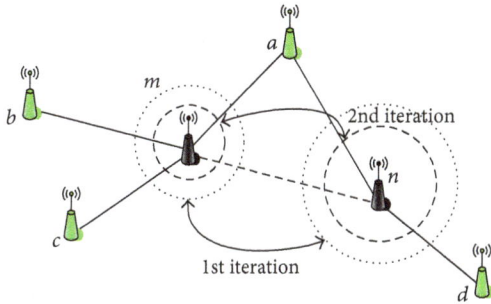

FIGURE 2: Iterative multilateration.

features such as the localization accuracy being high for the relatively near distance of nodes and the localization error being large for the relatively far distance of nodes.

Hence, in order to make the beacon nodes closer to the unknown nodes have greater weight and improve the localization accuracy of unknown node, the weighting function $(w(n) = 1/d_{ij}^2$, with d_{ij} being the distance from node i to node j) is constructed based on the distance between the nodes in this paper.

2.2. Iterative Multilateration Algorithm. Range-based localization techniques exploit multiple pairwise range measurements to estimate the locations of unknown nodes. Generally, three mathematical techniques are used for calculating the position of a receiver from signals received from several transmitters: triangulation, trilateration, and maximum likelihood multilateration. The triangulation allows the unknown node to calculate its position by measuring two directions towards two beacon nodes. Since the positions of beacon nodes are known, it is therefore possible to construct a triangle where one of the sides and two of the angles are known, with the unknown node at the third point. This information is enough to define the triangle completely and hence deduces the position of unknown node. The trilateration requires the distance between the receiver and transmitter to be measured. This can be done using a received signal strength indicator (RSSI), ToA, and so forth. The position of unknown node is estimated by three pairwise distances from the unknown node to different beacon nodes. Multiple range measurements between a node and its different neighbors can be used to improve the localization accuracy. The trilateration technique fails to provide an accurate estimate of position if the distance measurements are noisy. Instead, maximum likelihood estimation method is necessary to incorporate distance measurements from multiple neighbor nodes. However, usually, WSNs are deployed with a limited number of beacon nodes. Furthermore, because of ranging errors, more than three beacon nodes are required for accurate location estimation. Hence, the unknown node location can be estimated through the location estimates of neighbor nodes. Therefore, a multihop network of sensors can be localized with the help of a subset of beacon nodes. This procedure is called iterative multilateration.

Iterative multilateration is illustrated in Figure 2, where two unknown nodes m and n are surrounded by four beacon nodes, a, b, c, and d. Note that node m is surrounded by three beacon nodes (a, b, and c), whereas node n has only two beacon nodes (a and d) as neighbors. Since node m can communicate with at least three beacon nodes, the multi-lateration localization can be used to estimate its position. However, multilateration cannot be applied to node n, which requires additional distance information to a node with known location. In this case, iterative multilateration is used, where node m first estimates its position and becomes a beacon node. Using the additional information from node m, node n can perform multilateration and calculate its position. This information can in turn be utilized by node m to further improve the accuracy of its estimation. As a result, the uncertainty in location estimation as shown by the circles around nodes m and n can be decreased at each iteration. Hence, for improving localization accuracy, iterative multilateration algorithm is used in this paper.

3. Path Planning Algorithm Based on Grid Scan

The range-based localization algorithms are discussed so far by utilizing ranging measurements from nodes with fixed locations. Consequently, the localization accuracy is inherently limited depending on the network topology and the deployment strategy. Since a uniform deployment of beacon nodes is not feasible in practice, some portions of the network may have a lower density of beacon and the neighbors of a node may not be sufficient.

For improving the localization accuracy and reducing the cost of network construction, a localization technique based on a mobile beacon is proposed in [24, 28], but the mobile beacon path planning is not referred to. Next, a path planning algorithm based on grid scan which is the entire traverse in sensor field is proposed in this paper.

The sensor field is divided by the grid area which is shown in Figure 3. And the area of a grid cell $s_g = l_g^2$, where l_g denotes the edge length of a grid cell. The virtual beacon nodes are deployed when the mobile beacon node is moving at each vertex of cell grid. Hence, the mobile beacon path planning based on grid scan meets the requirements of complete coverage of sensor field. When the maximum communication radius of mobile beacon is defined, the quantity demand of virtual beacon nodes around the unknown node can be met by adjusting the edge length of a grid cell.

For the path planning based on grid scan, the mobile beacon is required through all vertexes of the grid cell. Here, the entire traverse path of the mobile beacon is denoted by the undirected graph TG = (V, E),

$$V = \{v_i\}, \quad v_i = \{v_{ij}\};$$
$$E = \{e^{ij}\}, \quad e^{ij} = \{e_{uv}^{ij}\}, \tag{2}$$

where $i = 1, \ldots, m, j = 1, \ldots, n, u = 1, \ldots, m, v = 1, \ldots, n, V$ denotes the traversing of the vertex set and the edge set, and E denotes the distance set of the current vertex to other vertexes. The entire traverse path depended on the deploy cycle of the virtual beacon nodes and the localization speed.

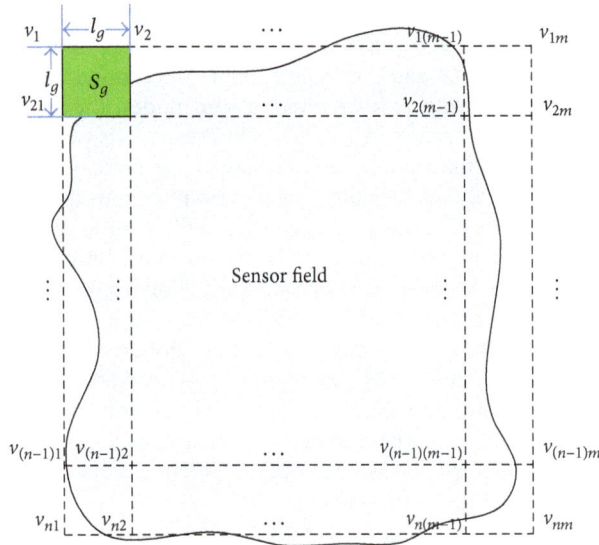

FIGURE 3: Grid cells of sensor field.

In this paper, $P_{\min}(V)$ denotes the shortest entire traverse path of the mobile beacon node; then

$$\text{dis}\left(P_{\min}(V)\right) = \sum_{e \in E} w(e), \tag{3}$$

where $w(e)$ denotes the distance between the two grid vertexes where the mobile beacon node has been passed.

Theorem 1. *For the mobile beacon path planning based on grid scan, with the shortest traverse path length of the mobile beacon node the following holds:*

$$\text{dis}\left(P_{\min}(V)\right) = (m \cdot n - 1)\, l_g. \tag{4}$$

Proof. if $P(V)$ denotes one of some entire traverse paths, then this path can be taken as one path tree, and the following holds:

$$\xi(P(V)) = v(P(V)) - 1, \tag{5}$$

where $\xi(P(V))$ is the number of edges, $v(P(V))$ is the number of vertexes, and $v(P(V)) = m \cdot n$. Then $\text{dis}(P(V))$ can be denoted as

$$\text{dis}(P(V)) = \sum_{k=1}^{\xi(P(V))} w(e_k) \geq \xi(P(V))(w_{\min}(e_k)), \tag{6}$$

where $e_k \in E$ and $w_{\min}(e_k)$ denotes the minimum value of $e_k (w_{\min}(e_k) = l_g)$. By substituting (5) and $w_{\min}(e_k)$ into (6), we can obtain

$$\text{dis}(P(V)) = (m \cdot n - 1)\, l_g; \tag{7}$$

then $\text{dis}(P_{\min}(V)) = (m \cdot n - 1)l_g$; the proof is completed.

There are many ways to obtain entire traverse path to hold in formula (4). However, a shortest entire traverse path based on grid scan can reduce the complexity of mobile

beacon node's implementation. For the entire traverse path of the mobile beacon in sensor field, this algorithm has adaptability to irregular sensor field and the better effectiveness of location. At the same time, the realization of algorithm is easy by adjusting fewer parameters. Therefore, a shortest entire traverse path based on grid scan is proposed in this paper. However, in the process of positioning, due to the longer deployment period of virtual beacon node, the localization time is longer than that of the static beacon node localization algorithm. Because the unknown nodes are likely to not receive timely all the reference localization information from some neighbor beacons, the unknown nodes begin to estimate positions for themselves, which can cause some loss of localization accuracy. To receive the more reference localization information from the neighbor beacon nodes and improve the localization accuracy, we set the following two conditions of localization start.

Condition 1. Consider

$$n_{\text{vitual beacon}} \geq k \cdot n_{\text{vitual beacon\_threshold}} \quad (k > 1). \tag{8}$$

Condition 2. Consider

$$n_{\text{all}} \geq n_{\text{threshold}},$$

$$n_{\text{vitual beacon}} \geq n_{\text{vitual beacon\_threshold}}, \tag{9}$$

where $n_{\text{vitual beacon}}$ is the number of virtual beacon nodes to be received by unknown nodes, $n_{\text{vitual beacon\_threshold}}$ is the threshold of virtual beacon to be set, $n_{\text{all}} = n_{\text{vitual beacon}} + n_{\text{unknown node}}$ is the total number of reference nodes, $n_{\text{threshold}}$ is the threshold of reference nodes to be set, and $n_{\text{unknown node}}$ is the total numbers of unknown nodes which have been positioned and received by the nodes to be positioned.

If the total number of neighbor beacon nodes is more than $n_{\text{vitual beacon\_threshold}}$, the unknown nodes begin to estimate their location according to Condition 1. Then after the location of those nodes is obtained, they begin to send the packets including their locations. After the unknown nodes which do not meet the conditions of localization having received previous packets including location, they begin to estimate their location according to Condition 2.

4. Simulation Experiments

In this section, simulation experiments are conducted to study the accuracy of path planning algorithm based on grid scan in Matlab. For this comparison, we illustrate the results of two localization approaches in terms of the average number of neighbor beacon nodes, the number of unresolved localization nodes, and localization error, that is, the approach of static beacon randomly deployed and the approach of RWP mobile model. Unless otherwise noted, a rectangular 100 m × 100 m sensor field is selected, where nodes are deployed randomly in the simulation experiments.

4.1. Localization Experiments by Beacons Randomly Deployed. In the following, we present the impact of beacon node's number on the positioning performance. The neighbor relationship of nodes is shown in Figures 4(a), 4(c), and 4(e) and

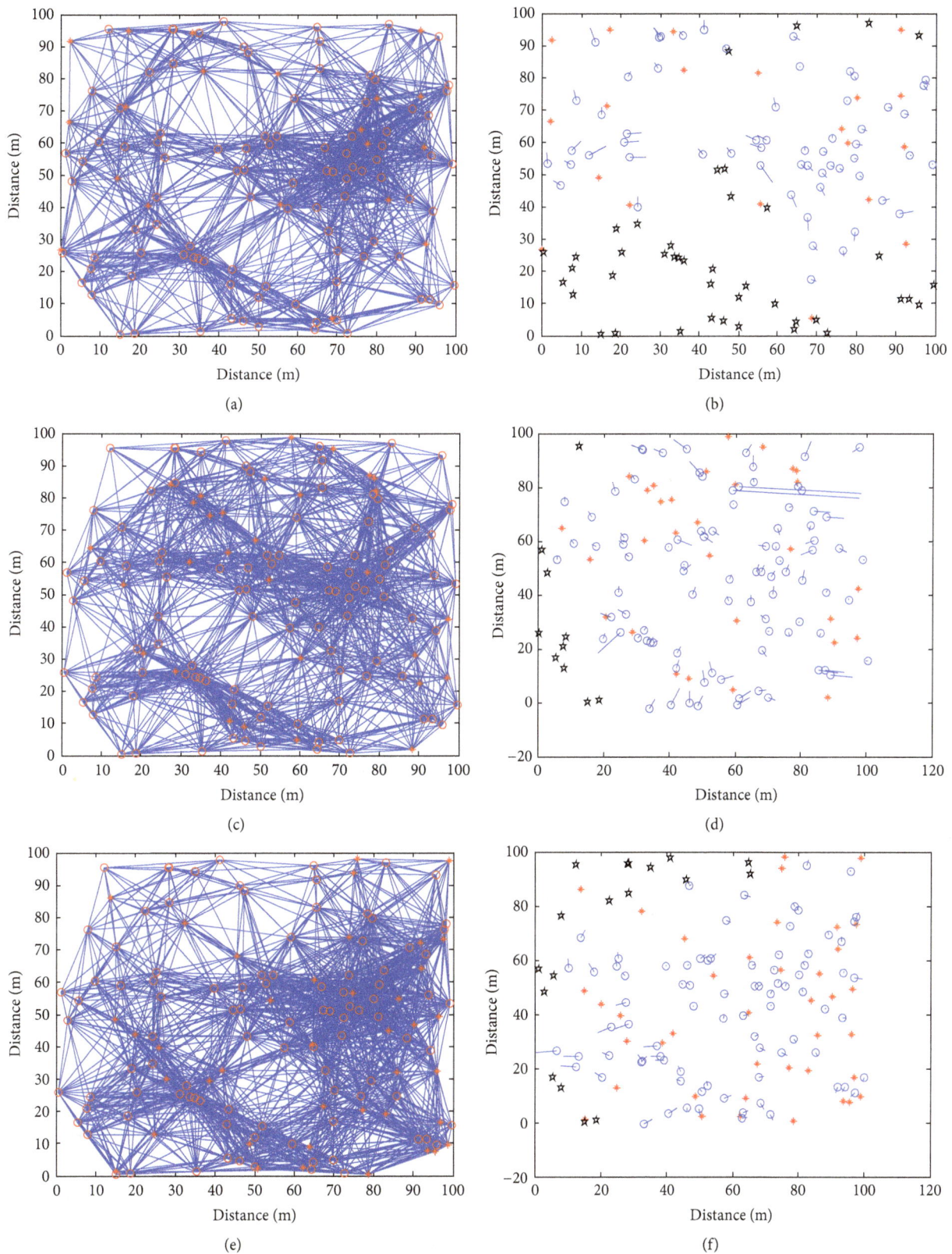

FIGURE 4: Connectivity and localization effect ((a), (c), and (e) are the neighbor relationship of nodes for the beacon node's number of 20, 30, and 40, resp. (b), (d), and (f) are the localization error for the beacon node's number of 20, 30, and 40, resp. The communication radius is always 30 m).

TABLE 1: Contrast of two kinds of localization scheme.

Deployment pattern of beacon node	Average connectivity of network	Average number of neighbor beacon nodes	Nodes of unresolved localization	Localization error
Case 1: beacon randomly deployed (communication radius = 30 m; beacon nodes = 25)	25.008	4.216	28	0.12441
Case 2: mobile beacon node (communication radius = 30 m; number of beacon nodes = 25)	26.208	4.992	1	0.08767

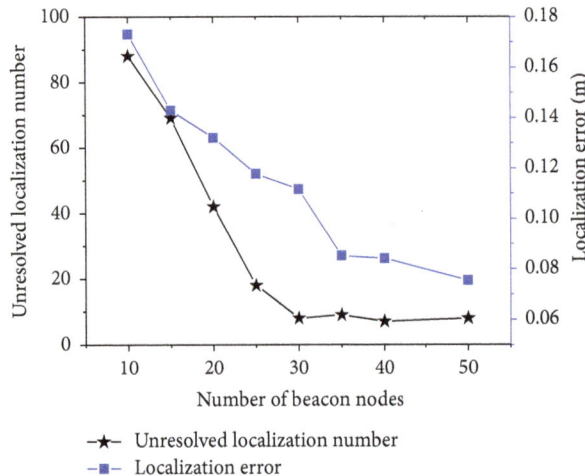

FIGURE 5: Relationship between localization effect and number of beacon nodes.

the localization error in Figures 4(b), 4(d), and 4(f) for the beacon node's number of 20, 30, and 40, respectively. The node's communication radius is set to 30 m. In this paper, unless otherwise noted, the dot "O" represents the actual position of unknown node, the asterisk "∗" represents the position of beacon node, and the pentagram "☆" represents the unresolved localization node. Those figures clearly show that the beacon node's number affects significantly the connectivity and localization effect.

The relationship of the localization error and the unresolved localization number at different numbers of beacon is shown in Figure 5. As shown in Figure 5, at 10 beacon nodes, the 88 unknown nodes are not located successfully and the localization error is also above 0.17. Increasing the beacon nodes number results in a rapid decrease in localization error and the unresolved localization number. However, when the beacon nodes exceed 30, the unresolved localization number tends to be stabilized. Even though the beacon nodes reach 50, 10% nodes are not located successfully still. The reason is that the beacon nodes are not deployed uniformly among the sensor fields. Hence, in those simulation conditions, to illustrate the advantages of mobile beacons, we can set 25 virtual beacon nodes in the proposed mobile path.

We present the impact of beacon node's number on the positioning performance.

Second, we present the impact of communication radius on the positioning performance. The localization error and the neighbor relationship of nodes for different communication radius are shown in Figure 6. The neighbor relationship of nodes is shown in Figures 6(a), 6(c), and 6(e) and the localization error in Figures 6(b), 6(d), and 6(f), for the communication radius of 20, 30, and 40, respectively. The number of beacon nodes is set to 30 m. It is observed that the communication radius is heavily affected by the connectivity and localization effect.

The relationship of the localization error and the number of unresolved localization nodes at different communication radius is shown in Figure 7. As shown in Figure 7, at 10 m communication radius, all the unknown nodes are not located successfully. Increasing the communication radius results in a rapid decrease in localization error and the unresolved localization number. However, when the communication radius exceeds 40 m, all the unknown nodes are located successfully and the localization error tends to be stabilized. Therefore, in these simulation conditions, the communication radius is set 30 m in the proposed mobile path in order to reduce power consumption of node.

4.2. Localization Experiments by a Mobile Beacon. In this section, according to preliminary experiments by beacons randomly deployed, the advantage of path planning algorithm based on a mobile beacon is illustrated.

Firstly, to illustrate the localization performance of proposed scheme by a mobile beacon, the localization method by beacon deployment randomly is also presented in simulation. In both cases, we all set 25 virtual beacon nodes and 30 m communication radius with the same unknown nodes in the same size sensor field. According to Section 3, the scan line divides the square deployment area into n by n subsquares ($n = 4$, $l_g = 5$ m in our case) and connects their vertexes using straight lines. Therefore, the 25 virtual beacon nodes are deployed uniformly in the sensor field. The node deployment is shown in Figures 8(a) and 8(b), the neighbor relationship of nodes is shown in Figures 8(c) and 8(d), and the localization error is shown in Figures 8(e) and 8(f).

The results are summarized in Table 1. In both cases, the average connectivity of network is little different. The average neighbor beacon number (Case 2) is obviously better than localization scheme by beacon deployment randomly. The nodes number of unresolved localization is only one

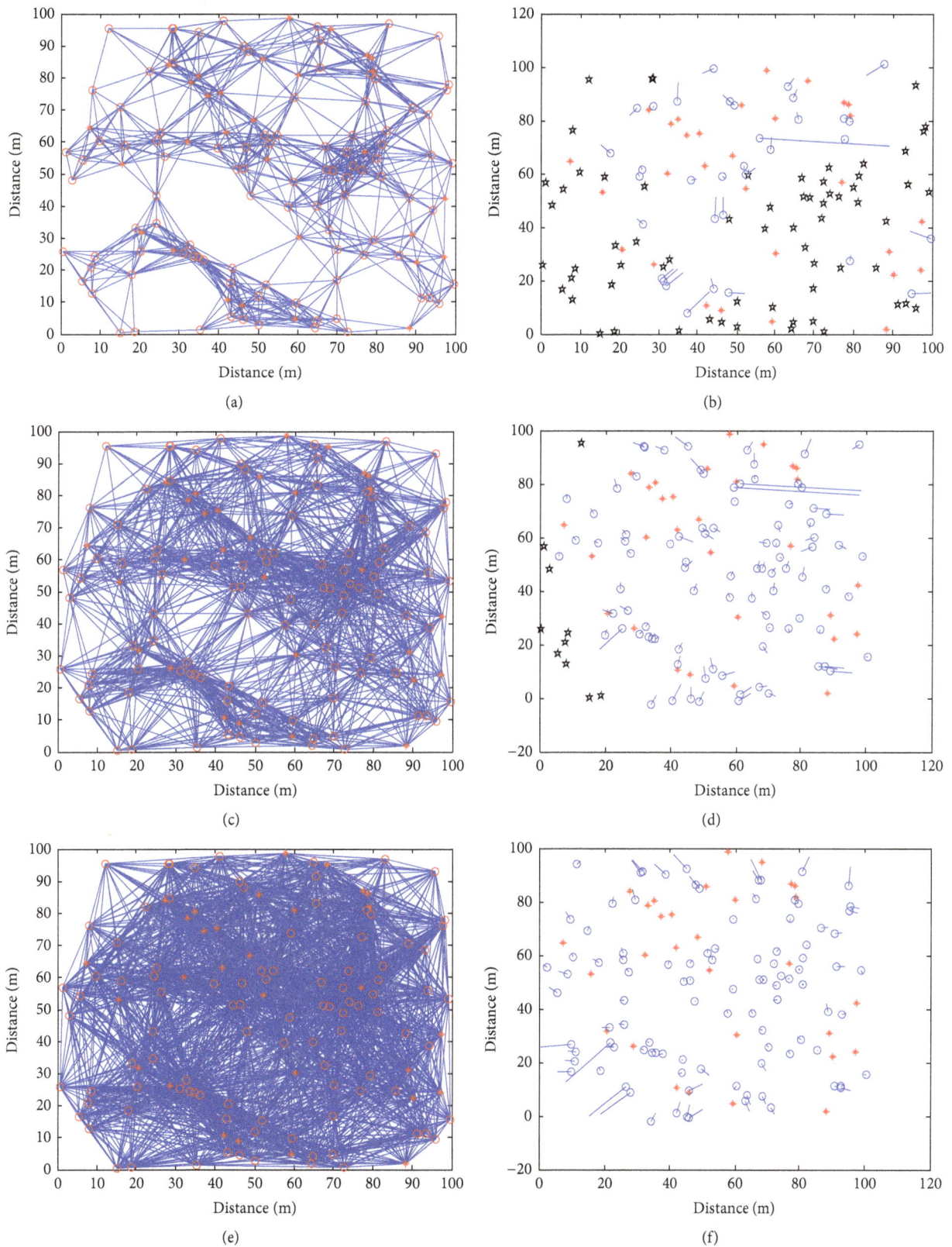

FIGURE 6: Connectivity and localization effect ((a), (c), and (e) are the neighbor relationship of nodes for the communication radius of 20 m, 30 m, and 40 m, resp. (b), (d), and (f) are the localization error for the communication radius of 20 m, 30 m, and 40 m, resp. The number of beacon nodes is always 30).

TABLE 2: Contrast of difference mobile paths.

Different mobile path planning	Distance of movement	Nodes of unresolved localization	Localization error
Case 3: path planning algorithm of grid scan (communication radius = 30 m; number of beacon nodes = 25)	600	1	0.08767
Case 4: random way point mobile path (communication radius = 30 m; number of beacon nodes = 25)	329	42	0.38701
Case 5: random way point mobile path (communication radius = 30 m; number of beacon nodes = 50)	783	7	0.16785

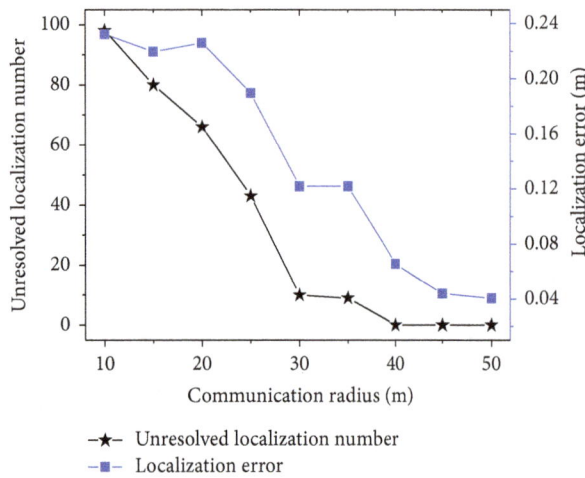

FIGURE 7: Relationship between localization effect and communication radius.

for scheme proposed, whereas there are 28 unresolved localization nodes for (Case 1). Localization error of proposed scheme (Case 2) is obviously lower than localization by beacon deployment randomly. Therefore, it is observed that proposed scheme by a mobile beacon is significantly better than localization scheme by beacon deployment randomly in the connectivity and localization effect in this paper.

Second, to illustrate the localization performance of proposed scheme, we also study the localization scheme by a mobile beacon, compared to the RWP mobile path. For Cases 3 and 4 of Table 2, we set 25 virtual beacon nodes and 30 m communication radius with the same unknown nodes in the same size sensor field. Scan line divides the square deployment area into n by n subsquares ($n = 4$, $l_g = 5$ m in our case) and connects their vertexes using straight lines as in Figure 9(a). Therefore, the 25 virtual beacon nodes are deployed uniformly in sensor field. To illustrate the advantages of grid scan, we also set 25 nodes (Case 4) and 50 nodes (Case 5) in the RWP mobile path. The beacon mobile path is shown in Figures 9(d) and 9(g), the neighbor relationship of nodes is shown in Figures 9(b), 9(e), and 9(h), and the localization error is shown in Figures 9(c), 9(f), and 9(i).

The results are summarized in Table 2. The nodes number of unresolved localization is only one for scheme proposed (Case 3), whereas there are 42 and 7 unresolved localization nodes for Cases 4 and 5, respectively. Although RWP scheme can reduce the number of unresolved localization nodes by improving mobile steps, it will cause an increase in the localization time and more network energy cost. The reason is that the virtual beacon nodes are not deployed uniformly among the sensor fields. Localization error of proposed scheme (Case 3) is obviously lower than RWP scheme (Cases 4 and 5).

5. Conclusions

In this work, a path planning algorithm based on grid scan which is the entire traverse in sensor field is proposed. In order to improve the localization accuracy, the weighting function is constructed based on the distance between the nodes. Furthermore, an iterative multilateration algorithm is also proposed to avoid decrease in the localization accuracy. At the same time, the start conditions of localization algorithm are also proposed. To evaluate the proposed path planning algorithm, the results of the static beacon randomly deployed and RWP mobile path in sensor field are also provided. It is obtained that proposed scheme by a mobile beacon is significantly better than localization scheme by beacon deployment randomly in localization effects.

Competing Interests

The authors declare that they have no competing interests.

Acknowledgments

This research was supported by the National Natural Science Foundation of China (no. 31101080), the China Postdoctoral Science Foundation (no. 2015M580254), the Heilongjiang Postdoctoral Science Foundation (no. LBH-Z15011), Harbin Science and Technology Innovation Youth Talents Special Fund (no. 2015RQQXJ094), "Academic Backbone" Project of Northeast Agricultural University (no. 15XG12), and Northeast Agricultural University Doctoral Start-Up Fund (no.

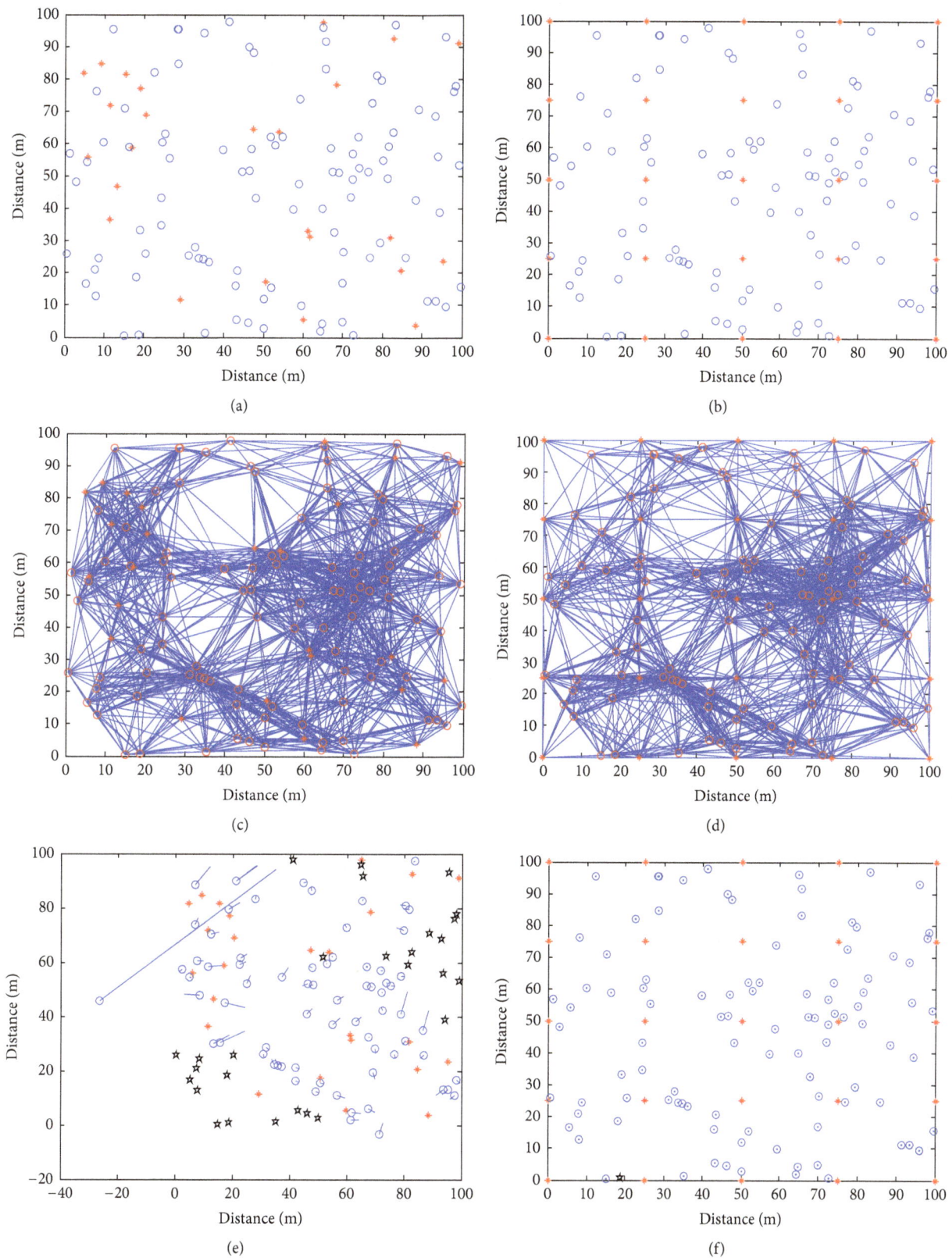

FIGURE 8: Connectivity and localization effect ((a) and (b) are the node deployment; (c) and (d) are the neighbor relationship of nodes; (e) and (f) are the localization error, in conditions of random deployment and grid scan, resp.).

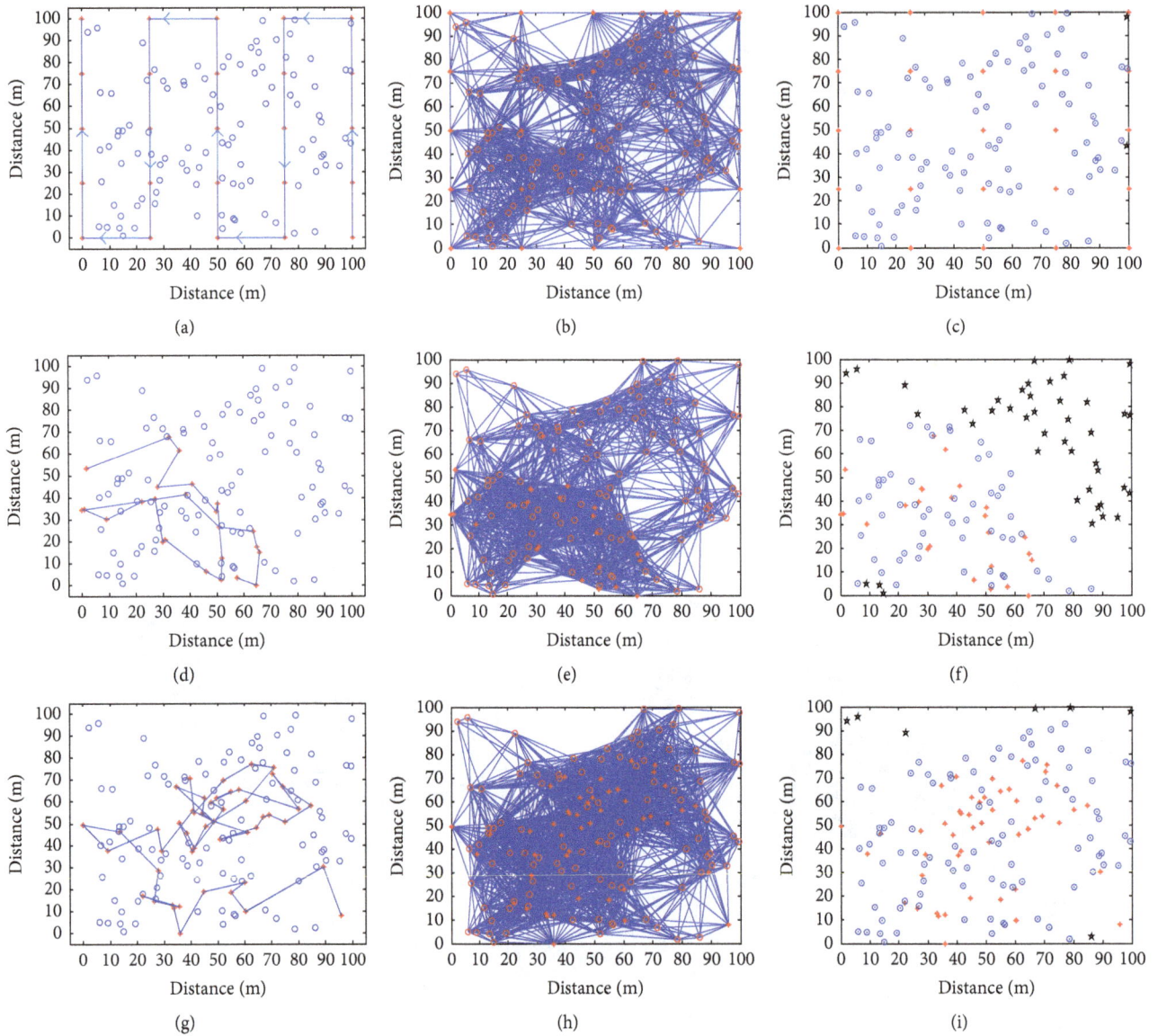

FIGURE 9: Contrast of difference mobile paths ((a) is the mobile path of grid scan in 25 virtual nodes. (d) and (g) are the mobile path with RWP in 25 and 50 virtual nodes, resp. (b), (e), and (h) are the neighbor relationship of nodes in conditions of grid scan and RWP, resp. (c), (f), and (i) are the localization error in conditions of grid scan and RWP resp. The communication radius is always 30 m).

2012RCB51). The authors thank Chengguo Fan, Jiali Du, and Dong Liu of Northeast Agricultural University for their work in revising the paper.

References

[1] K. Sohraby, D. Minoli, and T. Znati, *Wireless Sensor Networks: Technology, Protocols, and Applications*, John Wiley & Sons, London, UK, 2007.

[2] P. Bahl and V. N. Padmanabhan, "Radar: an in-building RF-based user location and tracking system," in *Proceedings of IEEE INFOCOM*, pp. 775–784, March 2000.

[3] A. Vempaty, O. Ozdemir, K. Agrawal, H. Chen, and P. K. Varshney, "Localization in wireless sensor networks: byzantines and

mitigation techniques," *IEEE Transactions on Signal Processing*, vol. 61, no. 6, pp. 1495–1508, 2013.

[4] N. Bulusu, J. Heidemann, and D. Estrin, "GPS-less low-cost outdoor localization for very small devices," *IEEE Personal Communications*, vol. 7, no. 5, pp. 28–34, 2000.

[5] J. M. Pak, C. K. Ahn, Y. S. Shmaliy, and M. T. Lim, "Improving reliability of particle filter-based localization in wireless sensor networks via hybrid particle/FIR filtering," *IEEE Transactions on Industrial Informatics*, vol. 11, no. 5, pp. 1089–1098, 2015.

[6] A. Maddumabandara, H. Leung, and M. Liu, "Experimental evaluation of indoor localization using wireless sensor networks," *IEEE Sensors Journal*, vol. 15, no. 9, pp. 5228–5237, 2015.

[7] A. Savvides, C. C. Han, and M. B. Strivastava, "Dynamic fine-grained localization in ad-hoc networks of sensors," in *Proceedings of the 7th Annual International Conference on Mobile

Computing and Networking, pp. 166–179, ACM, Rome, Italy, July 2001.

[8] Y. Shang and W. Ruml, "Improved MDS-based localization," in *Proceedings of the 3rd Annual Joint Conference of the IEEE Computer and Communications Societies (INFOCOM '04)*, pp. 2640–2651, IEEE, Hong Kong, China, March 2004.

[9] P. N. Pathirana, N. Bulusu, A. V. Savkin, and S. Jha, "Node localization using mobile robots in delay-tolerant sensor networks," *IEEE Transactions on Mobile Computing*, vol. 4, no. 3, pp. 285–296, 2005.

[10] C.-H. Ou, "A localization scheme for wireless sensor networks using mobile anchors with directional antennas," *IEEE Sensors Journal*, vol. 11, no. 7, pp. 1607–1616, 2011.

[11] B. A. Galstyan, B. Krishnamachari, K. Lerman, and S. Pattem, "Distributed online localization in sensor networks using a moving target," in *Proceedings of the 3rd International Symposium on Information Processing in Sensor Networks (IPSN '04)*, pp. 61–70, Berkeley, Calif, USA, April 2004.

[12] M. L. Sichitiu and V. Ramadurai, "Localization of wireless sensor networks with a mobile beacon," in *Proceedings of the Mobile Ad-Hoc and Sensor Systems (MASS '04)*, pp. 174–183, IEEE, Fort Lauderdale, Fla, USA, October 2004.

[13] J. Rezazadeh, M. Moradi, A. S. Ismail, and E. Dutkiewicz, "Superior path planning mechanism for mobile beacon-assisted localization in wireless sensor networks," *IEEE Sensors Journal*, vol. 14, no. 9, pp. 3052–3064, 2014.

[14] P. Corke, R. Peterson, and D. Rus, "Networked robots: flying robot navigation using a sensor net," *Springer Tracts in Advanced Robotics*, vol. 15, pp. 234–243, 2005.

[15] H. Q. Cui and Y. L. Wang, "Four-mobile-beacon assisted localization in three-dimensional wireless sensor networks," *Computers and Electrical Engineering*, vol. 38, no. 3, pp. 652–661, 2012.

[16] J. Xu and H.-Y. Qian, "Localization of wireless sensor networks with a mobile beacon," *Information Technology Journal*, vol. 12, no. 11, pp. 2251–2255, 2013.

[17] W. S. Wen and L. Wang, "Path planning of mobile beacon for localization based on distribution of unknown nodes," *Advanced Materials Research*, vol. 712–715, pp. 1933–1937, 2013.

[18] J.-F. Huang, G.-Y. Chang, and G.-H. Chen, "A historical-beacon-aided localization algorithm for mobile sensor networks," *IEEE Transactions on Mobile Computing*, vol. 14, no. 6, pp. 1109–1122, 2015.

[19] Z. Xia and C. Chen, "A localization scheme with mobile beacon for wireless sensor networks," in *Proceedings of the 6th International Conference on ITS Telecommunications*, pp. 1017–1020, IEEE, Chengdu, China, June 2006.

[20] G. D. Teng, K. G. Zheng, and G. Yu, "A mobile-beacon-assisted sensor network localization based on RSS and connectivity observations," *International Journal of Distributed Sensor Networks*, vol. 2011, Article ID 487209, 14 pages, 2011.

[21] K.-F. Ssu, C.-H. Ou, and H. C. Jiau, "Localization with mobile anchor points in wireless sensor networks," *IEEE Transactions on Vehicular Technology*, vol. 54, no. 3, pp. 1187–1197, 2005.

[22] S. Lee, E. Kim, C. Kim, and K. Kim, "Localization with a mobile beacon based on geometric constraints in wireless sensor networks," *IEEE Transactions on Wireless Communications*, vol. 8, no. 12, pp. 5801–5805, 2009.

[23] L. Dong and F. L. Severance, "Position estimation with moving beacons in wireless sensor networks," in *Proceedings of the IEEE Wireless Communications and Networking Conference (WCNC '07)*, pp. 2317–2321, IEEE, Hong Kong, March 2007.

[24] K. Kim and W. Lee, "MBAL: a mobile beacon-assisted localization scheme for wireless sensor networks," in *Proceedings of the 16th International Conference on Computer Communications and Networks (ICCCN '07)*, pp. 57–62, IEEE, Honolulu, Hawaii, USA, August 2007.

[25] B. Xiao and H. Chen, "A walking beacon-assisted localization in wireless sensor networks," in *Proceedings of the IEEE International Conference on Communications*, pp. 3070–3075, Glasgow, UK, June 2007.

[26] G. Teng, K. Zheng, and W. Dong, "MA-MCL: mobile-assisted monte carlo localization for wireless sensor networks," *International Journal of Distributed Sensor Networks*, vol. 2011, Article ID 671814, 8 pages, 2011.

[27] I. F. Akyildiz and M. C. Vuran, *Wireless Sensor Network*, John Wiley & Sons, London, UK, 2010.

[28] N. Patwari, J. N. Ash, S. Kyperountas, A. O. Hero III, R. L. Moses, and N. S. Correal, "Locating the nodes: cooperative localization in wireless sensor networks," *IEEE Signal Processing Magazine*, vol. 22, no. 4, pp. 54–69, 2005.

5

A Sequential Compressed Spectrum Sensing Algorithm against SSDH Attack in Cognitive Radio Networks

Zhuhua Hu,[1,2] **Yong Bai** ⓘ,[1,2] **Lu Cao** ⓘ,[1] **Mengxing Huang** ⓘ,[1,2] **and Mingshan Xie** ⓘ[1]

[1]*College of Information Science & Technology, Hainan University, Haikou 570228, China*
[2]*State Key Laboratory of Marine Resource Utilization in South China Sea, Hainan University, Haikou 570228, China*

Correspondence should be addressed to Yong Bai; bai@hainu.edu.cn

Academic Editor: George S. Tombras

Spectrum sensing is one of the key technologies in wireless wideband communication. There are still challenges in respect of how to realize fast and robust wideband spectrum sensing technology. In this paper, a novel nonreconstructed sequential compressed wideband spectrum sensing algorithm (NSCWSS) is proposed. Firstly, the algorithm uses a sequential spectrum sensing method based on history memory and reputation to ensure the robustness of the algorithm. Secondly, the algorithm uses the strategy of compressed sensing without reconstruction, which thus ensures the sensing agility of the algorithm. The algorithm is simulated and analyzed by using the centralized cooperative sensing. The theoretical analysis and simulation results reveal that, under the condition of ensuring the certain detection probability, the proposed algorithm effectively reduces complex computation of signal reconstruction, significantly reducing the wideband spectrum sampling rate. At the same time, in the cognitive wideband communication scenarios, the algorithm also achieves a better defense against the SSDF attack in spectrum sensing.

1. Introduction

Cognitive radio (CR) technology can increase the efficiency of spectrum utilization for wideband wireless communications [1]. Aiming at detecting spectrum holes, spectrum sensing is the precondition for the implementation of CR [2].

To detect the spectrum holes more effectively over a wide bandwidth in Cognitive Radio Network (CRN), traditional wideband spectrum sensing acquires the wideband signals with a high-speed analog-to-digital converter (ADC) and then uses digital signal processing techniques to detect spectral opportunities. However, it is very often limited by the capability of ADC hardware and unable to meet high-speed sampling rate stated in Nyquist sampling theorem for wideband spectrum sensing [3]. An effective solution to address the challenge is spectrum sensing technology based on the compressed sensing (or called compressed sampling, CS) theory. CS can maintain the structure and information of the original sparse signal far below the Nyquist sampling rate. There have been many research achievements in this field. Tian and Giannakis applied the CS to the research of wideband spectrum sensing and verified its

effectiveness [4]. Collaborative spectrum sensing from sparse observations in CRNs is studied in [5] by applying matrix completion and joint sparsity recovery to reduce sensing and transmission requirements. In order to reduce computational cost, spectrum holes can be obtained by means of partial reconstruction. Hong [6] presented a method of detecting the primary users (PUs) based on Bayesian compressed sensing, which can estimate important parameters of the primary user's signal directly from compressed sampling without the need for complete reconstruction. Thus it greatly reduces the computational complexity, but it still needs some parameters' distribution information of the original sparse signal. Actually, partial and complete signal reconstruction may not be required in many spectrum sensing applications. A method was proposed in [7] for nonreconstruction compressed detection of random signal under the maximum likelihood criterion. A new method for blind detection of signals by using nonreconstruction compressed sampling without prior knowledge was proposed in [8].

In compressive observation, the redundant observation is usually contained. In order to effectively reduce the number of observations, the sequential theory was introduced into the wideband spectrum sensing, and sequential compressive spectrum sensing algorithm was proposed in [9, 10]. The algorithm can make the compressive ratio adjusted adaptively according to the signal sparsity, which can reduce the sampling number, but this method still needs to reconstruct the original signal. In [11], according to the maximum interference endured for PU, the optimal false-alarm probability is set, which can obtain the largest throughput for second users (SUs) by improving the threshold method. However, the detected signal is still the deterministic signal, whose sparsity is known in [9–11]. In [12], nonreconstruction compressed sampling method combined with sequential testing for random signals was presented. In [12], only AWGN (Additive White Gaussian Noise) channel is considered, and the robustness of the algorithm was not guaranteed with malicious users (MUs) under the SSDF (Spectrum Sensing Data Falsification) attack [13].

When a priori knowledge of signal and sparsity is unknown, in order to agilely and robustly find out the spectrum holes not occupied by PUs, this paper presents an enhanced novel nonreconstructed sequential compressed wideband spectrum sensing (NSCWSS) algorithm. Firstly, a weighted sequential spectrum sensing method based on history memory model is designed. The method assumes that, in collaborative sensing environment, the initial credibility value of each secondary user is the same. After the end of each sensing, the value of credibility of corresponding users will be updated. Then the sensing results of secondary user are assigned with different weights corresponding to its credibility and history accuracy in the final fusion, which can improve the antijamming ability of the algorithm to launch SSDF attacks. In addition, the algorithm uses a compression sensing method without reconstruction, which combined with the theory of sequential detection; thus it can ensure the least average computation for signal detection.

2. The Process of Nonreconstructed Compressed Detection

In Cognitive Radio Network, SUs need to judge whether PUs exist or not for dynamic access without affecting the normal communications of the PUs. The compressed detection model can be expressed as

$$y = \Phi x. \tag{1}$$

Under the fading channel, Corresponding to the cases that the PU does not exist or exist, a binary hypothesis testing model can be established as follows:

$$H_0 : y = \Phi \omega_{SL},$$
$$H_1 : y = \Phi \left(h_{SLi} s + \omega_{SL} \right), \tag{2}$$

where $s = [s(0), \ldots, s(n-1)]^T$ is the sequence of samples of random sparse signals, h_{SLi} is the channel fading gain of the ith sensing link, and $\omega_{SL} = [\omega_{SL}(0), \ldots, \omega_{SL}(n-1)]^T$ is the sequence of the samples of noise of the sensing link, which is independent and identically distributed (i.i.d) with one-sided

power spectral density N_0. Here, we assume that s and ω_{SL} are independent of each other; n is the sampling number, and Φ is the observation matrix with dimension $m \times n$. Normally, $m = O(K \log(n/K))$ is required, where K is the signal sparsity.

Considering fading channels, Rayleigh and Rician distributions do, in many cases, model the envelope of the signal through the fading channel very well. However, in the actual wireless communication environments, it is found that the Nakagami distribution provides better matching degree with the actual test [14]. In comparison with the Rician distribution, the Nakagami distribution does not need to be assumed in terms of direct conditions. Thus, the sensing links are assumed to be subject to a Nakagami-m fading; then the received power $Y_{SLi} = v_i(t)^2$ of ith CR user obeys a Gamma distribution. The probability density function (PDF) of Y_{SLi} is given by [14–17]

$$f_{Y_{SLi}} = \left(\frac{m_i}{\sigma_{Y_{SLi}}} \right)^{m_i} \frac{Y_{SLi}^{m_i-1}}{\Gamma(m_i)} \exp\left(-\frac{m_i}{\sigma_{Y_{SLi}}} Y_{SLi} \right) \tag{3}$$

$$Y_{SLi} \geq 0,$$

where $m_i \geq 1/2$ denotes the severity of fading, $\Gamma(\cdot)$ is the Gamma function, $\sigma_{Y_{SLi}} = E[Y_{SLi}]$ is the local-mean power of $v_i(t)$, and $v_i(t) = h_{SLi} s + \omega_{SL}$ denotes the received instantaneous signal level. As m_i increases, the level of fading decreases. When $m_i = 1/2$ or $m_i = 1$, the sensing channel is subject to one-sided Gaussian fading and the Rayleigh fading, respectively. When $m_i > 1$, the Nakagami fading can be approximately equivalent to the Rician fading. $\sigma_{Y_{SLi}}$ is given by [16]

$$E[Y_{SLi}] = \begin{cases} \sigma_{Y_{SLi0}} = N_0 & H_0 \\ \sigma_{Y_{SLi1}} = E_i + N_0 = N_0 \left(1 + \overline{\gamma}_{SLi} \right) & H_1, \end{cases} \tag{4}$$

where $\overline{\gamma}_{SLi} = E_i/N_0$ is the average SNR of the ith sensing channel and $E_i = E[(h_{SLi} s)^2]$ is the expected local-mean power of the PUs under H_1 assumption.

In the nonreconstruction approach we can test the two assumptions in (2) by directly processing the observation vector $y = [y(0), \ldots, y(m-1)]^T$. Here, m denotes the number of observations, and we assume that each observation is statistically independent of each other. Then, with H_0 assumption, the observation sequence y is Gauss random vector with mean 0 and variance $N_0 \Phi \Phi^T$. While the observation sequence y is Gauss random vector under H_1 condition, its mean is $E(\Phi h_{SLi} s)$, and the variance is $(N_0(1 + \overline{\gamma}_{SLi})) \Phi \Phi^T$.

According to the Neyman-Pearson theorem [18], given a false-alarm probability $P_f = \alpha$, the detection probability P_d is maximized with the test

$$\Lambda(x) = \frac{p(y, H_1)}{p(y, H_0)} > \zeta. \tag{5}$$

The threshold ζ is determined by

$$P_f = \int_{\{y : \Delta(y) > \zeta\}} p(y \mid H_0) \, dy = \alpha. \tag{6}$$

Putting the probability density function of y into (5) and taking the logarithmic transformation [3], we can get

$$y^T \left(\Phi\Phi^T\right)^{-1} y > 2\frac{N_0\left(N_0 + E_i\right)}{E_i} \ln\left[\zeta\frac{\left(N_0 + E_i\right)^{m/2}}{N_0{}^m}\right] \quad (7)$$

$$:= \eta.$$

Thus, the detection statistic for random signal based on nonreconstructed compressed sampling is expressed as

$$t = y^T\left(\Phi\Phi^T\right)^{-1} y. \quad (8)$$

Formula (8) contains all the information needed for signal compressed detection. In addition, if $\Phi = I$, it shows no compression capacity. That is to say, it is converted into the convertional energy-detection based spectrum sensing. The detection statistics t follow χ_m^2 distribution [12]. Then we can get

$$P_f = \Pr\left\{t > \eta; H_0\right\} = \Pr\left\{\frac{t}{N_0} > \frac{\eta}{N_0}; H_0\right\}$$

$$= Q_{\chi_m^2}\left(\frac{\eta}{N_0}\right) = \alpha, \quad (9)$$

$$\eta = N_0 Q_{\chi_m^2}^{-1}(\alpha), \quad (10)$$

$$P_d = \Pr\left\{t > \eta; H_1\right\} = \Pr\left\{\frac{t}{N_0 + E_i} > \frac{\eta}{N_0 + E_i}\right\}$$

$$= Q_{\chi_m^2}\left(\frac{\eta}{N_0 + E_i}\right), \quad (11)$$

where the right tailed distribution of the random variables in χ_m^2 distribution is $Q_{\chi_m^2}(x) = \int_x^\infty p(t)dt$.

Further, to simplify calculation, by using approximate expression method of χ_m^2 distribution, we can get

$$P_f = Q_{\chi_m^2}\left(\frac{\eta}{N_0}\right) \approx Q\left(\frac{\eta/N_0 - m}{\sqrt{2m}}\right). \quad (12)$$

From the above analysis, the relation equation $P_d(\alpha)$ between P_d and P_f can be given by

$$P_d(\alpha) = Q\left(\frac{Q^{-1}(\alpha) - \left(E_i/N_0\right)\sqrt{m/2}}{E_i/N_0 + 1}\right). \quad (13)$$

In (13), $Q(z) = \int_z^\infty \exp(-u^2/2)du$.

$\bar{\gamma}_{SLi} = E_i/N_0$ is the average SNR related to the ith sensing channel, and then

$$P_d(\alpha) = Q\left(\frac{Q^{-1}(\alpha) - \bar{\gamma}_{SLi}\sqrt{m/2}}{\bar{\gamma}_{SLi} + 1}\right). \quad (14)$$

Obviously, Because the $Q(\cdot)$ function is strictly monotonically decreasing, P_d will increase with the increase of SNR $\bar{\gamma}_{SLi}$ and observation number m under given P_f. From (9), the theoretical observation number m required by the compressed detection algorithm can be given as

$$m = 2\left(\frac{N_0 Q^{-1}(\alpha) - Q^{-1}(P_d)\left(N_0 + E_i\right)}{E_i}\right)^2. \quad (15)$$

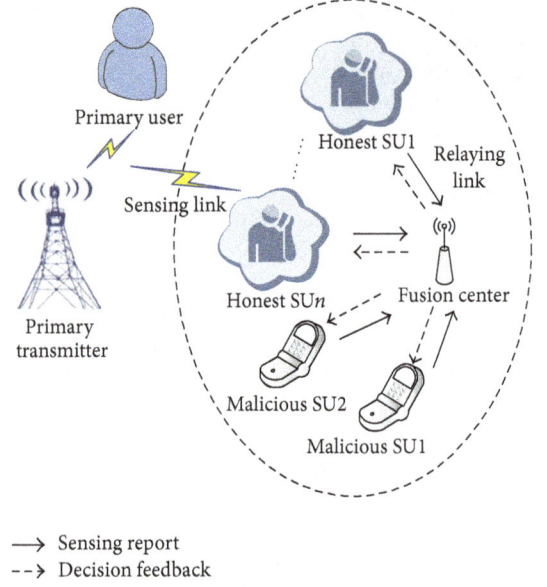

→ Sensing report
--→ Decision feedback

FIGURE 1: System network model for centralized collaborative spectrum sensing with SSDF attack.

3. System Modeling

Centralized collaborative spectrum sensing can be seen as a series of processes of voting, decision, and fusion of binary decisions. Such a system is mainly composed of PU, SUs, malicious users, and fusion center (FC). The system model is shown in Figure 1.

In centralized collaborative spectrum sensing, the final result is obtained by the centralized fusion of the sensed information from each secondary user. Cognitive users and primary users coexist. We adopt the detect-forward (DF) relaying strategy between the SU and FC. All honest cognitive users need to submit their local-decsion, according to their independent observations, to the fusion center within a sampling period, respectively, meanwhile the SSDF attackers probably send false spectrum sensing reports, trying to cause a wrong global decision about spectrum availability at the FC [19]. The fusion center can determine the presence of primary user based on all collected local-decsions from SUs. On the basis of sequential sensing theory, when SU cannot make a decision at the CR node, an additional observation should be added.

In reality, it is difficult to predict the sparsity of the signals. When the predetermined value of the signal sparsity is larger than the actual one, it is inevitable to produce redundant observation sequence. These extra observation sequences increase the number of unnecessary compression measurements without improving the detection performance. On the other hand, if the predetermined value of the signal sparsity is set too low, the effective sample values of observation sequence are not enough, and it will be difficult to meet the requirements of sensing accuracy. Therefore, this paper adopts nonreconstruction method for spectrum sensing. A single SU detects signal based on sequential compressive sensing [9]. One row of $1 \times N$ vector Φ_i ($i = 1, 2, \ldots$) is

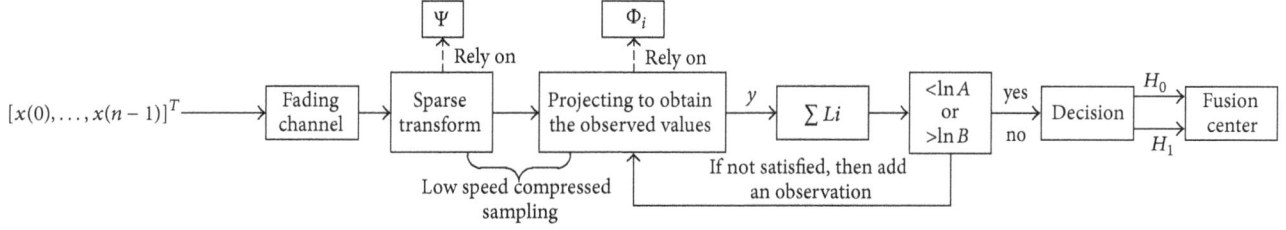

FIGURE 2: Spectrum sensing based on nonreconstructed sequential compression.

selected from the original observation matrix Φ. SUs collect a fixed number of data $x = [x(0), \ldots, x(n-1))]^Y$ and project on the observation vector Φ_i to get low dimensional observation value y_i. Then, the likelihood ratio calculation, denoted as Li, is calculated and it is compared with the thresholds $\ln(A)$ and $\ln(B)$ and then sends feedback of the final results to the node. If we can make the decision with Φ_i, then the test is over; otherwise, the observation vector Φ_{i+1} needs to be selected and x is projected onto it. Then joint judgment is made together with previous observed values. Such a process is repeated until the decision can be made. In Figure 2, Ψ is the sparse transformation matrix.

As mentioned in the Introduction, an effective solution for the technological challenge of ADC hardware is the wideband spectrum sensing technique based on compressed sensing theory. In fact, users just want to extract useful information from the observation sequence or filter out the information that they are not interested in subsequent processing. With respect to the sensing efficiency, it is not wise to completely reconstruct the original signal by using compressed sampling data.

The sequential nonreconstructed compressed sensing procedure, based on single CR node detection, is shown in Figure 2. The thresholds A and B are determined by P_d and P_f, respectively, as

$$A = \frac{1 - P_d}{1 - P_f},$$
$$B = \frac{P_d}{P_f}. \tag{16}$$

From the above, we can rewrite (8) and get

$$t_i = y_i^T \left(\Phi_i \Phi_i^T\right)^{-1} y_i. \tag{17}$$

For the kth CR user, we get a sampling sequence $X_k = [x_k(0), \ldots, x_k(n-1)]^T$, $k = 1, 2, \ldots, N$, within the sampling period T. Then the data received at the N CR nodes can be expressed in the matrix $\kappa = [X_1^T, \ldots, X_N^T]^T$. After the projection transformation between κ and Φ_i is carried out, we can obtain $y_i = [y_{1i}, \ldots, y_{Ni}]^T$.

Putting each element of y into (17), we can obtain the test value

$$t_i = [t_{1i}, \ldots, t_{ki}, \ldots, t_{Ni}]^T \quad k = 1, 2, \ldots, N. \tag{18}$$

Obviously, if we take the kth CR user as an example, t_{ki} satisfies the χ^2 distribution with the degree of freedom 1 [12].

$$H_1 : \frac{t_{ki}}{(N_0 + E_i)\,\Phi_i\Phi_i^{\,T}} \sim \chi^2,$$
$$H_0 : \frac{t_{ki}}{N_0\Phi_i\Phi_i^{\,T}} \sim \chi^2. \tag{19}$$

Furthermore, we can obtain the following formula by means of the log-likelihood ratio:

$$\sum_{m=1}^{M} Li = \sum_{m=1}^{M} \ln \frac{P_m\left(t_{ki} \mid H_1\right)}{P_m\left(t_{ki} \mid H_0\right)}$$
$$= \sum_{m=1}^{M} \left(\ln \frac{P_m\left(t_{ki} \mid H_1\right)}{P_m\left(t_{ki} \mid H_0\right)} \right) \tag{20}$$
$$= \sum_{m=1}^{M} \left(\ln \left(\frac{N_0}{E_i + N_0} \right)^{1/2} + \frac{t_{ki}E_i}{2N_0\left(N_0 + E_i\right)} \right)_m,$$

where M is the number of sequential observations. $\sum Li$ denotes the sum of the current value of Li and the value of previous likelihood ratio. Comparing $\sum Li$ with $\ln(A)$ and $\ln(B)$ continually until the decision can be made, then we can also get the theoretical value of the total average number of observations:

$$E[M] = \frac{\left(\sum_{n=1}^{N} E_n[M]\right)}{N}$$
$$= \frac{\sum_{n=1}^{N} \left(P\left(H_1\right) \left(\left(P_d \ln B + (1 - P_d) \ln A\right) / \left(\ln \sqrt{N_0/(E_i + N_0)} + E_i/2N_0 \right) \right) \right)_n}{N}$$
$$+ \frac{\sum_{n=1}^{N} \left(P\left(H_0\right) \left(\left(P_f \ln B + (1 - P_f) \ln A\right) / \left(\ln \sqrt{N_0/(E_i + N_0)} + E_i/2(N_0 + E_i) \right) \right) \right)_n}{N}. \tag{21}$$

From (21), we can know that the number of observations is closely related to the signal-to-noise ratio.

4. NSCWSS Algorithm

4.1. Traditional Sequential Detection [9]

$$S_n = \prod_{i=0}^{n} \frac{P(S_i \mid H_1)}{P(S_i \mid H_0)}, \tag{22}$$

where the sampling value n is arbitrary value of the number of all collaborative nodes and we take the double threshold detection method for detection. With given P_f and P_{md}, the two thresholds can be expressed as

$$\xi_0 = \frac{P_f}{1 - P_{md}},$$
$$\xi_1 = \frac{1 - P_f}{P_{md}}. \tag{23}$$

The rules of double threshold decision are as follows:

$$S_n \geq \xi_1 \longrightarrow H_1,$$
$$S_n \leq \xi_0 \longrightarrow H_0, \tag{24}$$
$$\xi_1 \geq S_n \geq \xi_0 \longrightarrow \text{continue}.$$

4.2. Algorithm Idea. In traditional sequential detection, each data fusion does not need to traverse all the CR user's sensing results, which improves the sensing speed. Nevertheless, the deficiency is that it treats all the nodes equivalently, which affects the effectiveness and accuracy of each sensing for different CR nodes.

In the process of centralized collaborative spectrum sensing detection, there might be a secondary user simulating the authorized signal to send false information to gain a higher access authority, which makes the fusion center to make a false judgment and thus seriously affects the efficiency and accuracy of overall sensing. In order to ensure that the collaborative spectrum sensing under SSDF attacks to have a high degree detection accuracy, the aggregation center needs to identify and try to eliminate the interference of false spectrum detection information sent by those malicious users. In addition, the occurrences of these disturbances are random and uncertain, so we need a better data fusion scheme. A weighted sequential probability ratio test (WSPRT) is presented in [20]. An improved WSPRT algorithm with advanced weight sequential log-likelihood ratio detection (AWSPRT) is presented in [21]. Those proposed methods apply information of trust degree to data fusion, and the fusion center distinguishes the credibility of spectrum sensing results by trust value, which can be against SSDF attack to a certain extent. However, because these two algorithms have no combination with compressed sampling and the computational complexity is high, they are hard to be used for wideband spectrum sensing.

For the problems mentioned above, this paper proposes a robust weighted sequential spectrum sensing algorithm based on history information and compressed non-reconstruction sampling technique. The proposed algorithm assigns weights to SUs by building weighted vector W. Meanwhile, the fusion center will efficiently record each sensing decision result. When there exists malicious signal, FC gives them lower weights to decrease the trust degree of sensor nodes and carries on the record. Therefore, the interference of the malicious nodes on the whole sensing network can be reduced. Another benefit, by the introduction of the nonreconstructed compressed sensing approach, is that our proposed algorithm (NSCWSS) can greatly reduce the sensing time by the lower computing complexity. Structure description of the NSCWSS algorithm, based on DF strategy, is as shown in Figure 3.

In FC, each sensing result of all SUs is recorded by using a struct. All record nodes are stored by using data struct of block linked list, which is described as Figure 4. Struct fields are all initially set to 0. Figure 4 records N times' spectrum sensing results for each secondary user. In Figure 4, b_{ji} denotes difference between decision result of SU_j and FC in the ith time, and l_{ji} is the concept of correct rate, which is the ratio of the times that the spectrum sensing decision results of SU_j are the same as the final decision of the fusion center to the first i times.

The value of b_{ji} and l_{ji} is calculated as follows:

$$b_{ji} = \begin{cases} 0, & H_{ji} = H_{FC} \\ 1, & H_{ji} \neq H_{FC} \end{cases} \tag{25}$$
$$l_{ji} = 1 - \frac{\sum_{m=1}^{i} b_{jm}}{i}.$$

In [20], if the user is judged to be a malicious user, its weight is set to 0. That is, the data of malicious users is abandoned and will not participate in the process of data fusion. When there are many malicious users, the data sent to the fusion center from secondary users may not reach decision condition. Then the detection performance of the system will decline, which wastes information resources. If we give a negative value to the data from less trusted users, this will effectively reduce the waste of the information resources.

We set the reputation vector to $R = [r_1, r_2, \ldots, r_i, \ldots, r_{M-1}, r_M]^T$, and initial value of elements is set to 0, which is corresponding to the reputation value of each CR user. After each round of sensing, the corresponding user's reputation value will be updated. In the real environment, once there is a malicious user, detection probability will drop sharply. So, the cumulative rate of accuracy is the 2/3 of attenuation rate. The update of credibility is performed as follows:

$$r_j = r_j + 1 \quad H_{ji} = H_{FC},$$
$$r_j = r_j - 1.5 \quad H_{ji} \neq H_{FC}. \tag{26}$$

Here we assume that the sensing weight of each CR node is w_i, and its initial value is set to 1. Constant $-g$ is as the lower limit of credibility. g is usually assigned the value 5.

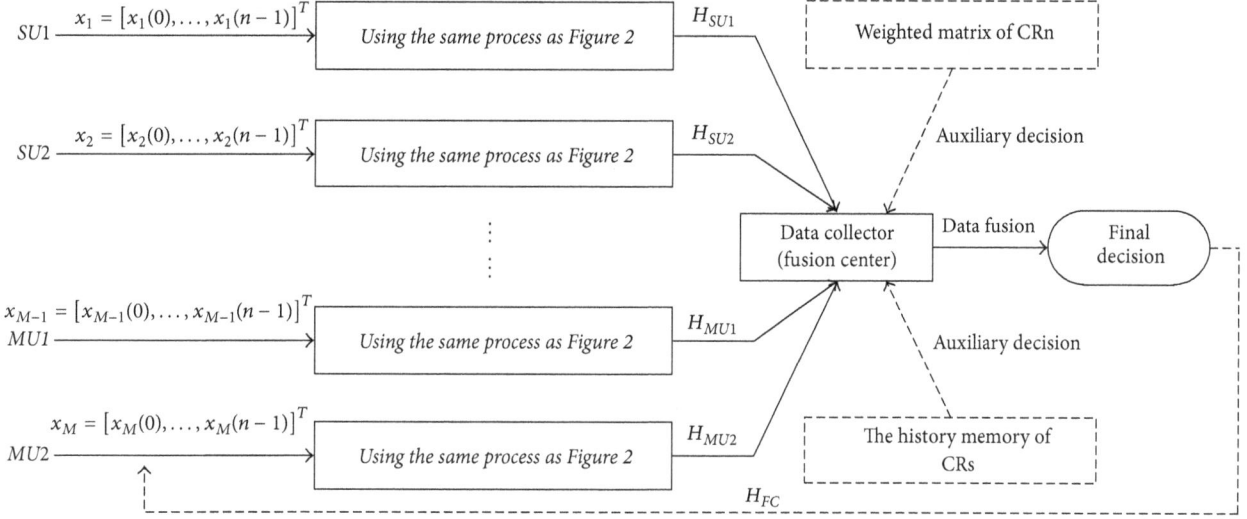

FIGURE 3: The flow chart of NSCWSS algorithm.

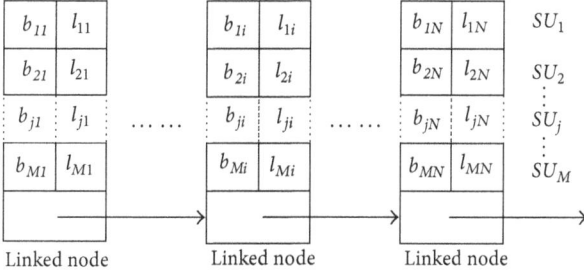

FIGURE 4: History memory model.

Then we set weighted coefficient to $W_i = [w_{1i}, w_{2i}, \ldots, w_{ji}, \ldots, w_{(M-1)i}, w_{Mi}]^T$ in the ith sensing cycle:

$$w_{ji} = f\left(r_{ji}\right) \begin{cases} \dfrac{r_{ji} \times l_{ji}}{\max\left(r_i\right) + g}, & r_{ji} \leq -g \\[3mm] \dfrac{r_{ji} \times l_{ji} + g}{\max\left(r_i\right) + g}, & r_{ji} > -g. \end{cases} \tag{27}$$

Normalization of w_{ji} is required in applications. Then the new decision variables can be expressed as

$$S_M = \prod_{i=0}^{M} \left(\frac{P\left[S_i \mid H_1\right]}{P\left[S_i \mid H_0\right]}\right)^{w_{ji}}. \tag{28}$$

The NSCWSS algorithm compares the result S_M with the two threshold values of ξ_0 and ξ_1 to get the final decision results. The fusion center needs to analyze the reliability degree of each user's information and then compares the test results of secondary user with the final fusion results to evaluate the accuracy of CR user's detection and improve the capability of anti-interference of system on SSDF attack. Compared to the previous algorithms, our proposed algorithm has strong robustness in the situation of collaborative spectrum sensing where there are malicious users. Thus it can also effectively resist the SSDF attack and improves the detection probability of system and significantly reduces the wrong judgment of the whole system.

5. Simulation Results and Discussion

We verify the detection performance of the proposed algorithm by means of computer simulation and numerical calculation. In the simulation, for simplicity, we set the original PU signal as a random sparse signal obeying the Gauss distribution, which is sparse in the time domain. The degree of sparsity is $K = 10$. SNR $= E/\sigma_0^2$, where E is the average power of $s(n)$. In the simulation, we assume that the total length of the signal $n = 400$ and carry out 1000 times Monte Carlo simulation to get statistical average results.

It is assumed that there is no exchange of information between the CR nodes. That is, the detection results are independent of each other. For a single node user, we assume the detection probability of $P_d = 0.9$ and $m_i = 0.6$. Under different false-alarm probability P_f, the number of required observations for sequential compressed detection is shown in Figure 5.

The curves marked with "o"; "△"; and "+" correspond to simulation results when SNR is 21 dB, 24 dB, and 27 dB, respectively. It can be seen that the theoretical value is in good agreement with the simulation results under different SNRs. It can also be seen from Figure 5 that the number of observed values will be reduced with the increase of the SNR.

In the centralized collaborative sensing environment, Figure 6 shows, when $P_d = 0.9$ and $m_i = 0.6$, the required observations with NSCWSS and nonsequential compressed detection algorithm (CD) under different false-alarm probability. It can be seen from Figure 6 that the NSCWSS algorithm can reduce the number of observations from 35% to 50%. However, since the DF strategy is adopted, there is no little variation in the average number of observations in a cooperative sensing environment compared to a single node.

Figure 7 is the number of sample values for $m_i = \{0.5, 1.0, 1.5, 2.0, 2.5\}$ when SNR is from 15 dB to 25 dB in

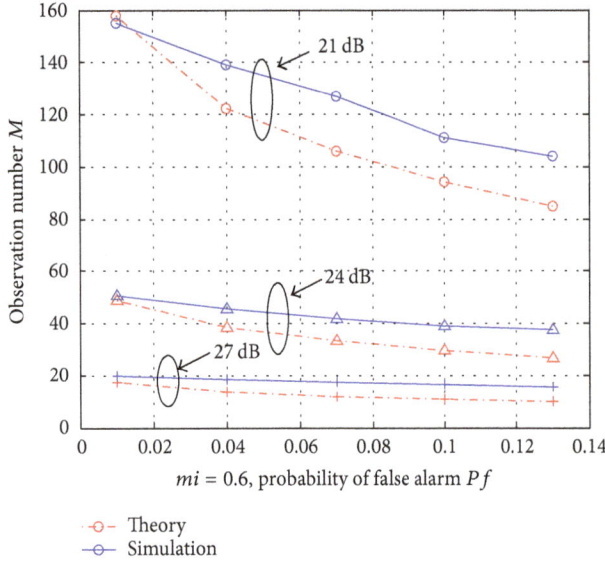

FIGURE 5: Observation number for different false-alarm probability.

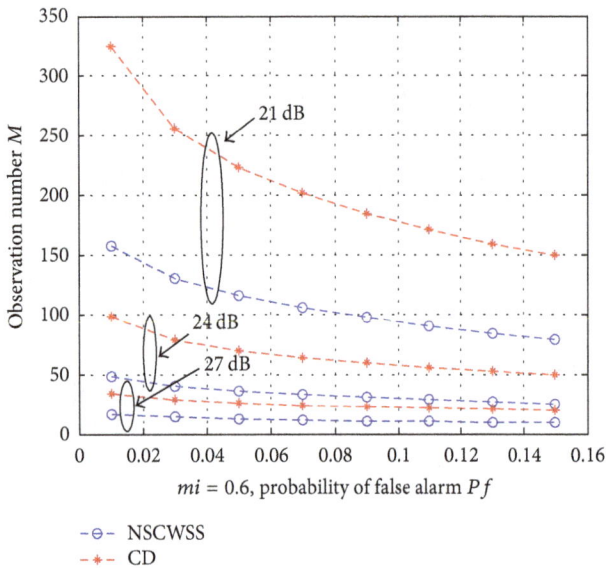

FIGURE 6: Performance of NSCWSS and CD at different SNR.

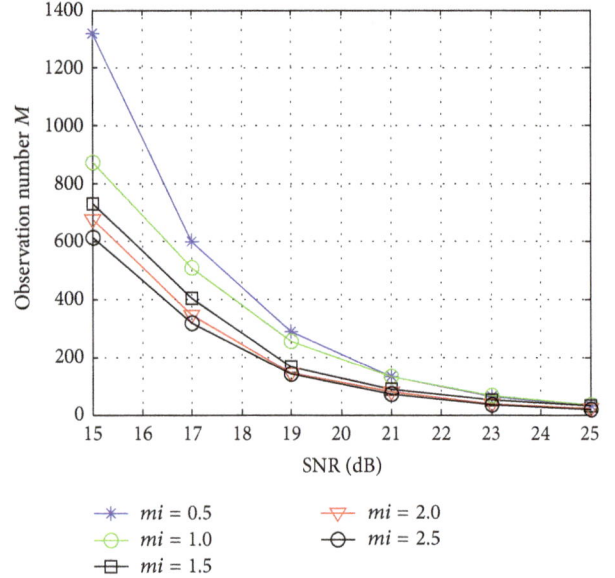

FIGURE 7: Required detection number M versus SNR under different m_i.

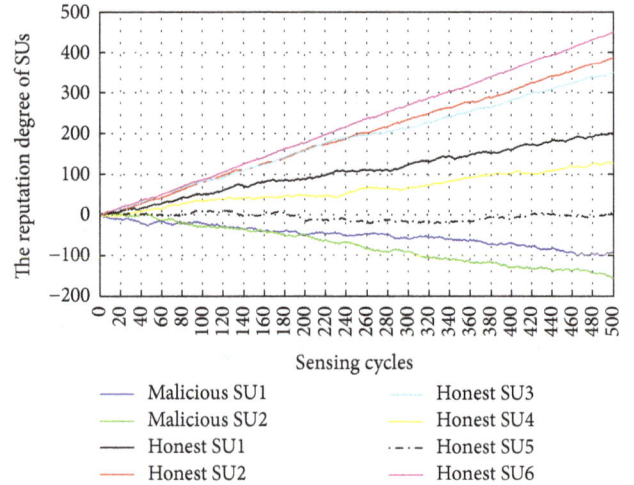

FIGURE 8: The changes of reputation for SUs.

the collaborative sensing environment. We can see that the number of observations required for detection increases as the degree of channel fading increases.

In the case of Figures 8 and 9, there have been two malicious users. One of them implements an "Always-1" attack, and another one implements the "Always-0" attack. The reputation and weighted value of the malicious users are significantly reduced with the increase of number of sensing, which shows that algorithm can well identify malicious users to attack.

Assume that the number of SU is 50. Figure 10 shows that NSCWSS algorithm has the better system sensing performance than "AND" algorithm, CD algorithm, and traditional WSPRT fusion algorithm with different proportion of malicious users under the "Always-1" or "Always-0" attack.

Frome Figure 10, we can see that NSCWSS algorithm can maintain higher detection success probability, even if there have been more malicious users to attack.

6. Conclusion

In this paper, we propose a new robust weighted sequential nonreconstruction spectrum sensing algorithm based on history memory for wideband communications and analyze its detection performance. The simulation analysis shows that, under the prerequisite of guaranteeing certain detection probability, the proposed algorithm can greatly reduce the number of sampled observations and reduce the computation of sensing. Compared with the conventional sequential detection techniques, the proposed algorithm can better eliminate

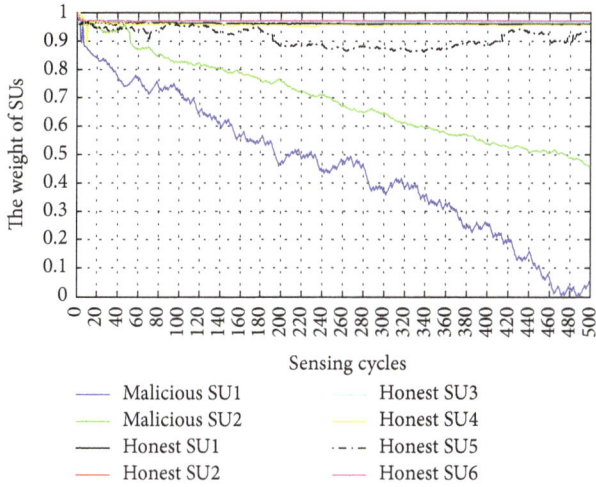

FIGURE 9: The changes of weighting for SUs.

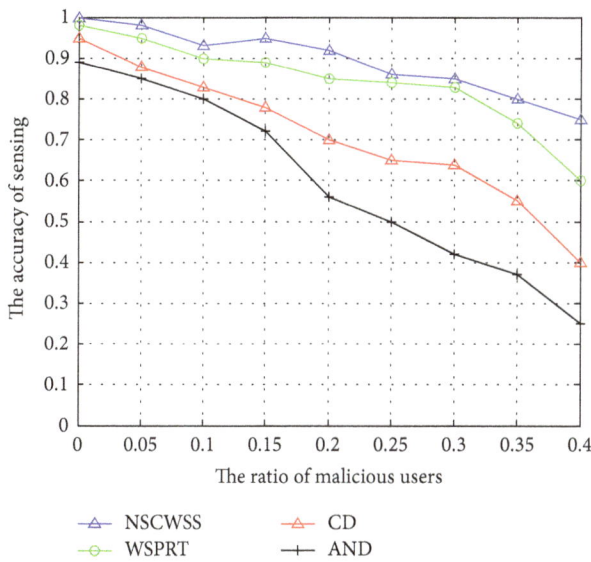

FIGURE 10: The performance analysis for different algrithms.

sample parameter uncertainty effect and improve the anti-interference ability of the SSDF attack, which can enhance the effectiveness and stability of the overall sensing of the spectrum, and it can satisfy the requirement of robustness for wideband spectrum sensing.

Conflicts of Interest

The authors declare that there are no conflicts of interest regarding the publication of this paper.

Acknowledgments

This project is supported by the National Natural Science Foundation of China (no. 61561017), the Hainan Province Natural Science Foundation of China (no. 617033), the Open Subproject of State Key Laboratory of Marine Resource Utilization in South China Sea (no. 2016013B), Oriented Project of State Key Laboratory of Marine Resource Utilization in South China Sea (no. DX2017012), the Postgraduate Practice and Innovation Project of Hainan University, and the Major Science and Technology Project of Hainan Province (no. ZDKJ2016015).

References

[1] H. Sun, A. Nallanathan, C. Wang, and Y. Chen, "Wideband spectrum sensing for cognitive radio networks: a survey," *IEEE Wireless Communications Magazine*, vol. 20, no. 2, pp. 74–81, 2013.

[2] G. I. Tsiropoulos, O. A. Dobre, M. H. Ahmed, and K. E. Baddour, "Radio resource allocation techniques for efficient spectrum access in cognitive radio networks," *IEEE Communications Surveys & Tutorials*, vol. 18, no. 1, pp. 824–847, 2014.

[3] M. A. Davenport, P. T. Boufounos, M. B. Wakin, and R. G. Baraniuk, "Signal processing with compressive measurements," *IEEE Journal of Selected Topics in Signal Processing*, vol. 4, no. 2, pp. 445–460, 2010.

[4] Z. Tian and G. B. Giannakis, "Compressed sensing for wideband cognitive radios," in *Proceedings of the IEEE International Conference on Acoustics, Speech and Signal Processing (ICASSP '07)*, pp. IV1357–IV1360, Honolulu, HI, USA, April 2007.

[5] J. Meng, W. Yin, H. Li, E. Hossain, and Z. Han, "Collaborative spectrum sensing from sparse observations in cognitive radio networks," *IEEE Journal on Selected Areas in Communications*, vol. 29, no. 2, pp. 327–337, 2011.

[6] S. Hong, "Multi-resolution bayesian compressive sensing for cognitive radio primary user detection," in *Proceedings of the 53rd IEEE Global Communications Conference (GLOBECOM '10)*, pp. 1–6, December 2010.

[7] B. W. Li, Y. G. Li, and Y. G. Zhu, "Non-reconstruction compressive detection of random signal using maximum likelihood criterion and its analysis," *Journal of Signal Processing*, vol. 29, no. 8, pp. 996–1002, 2013.

[8] K.-T. Cao, X.-Q. Gao, and D.-L. Wang, "Wideband compressive spectrum sensing without reconstruction based on random matrix theory," *Journal of Electronics and Information Technology*, vol. 36, no. 12, pp. 2828–2834, 2014.

[9] D. M. Malioutov, S. R. Sanghavi, and A. S. Willsky, "Sequential compressed sensing," *IEEE Journal of Selected Topics in Signal Processing*, vol. 4, no. 2, pp. 435–444, 2010.

[10] H. Zheng, S. Xiao, and X. Wang, "Sequential compressive target detection in wireless sensor networks," in *Proceedings of the 2011 IEEE International Conference on Communications (ICC '11)*, vol. 6, pp. 1–5, June 2011.

[11] L. Lu, X. Zhou, and G. Y. Li, "Optimal sequential detection in cognitive radio networks," in *Proceedings of the 2012 IEEE Wireless Communications and Networking Conference (WCNC '12)*, pp. 289–293, April 2012.

[12] S. Y. Tu, X. Q. Song, Y. G. Zhu et al., ""Detection of random signal based on unreconstructed sequential compressive sensing and its analysis in cognitive wireless network," *Journal of Signal Processing*, vol. 30, no. 2, pp. 205–213, 2014.

[13] S. Althunibat, B. J. Denise, and F. Granelli, "Identification and punishment policies for spectrum sensing data falsification attackers using delivery-based assessment," *IEEE Transactions on Vehicular Technology*, vol. 65, no. 9, pp. 7308–7321, 2016.

[14] M. Nakagami, "The m-distribution-A general formula of intensity distribution of rapid fading," *Statistical Methods in Radio Wave Propagation*, pp. 3–34, 1960.

[15] M. Abdel-Hafez and M. Cafak, "Performance analysis of digital cellular radio systems in Nakagami fading and correlated shadowing environment," *IEEE Transactions on Vehicular Technology*, vol. 48, no. 5, pp. 1381–1391, 1999.

[16] S. Hussain and X. N. Fernando, "Performance analysis of relay-based cooperative spectrum sensing in cognitive radio networks over non-identical nakagami-m channels," *IEEE Transactions on Communications*, vol. 62, no. 8, pp. 2733–2746, 2014.

[17] S. Hussain and X. N. Fernando, "Closed-form analysis of relay-based cognitive radio networks over Nakagami-m fading channels," *IEEE Transactions on Vehicular Technology*, vol. 63, no. 3, pp. 1193–1203, 2014.

[18] N. Kundargi and A. Tewfik, "A performance study of novel sequential energy detection methods for spectrum sensing," in *Proceedings of the 2010 IEEE International Conference on Acoustics, Speech, and Signal Processing (ICASSP '10)*, pp. 3090–3093, Dallas, TX, USA, March 2010.

[19] A. Attar, H. Tang, A. V. Vasilakos, F. R. Yu, and V. C. M. Leung, "A survey of security challenges in cognitive radio networks: solutions and future research directions," *Proceedings of the IEEE*, vol. 100, no. 12, pp. 3172–3186, 2012.

[20] R. Chen, J.-M. Park, and K. Bian, "Robust distributed spectrum sensing in cognitive radio networks," in *Proceedings of the 27th Conference on Computer Communications (INFOCOM '08)*, pp. 31–35, April 2008.

[21] J.-H. Zhao, F. Li, and T. Yang, "Weight sequential log-likelihood ratio detect algorithm with malicious users removing," *Journal of China Universities of Posts and Telecommunications*, vol. 20, no. 2, pp. 60–65, 2013.

6

Efficient Cross-Layer Optimization Algorithm for Data Transmission in Wireless Sensor Networks

Chengtie Li, Jinkuan Wang, and Mingwei Li

School of Information Science and Engineering, Northeastern University, Shenyang, Liaoning 110819, China

Correspondence should be addressed to Chengtie Li; neuqlct@hotmail.com

Academic Editor: Jun Bi

In this paper, we address the problems of joint design for channel selection, medium access control (MAC), signal input control, and power control with cooperative communication, which can achieve tradeoff between optimal signal control and power control in wireless sensor networks (WSNs). The problems are solved in two steps. Firstly, congestion control and link allocation are separately provided at transport layer and network layer, by supply and demand based on compressed sensing (CS). Secondly, we propose the cross-layer scheme to minimize the power cost of the whole network by a linear optimization problem. Channel selection and power control scheme, using the minimum power cost, are presented at MAC layer and physical layer, respectively. These functions interact through and are regulated by congestion rate so as to achieve a global optimality. Simulation results demonstrate the validity and high performance of the proposed algorithm.

1. Introduction

Wireless sensor networks find many applications in military areas detection, habitat monitoring, and so on. Performance degenerate analysis has gained much interest due to unbalanced power allocation in physical layer, excessive contention wireless channel in MAC layer, unfair link capacity allocation in network layer, and the inappropriate transport protocol in transport layer. WSNs suffer from several restrictions of the sink nodes and sensor nodes on account of the battery powered, computation complexity, communication, and storage capabilities [1–3].

MAC (medium access control), which is critical technology concerning net performance [4], is in charge of allocation wireless communication resources for contention nodes in protocol stack.

Scheduling effectively achieves resource allocation for wireless communication. To reach a satisfying scheduling scheme, scheduling cost, and scheduling objective regularly compromised, the scheduling problem is transferred to multiobjective optimization. Recently, scholars integrated scheduling and power control for better performance and obtained certain achievements [5, 6].

Reference [7] provided an in-depth analysis on the CS-based medium access control schemes and revealed the impact of communication signal-to-noise ratio on the reconstruction performance. Authors showed the process of the sensor data converted to the modulated symbols for transmission and how the modulated symbols are recovered via compressed sensing. Reference [8] studied the optimal flow control by a multiobjective linear programming problem, which achieved the optimization between utility and lifetime in WSNs. Reference [9] jointly designed rate control, scheduling, and power control with stochastic optimization problems to achieve cross-layer optimization protocol designing. Reference [10] investigated optimal power control, rate adaptation, and scheduling for an ultrawideband-based Intravehicular Wireless Sensor Network for one-electronic-control-unit (ECU) and multiple-ECU cases. These methods are widely applied into many-to-one data transmission control protocols and solved the problems in every aspect [11–14]. However, the above algorithms did not give a comprehensive analytic process for considering energy consumption, channel selection, link capacity, and congestion control to ensure better performance.

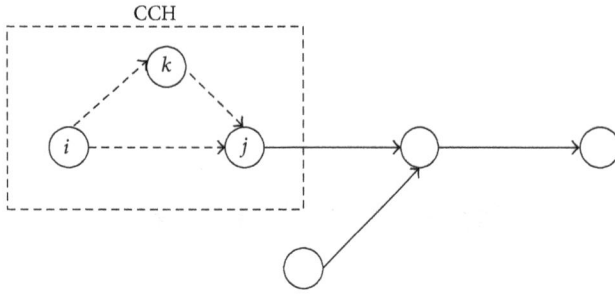

FIGURE 1: Multihop routing cooperative communication model.

In this paper, cross-layer optimal design is presented, which considers the influence for congestion rate at physical layer, MAC layer, network layer, and transport layer to achieve minimum power cost in WSNs. The algorithm coordinates communication protocols by congestion rate in the several layers to solve lossy wireless channel, excessive contention and unfair access, disabled bandwidth allocation, and the fundamentally inappropriate mechanisms of TCP. First, we design a packet error rate control strategy based on congestion rate, which makes the packet error rate in the domain of validity. The power is exactly allocated on the basis of "to each according to his needs." Second, the optimal schemes of control input and link capacity with compressed sensing are discussed. Third, we construct minimum energy problem with power control function, congestion control function, link allocation function, and so on. The optimum solutions are the optimal power control and channel selection.

2. Network and Node Model

For energy efficiency purpose, the transmission power at every node is assumed to be adjustable resulting in congestion price: when the congestion rate is bigger, the power should be increased for data transmitted successfully; when the congestion rate is smaller, the power is only satisfied for data transmitted regularly. We assume that the sensor network consists of N nodes, of which the sink node is defined as node #N.

Communication from node i to node j contains directed communication hop (DCH) $i \rightarrow j$ [15] and cooperative communication hop (CCH) $i \rightarrow k \rightarrow j$, as shown in Figure 1. CCH consists of a sender, a relay, and a receiver (i, k, j in Figure 1, resp.), such that both direct and relayed copies of the information transmitted by the sender are received at the receiver.

3. Cross-Layer Optimization Design of Cooperative Communication

In the wireline networks, the protocols hold rigorous constraint at each layer, which is only responsible for oneself, not being infiltrated. Physical layer is mostly in charge of power allocation; channel selection is achieved in MAC layer; flow control is implemented in transport layer; network layer is usually responsible for scheduling design. However, the requirement like this does not occur in WSNs. Cooperative

communication is applied for protocols at each layer to make net performance better.

3.1. Power Control. Define packet error rate with white Gaussian noise [15] as follows:

$$\Omega_{ik,DC} = 1 - \left(1 - \frac{1}{2}e^{-\gamma_{ik}/2}\right)^L, \tag{1}$$

where $\gamma_{ik} = (P_{tx,tot,i}g_{ik}/P_n) \times (B_n/R)$ is energy per bit to noise power spectral density ratio, $P_{tx,tot,i}$ is the aggregated power for transmitting a packet at i, B_n is the noise bandwidth, g_{ik} is the path loss of link (i, k) in linear units, P_n is the average noise power, R is the maximum link data rate in bits per second, bits/s, and L is the length of the packet. The power consumption of a packet transmitted successfully is $P_{tx,tot,i} = (1 + p_i)P_{tx,i}$ if congestion occurs in the WSNs, $P_{tx,i}$ denotes transmitted power to a packet at ideal condition (congestion does not occur), and p_i is the congestion rate at node i; the average noise power $P_n = (1 + p_i)\overline{P}_n$, where $\overline{P}_n$ is the average noise power at ideal condition; the maximum link data rate $R = (1 + p_i)\overline{R}$, where $\overline{R}$ is the maximum link data rate at ideal condition. Thus,

$$\gamma_{ik} = \frac{(1 + p_i)P_{tx,i}g_{ik}}{(1 + p_i)\overline{P}_n} \frac{B_n}{(1 + p_i)\overline{R}} = \frac{P_{tx,i}g_{ik}}{(1 + p_i)\overline{P}_n} \frac{B_n}{\overline{R}}. \tag{2}$$

We can conclude that γ_{ik} is adjusted to be smaller, and $\Omega_{ik,DC}$ is bigger when congestion rate is increased from formula (2). The power, appropriately adjusted by congestion rate, makes the packet error rate constraint the domain of validity. The power is exactly allocated on the basis of "to each according to his needs," which can extensively highlight power efficiency.

3.2. Congestion Control. Recently, lots of researchers have addressed the signal function with cross-layer optimal design to achieve better performance due to the characteristic of higher communication consumption and lower data process consumption in WSNs [16–18]. However, almost all congestion control algorithms did not consider the data preprocessing before being transmitted [19]. In this section, we try to reduce the number of transmissions with a Toeplitz matrix in compressed sensing that enables the number of transmissions to be extensively decreased in WSNs. The transmission of compressed signal is not only relieving the congestion in data process overabundance but also economizing energy in the transmission. Compressed sensing model is expressed as follows:

$$y(t) = \Phi x_i(t) + \varepsilon, \tag{3}$$

where $x_i(t) \in \mathfrak{R}^{n \times 1}$ is input signal, $y(t) \in \mathfrak{R}^{m \times 1}$ is measurement vector, $\Phi \in \mathfrak{R}^{m \times n}$ is sensing matrix, $m \ll n$, and $\varepsilon \in \mathfrak{R}^{m \times 1}$ is unknown vector for measurement noise. Suppose that the input signal is sparse, and sensing matrix satisfies restricted isometry property (RIP) [20]. To speed up congestion control, we select Toeplitz matrix as sensing matrix, meeting $\|\Phi\|_2 \leq 1$.

The scheduling algorithm and signal control scheme are the same in each link or at each node. In this paper, only consider the scheduling in (i, k) and signal control at i. The linear function $A(f_{ik})$ of the link capacity f_{ik} in (i, k) denotes the service capacity, and the service requirement at node i by a linear function $H(x_i(t), x_i)$ of the control input signal $x_i(t)$ and original signal x_i indicates the service requirement. Consider

$$Z(f_{ik}, x_i(t), x_i) = (1 - p_i) A(f_{ik}) - H(x_i(t), x_i). \quad (4)$$

Formula (4) represents supply and demand function of the service. The valid service capacity achieves optimality if and only if it is exactly satisfied with service requirement. Otherwise, it will bring unnecessary energy consumption:

$$x_i(t) = \arg \min_x Z(f_{ik}, x_i(t), x_i) + p_i \binom{\varepsilon}{0}. \quad (5)$$

The most desired input feedback signal is $\arg \min_x Z(f_{ik}, x_i(t), x_i)$. Considering congestion in the WSNs, however, input signal should reach to expression in (5).

3.3. Scheduling Algorithm. The adjustment of the transmitted power in Section 3.1, being changed in the transmission range and the connectivity of the node, causes the changes in the requirement of the link capacity. In Section 3.2, the input signal satisfying supply and demand function minimum is generated. On this basis, we will discuss the link capacity allocation in the link (i, k) in this section. From analysis, the network optimal condition is that supply and demand function attains minimum. When function (4) is the actual minimum, the link capacity f_{ik} is the optimal link capacity. In this scheduling, on the one hand, unnecessary energy consumption with link capacity too big is effectively avoided; on the other hand, the new congestion with link capacity too small is suppressed. Thus, f_{ik} can be denoted as follows:

$$f_{ik} \in \arg \min Z(f_{ik}, x_i(t), x_i). \quad (6)$$

3.4. Channel Selection. Power control is applied to dynamically adjust transmission power, which can not only effectively reduce energy consumption in communication, but also prolong network lifetime in WSNs. In addition, power control could remarkably influence the topology control, connectivity, throughput, and real-time of message transmission. Moreover, MAC layer or cross-layer design with power control evidently optimizes network performance and highlights QoS. In this section, we will discuss how to select channel by cutting down unnecessary energy consumption [21]. Energy is mainly expended at communications, and computation brings the energy to be neglected. For a CCH with relay node k, power consumption chiefly includes three parts: the first part is transmission power of a packet sent at node i;

$$C_{S,k,ij} = P_{tx,i,tot}; \quad (7)$$

the second part indicates that, for relay node k, the power aggregate to receive a packet from node i and successfully transmit a packet to node j is

$$C_{R,k,ij} = P_{rx,k,cir} + (1 - \Omega_{ik,DC}) P_{tx,i,tot}. \quad (8)$$

The third part indicates that, for node j, the power aggregate to receive a packet from node i and node k is

$$C_{D,k,ij} = P_{rx,j,cir} + (1 - \Omega_{ik,cir}) P_{rx,j,cir}. \quad (9)$$

Suppose that $P_{rx,k}$ is the required power for receiving a packet for node k at ideal condition. Congestion should be considered for data valid transmission. To achieve energy optimization, let

$$\begin{aligned} C_{S,k,ij} &= P_{tx,i,tot} = (1 + p_i) P_{tx,i} \\ C_{R,k,ij} &= P_{rx,k,cir} + (1 - \Omega_{ik,DC}) P_{tx,tot,k} \\ &= (1 + p_k) P_{rx,k} + (1 - \Omega_{ik,DC}) (1 + p_k) P_{tx,k} \\ C_{D,ij,k} &= P_{rx,j,cir} + (1 - \Omega_{ik,DC}) P_{rx,j,cir} \\ &= (1 + p_j) P_{rx,j} \\ &\quad + (1 - \Omega_{ik,DC}) (1 - \Omega_{kj,DC}) (1 + p_j) P_{rx,j}. \end{aligned} \quad (10)$$

From formula (10), the power aggregate is as follows:

$$\begin{aligned} C_{ij,k} &= C_{S,k,ij} + C_{R,k,ij} + C_{D,k,ij} \\ &= (1 + p_i) P_{tx,i} + (1 + p_k) P_{rx,k} \\ &\quad + (1 - \Omega_{ik,DC}) (1 + p_k) P_{tx,k} + (1 + p_j) E_{rx,j} \\ &\quad + (1 - \Omega_{ik,DC}) (1 - \Omega_{kj,DC}) (1 + p_j) E_{rx,j}. \end{aligned} \quad (11)$$

The energy for the packet transmitted or sent is considered equal at each node (not considering congestion); thus (11) becomes

$$\begin{aligned} C_{ij,k} &= C_{S,k,ij} + C_{R,k,ij} + C_{D,k,ij} \\ &= (1 + p_i) P_T + (1 + p_k) (P_R + (1 - \Omega_{ik,DC}) P_T) \\ &\quad + [(1 - \Omega_{ik,DC}) (1 - \Omega_{kj,DC}) + 1] (1 + p_j) P_R. \end{aligned} \quad (12)$$

Suppose that the desired packet error ratio at hop is Ω_{obj} and power allocation constraint packet error ratio at hop can be denoted:

$$\begin{aligned} \min \quad & \{C_{ij,k}, \ \forall k \in [2, N], \ k \neq i, j\} \\ \text{s.t.} \quad & \max(\Omega_{ik,DC}, \Omega_{kj,DC}) \leq \Omega_{obj} \\ & (1 - p_u) A(f_{uv}) - H(x_u(t), x_u) \geq 0 \\ & 0 \leq p_u \leq 1 \\ & x_u(t) = \arg \min_x Z(f_{uv}, x_u(t), x_u) + p_u \binom{\varepsilon}{0} \\ & f_{ik} = f_{uv} \in \arg \min_f Z(f_{uv}, x_u(t), x_u) \\ & p_u(t+1) \\ & = p_u(t) + \gamma_t (H(x_u(t), x_u) - A(f_{uv})). \end{aligned} \quad (13)$$

Node k making (13) optimal is the optimal occupied node.

3.5. Algorithm Design

Algorithm 1.

    Step 1. Initialization

        Step 1.1. Initialize original signal $x_0 \in R^{n \times 1}$, and select appropriate $\Phi \in R^{m \times n}$, $m \ll n$, calculate L;
        Step 1.2. Given g_{ik}, B_n, $\overline{p}_n$, $\overline{R}$.

    Step 2. Given the original $p_u(0)$ and the expression of $H(x_u(t), x_u)$, $A(f_{uv})$,

        Step 2.1. $\forall t \in [0, N]$, calculate $p_u(t)$, $x_u(t)$ and f_{uv};
        Step 2.2. Given Ω_{obj},
        If $(1 - p_u)A(f_{uv}) - H(x_u(t), x_u) \geq 0$

            Then calculate $\Omega_{ik,\mathrm{DC}}$, $\Omega_{kj,\mathrm{DC}}$;

        Else go to Step 1.

    Step 3

    For $f_{uv} \in \arg\min_f Z(f_{uv}, x_u(t), x_u)$
    If $0 \leq p_u(t) \leq 1$, $\max(\Omega_{ik,\mathrm{DC}}, \Omega_{kj,\mathrm{DC}}) \leq \Omega_{\mathrm{obj}}$

        Then calculate $C_{ij,k}$;

    Else if $0 \leq p_u(t) \leq 1$, $\max(\Omega_{ik,\mathrm{DC}}, \Omega_{kj,\mathrm{DC}}) > \Omega_{\mathrm{obj}}$

        Then go to Step 2.2;
        Else if $p_u(t) > 1$

            Then go to Step 2.1;
            End

        End

    End

    End For.

The performance analysis of the algorithm is presented and proves its validity.

Theorem 2. *If $x_u(t) \in \mathfrak{R}^{n \times 1}$ is sparse signal satisfying formula (5), measurement vector with noise is expressed to $y_u(t) = \Phi x_u(t) + \varepsilon$, where the optimal vector is y_u^* constrained by $\|y_u^* - y_u(t)\| \leq \delta_u$. Let sensing matrix $\Phi \in \mathfrak{R}^{m \times n}$ obey restricted isometry property (RIP); then*

$$\|x_u(t) - x_u^*\|_2 \leq \frac{p_u \delta_u}{1 - p_u \|\Phi\|_2}, \tag{14}$$

where x_u^* is the optimal value of $x_u(t)$.

Proof. By formula (5), we have

$$\|x_u(t) - x^*\|_2 = \left\| \arg\min_x Z(f_{uv}, x_u(t), x_u) \right.$$
$$- \arg\min_x Z(f_{uv}, x_u^*, x_u) \tag{15}$$
$$+ \left. p_u \begin{pmatrix} y - \Phi x_u(t) - (y^* - \Phi x_u^*) \\ 0 \end{pmatrix} \right\|_2$$

$$\leq \left\| \arg\min_x Z(f_{uv}, x_u(t), x_u) \right.$$
$$\left. - \arg\min_x Z(f_{uv}, x_u^*, x_u) \right\|_2 \tag{16}$$
$$+ \left\| p_u \begin{pmatrix} y - \Phi x_u(t) - (y^* - \Phi x_u^*) \\ 0 \end{pmatrix} \right\|_2$$

$$\leq \left\| \arg\min_x Z(f_{uv}, x_u(t), x_u) \right.$$
$$\left. - \arg\min_x Z(f_{uv}, x_u^*, x_u) \right\|_2 + p_u \|y - y^*\|_2 \tag{17}$$
$$+ p_u \|\Phi x_u(t) - \Phi x_u^*\|_2.$$

The optimal value x_u^* is the minimum value of $x_u(t)$; formula (17) is transformed into

$$\left\| \arg\min_x Z(f_{uv}, x_u(t), x_u) \right.$$
$$\left. - \arg\min_x Z(f_{uv}, x_u^*, x_u) \right\|_2 = 0. \tag{18}$$

Thus, formula (17) becomes

$$= p_u \|y - y^*\|_2 + p_u \|\Phi x_u(t) - \Phi x_u^*(t)\|_2$$
$$\leq p_u \|y - y^*\|_2 + p_u \|\Phi\|_2 \|x_u(t) - x_u^*(t)\|_2. \tag{19}$$

Therefore formula (15) becomes

$$\|x_u(t) - x^*\|_2 \leq \frac{p_u}{1 - p_u \|\Phi\|_2} \|y - y^*\|_2$$
$$\leq \frac{p_u \delta_u}{1 - p_u \|\Phi\|_2}. \tag{20}$$

Thus, algorithm (5) converges statistically to within a small neighborhood of the optimal values x^*.

Since p_u is continuous, Theorem 2 implies that the input signal approaches the optimal x^* when δ_u is small enough.

4. Simulation Analysis

In this section, we exhibit numerical examples and simulation results for the proposed algorithm (cross-layer optimal design, CLOD). We conduct numerical experiments using NS2 to confirm the efficiency of the proposed algorithm. We also perform simulations using Lee and Lim [3] and DCH to validate our assumptions. In our numerical examples, we set

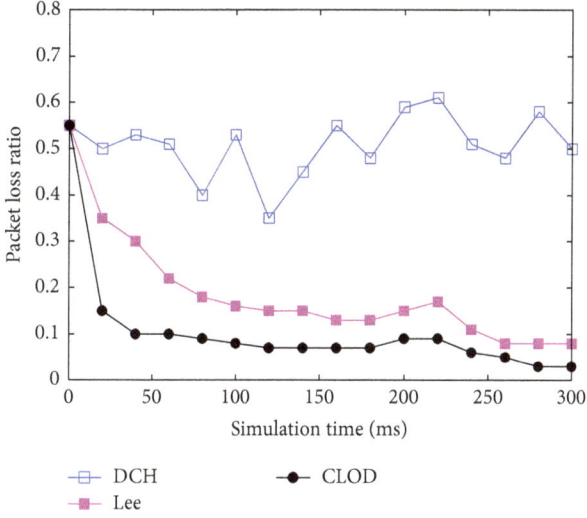

FIGURE 2: Comparison of dropped packet.

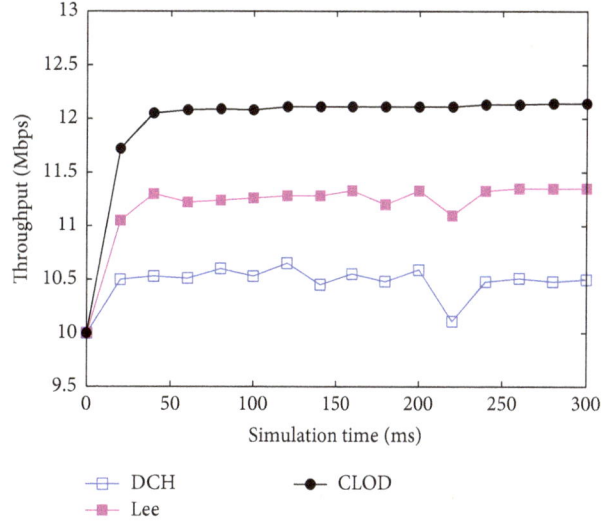

FIGURE 3: Comparison of throughput.

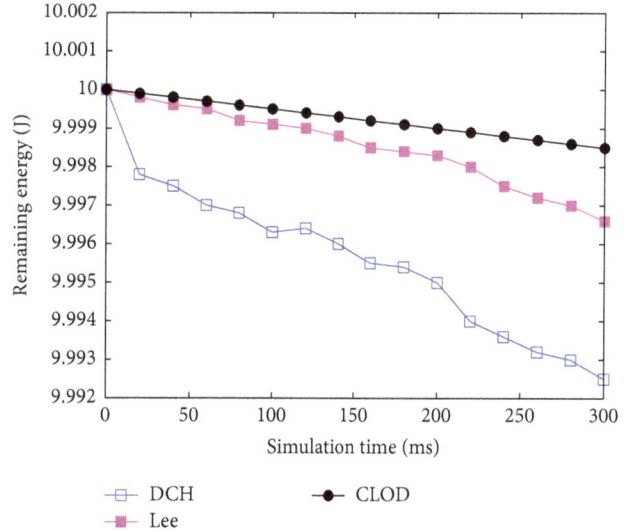

FIGURE 4: Comparison of remaining energy.

a net topology with 100 nodes placed randomly in a 10×10 field, where the distance of neighboring nodes is set to 20 m. The buffers are 10–100 packets. Packet size is 1024 bits, and simulation time is 300 ms.

Figure 2 indicates that the dropped packets of the Lee algorithm and CLOD algorithm are relatively lower. This illustrates that the algorithms could relieve congestion at a certain extent. However, there are some differences in the algorithms: CLOD dropped packet is the lowest when node-level congestion and link-level congestion are simultaneously existing. This verifies that the algorithm achieves better network control by signal compressed and channel selection; the dropped packet is lower in CLOD algorithm than Lee algorithm, which is owing to only dispose node-level congestion and ignore link-level congestion in Lee algorithm; DCH makes the congestion stay at the top because the coping mechanism for congestion is weak.

Figure 3 shows that CLOD makes transmitted packet remarkably increase per second for signal compressed and channel selection strategy. The other algorithms cannot achieve so high throughput for required transmitted data too large resulting in congestion. In addition, the Lee algorithm and DCH algorithm show incapability of channel contention triggered congestion.

Figure 4 displays comparison with energy in three algorithms. From Figure 4, it can be observed that the best results of CLOD can reach the lowest energy consumption. Such a phenomenon implies that compressed data makes the transmission traffic drastically decrease, which can effectively reduce energy consumption, and distinctly relieve node-level congestion. Channel selection makes certain positive contribution to save energy, and link capacity allocation can be used to take full advantage of limited energy. The others are to be considerably inferior to CLOD in this problem.

5. Conclusions

In this paper, we propose a cross-layer optimal protocol by integrating using "cross-layer design," "optimal theory," and "compressed sensing" to achieve higher throughput and power efficiency. In the power control protocol, by adjusting congestion rate to the reduction of power, we could attain lower power consumption than the former algorithms. In the proposed input signal control algorithm, the signal size is adjusted to stable state. Link capacity allocation is presented by supply and demand function of the service. Channel is selected by energy minimum optimal. Through the analysis, accuracy of signal transmission is guaranteed, although signal is to be lossy compression. In addition, simulation results show that the proposed algorithm offers better performance in terms of throughput and power consumption compared with the other protocols.

Conflict of Interests

The authors declare that there is no conflict of interests regarding the publication of this paper.

Acknowledgments

This work has been supported by the Fundamental Research Funds for the Central Universities of China no. N142303013 and the Program of Science and Technology Research of Hebei University no. QN2014326.

References

[1] J. Cheng, Q. Ye, H. Jiang, D. Wang, and C. Wang, "STCDG: an efficient data gathering algorithm based on matrix completion for wireless sensor networks," *IEEE Transactions on Wireless Communications*, vol. 12, no. 2, pp. 850–861, 2013.

[2] Y. Li, M. Sheng, C. Wang, X. Wang, Y. Shi, and J. Li, "Throughput-delay tradeoff in interference-free wireless networks with guaranteed energy efficiency," *IEEE Transactions on Wireless Communications*, vol. 14, no. 3, pp. 1608–1621, 2015.

[3] H.-J. Lee and J.-T. Lim, "Cross-layer congestion control for power efficiency over wireless multihop networks," *IEEE Transactions on Vehicular Technology*, vol. 58, no. 9, pp. 5274–5278, 2009.

[4] W. Chen and I. J. Wassell, "Optimized node selection for compressive sleeping wireless sensor networks," *IEEE Transactions on Vehicular Technology*, 2015.

[5] L. Orihuela, F. Gomez-Estern, and F. R. Rubio, "Scheduled communication in sensor networks," *IEEE Transactions on Control Systems Technology*, vol. 22, no. 2, pp. 801–808, 2014.

[6] D. Liao and A. K. Elhakeem, "A cross-layer joint optimization approach for multihop routing in TDD-CDMA wireless mesh networks," *European Transactions on Telecommunications*, vol. 23, no. 1, pp. 6–15, 2012.

[7] T. Xue, X. D. Dong, and Y. Shi, "Multiple access and data reconstruction in wireless sensor networks based on compressed sensing," *IEEE Transactions on Wireless Communications*, vol. 12, no. 7, pp. 3399–3411, 2013.

[8] J. Chen, S. He, Y. Sun, P. Thulasiraman, and X. M. Shen, "Optimal flow control for utility-lifetime tradeoff in wireless sensor networks," *Computer Networks*, vol. 53, no. 18, pp. 3031–3041, 2009.

[9] S. Y. Xiang and L. Cai, "Transmission control for compressive sensing video over wireless channel," *IEEE Transactions on Wireless Communications*, vol. 12, no. 3, pp. 1429–1437, 2013.

[10] Y. Sadi and S. C. Ergen, "Optimal power control, rate adaptation, and scheduling for UWB-based intravehicular wireless sensor networks," *IEEE Transactions on Vehicular Technology*, vol. 62, no. 1, pp. 219–234, 2013.

[11] S. B. He, J. M. Chen, D. K. Y. Yau, and Y. X. Sun, "Cross-layer optimization of correlated data gathering in wireless sensor networks," *IEEE Transactions on Mobile Computing*, vol. 11, no. 11, pp. 1678–1691, 2012.

[12] Y. B. Wang, M. C. Vuran, and S. Goddard, "Cross-layer analysis of the end-to-end delay distribution in wireless sensor networks," *IEEE/ACM Transactions on Networking*, vol. 20, no. 1, pp. 305–318, 2012.

[13] X. Zhang, H. V. Poor, and M. Chiang, "Optimal power allocation for distributed detection over MIMO channels in wireless sensor networks," *IEEE Transactions on Signal Processing*, vol. 56, no. 9, pp. 4124–4140, 2008.

[14] G. A. Shah, V. C. Gungor, and O. B. Akan, "A cross-layer QoS-aware communication framework in cognitive radio sensor networks for smart grid applications," *IEEE Transactions on Industrial Informatics*, vol. 9, no. 3, pp. 1477–1485, 2013.

[15] L. Shi and A. O. Fapojuwo, "Cross-layer optimization with cooperative communication for minimum power cost in packet error rate constrained wireless sensor networks," *Ad Hoc Networks*, vol. 10, no. 7, pp. 1457–1468, 2012.

[16] M. C. Vuran and I. F. Akyildiz, "XLP: a cross-layer protocol for efficient communication in wireless sensor networks," *IEEE Transactions on Mobile Computing*, vol. 9, no. 11, pp. 1578–1591, 2010.

[17] P. F. Zhang and S. Jordan, "Cross layer dynamic resource allocation with targeted throughput for WCDMA data," *IEEE Transactions on Wireless Communications*, vol. 7, no. 12, pp. 4896–4906, 2008.

[18] G. Athanasiou, T. Korakis, O. Ercetin, and L. Tassiulas, "A cross-layer framework for association control in wireless mesh networks," *IEEE Transactions on Mobile Computing*, vol. 8, no. 1, pp. 65–80, 2009.

[19] P. Yang and B. Chen, "To listen or not: distributed detection with asynchronous transmissions," *IEEE Signal Processing Letters*, vol. 22, no. 5, pp. 628–632, 2015.

[20] H. Mamaghanian, N. Khaled, D. Atienza, and P. Vandergheynst, "Compressed sensing for real-time energy-efficient ECG compression on wireless body sensor nodes," *IEEE Transactions on Biomedical Engineering*, vol. 58, no. 9, pp. 2456–2466, 2011.

[21] F. Fazel, M. Fazel, and M. Stojanovic, "Random access compressed sensing for energy-efficient underwater sensor networks," *IEEE Journal on Selected Areas in Communications*, vol. 29, no. 8, pp. 1660–1670, 2011.

7

Design of CPW-Fed Antenna with Defected Substrate for Wideband Applications

Amar Sharma,[1] Puneet Khanna,[1] Kshitij Shinghal,[2] and Arun Kumar[1]

[1]Electronics and Communication Engineering, IFTM University, Moradabad, India
[2]Electronics and Communication Engineering, MIT, Moradabad, India

Correspondence should be addressed to Amar Sharma; amar.charu@gmail.com

Academic Editor: John N. Sahalos

A CPW-fed defected substrate microstrip antenna is proposed. The proposed antenna shows wideband applications by choosing suitable defected crown shaped substrate. Defected substrate also reduces the size of an antenna. The radiating patch of proposed antenna is taken in the form of extended U-shape. The space around the radiator is utilized by extending the ground plane on both sides of radiator. Simulation of proposed antenna is done on Ansoft's High Frequency Structure Simulator (HFSS v. 14). Measured results are in good agreement with simulated results. The prototype is taken with dimensions 36 mm × 42 mm × 1.6 mm that achieves good return loss, constant group delay, and good radiation characteristics within the entire operating band from 4.5 to 13.5 GHz (9.0 GHz) with 100% impedance bandwidth at 9.0 GHz centre frequency. Thus, the proposed antenna is applicable for C and X band applications.

1. Introduction

In the rapid increasing technology of wireless communication, there is a great demand of compact, low profile, low cost, and light weight microstrip antenna [1]. Microstrip antennas are mostly used in military and commercial applications. However, the main disadvantage of microstrip antenna is narrow bandwidth that limits its applications. Enhancement in performance is necessary to cover the demand of wide impedance bandwidth. For obtaining wide impedance bandwidth, lots of techniques have been recommended such as defected structure shape, defected ground structure, slotted patch antenna, stacked patch antenna, and planar monopole antenna [2–7]. Several studies have been reported by the researchers on the microstrip patch antenna with defected ground structure for obtaining wideband/ultrawideband (UWB) [6, 8–11]. However, in these types of antenna, large space available on both sides of the radiator is not fully utilized and increases the cost of antenna. In these circumstances, coplanar waveguide (CPW) fed microstrip patch antennas play a vital role in utilizing the space available around the radiator [12–18]. In CPW-fed technique the radiating element and the ground plane are on the same side of the substrate. Small amount of work is also reported on defected substrate technique [19]. By making defected substrate, a new structure is formed that shows wideband characteristics.

In this paper, a defected crown shaped substrate wideband microstrip antenna is proposed and designed. The proposed antenna possesses a method to minimize the substrate size so that the overall size of the antenna can be minimized. The proposed antenna uses a crown shaped substrate over the conventional rectangular substrate for reducing the overall dimensions of patch antenna. The ground plane and the radiator plane are on single side of the substrate, so that the large space around the radiator can be fully utilized. In the next section the antenna geometry is discussed in detail. Section 3 covers the parametric study of proposed antenna in detail. Experimental results are discussed in Section 4. Section 5 covers all the discussion made in earlier sections.

2. Antenna Geometry

The geometry of the proposed antenna is shown in Figure 1. It has overall dimensions $L_s \times W_s$ and is fabricated on low cost FR4 substrate of thickness h = 1.6 mm, whose

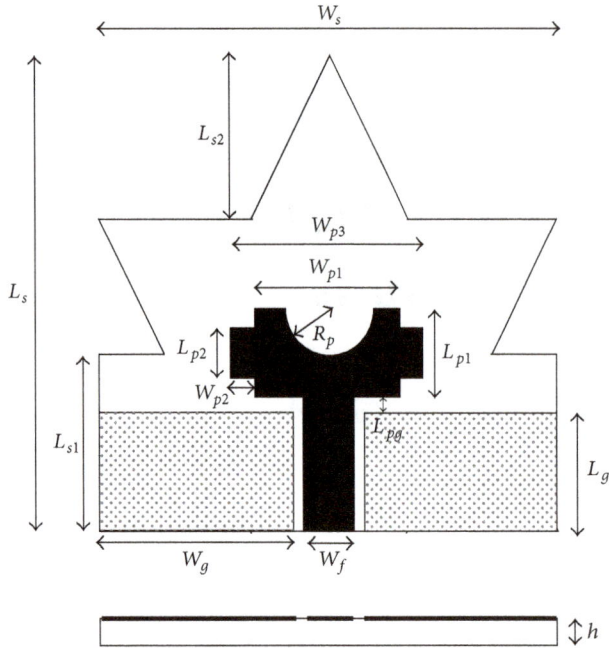

FIGURE 1: Schematic configuration of the proposed defected substrate CPW-fed wideband antenna.

FIGURE 2: Photograph of the fabricated defected substrate CPW-fed wideband antenna.

relative permittivity ϵ_r = 4.4 and loss tangent $\tan\delta$ = 0.0019. A photograph of fabricated defected substrate CPW-fed wideband antenna is shown in Figure 2. The size of the proposed antenna is obtained by using mathematical formulation for patch antennas. In the proposed antenna the mathematical modelling is based on rectangular patch, but due to five cutting slots in radiating patch the overall size calculation is not as simple as rectangular patch antenna. Therefore it has been optimized by using electromagnetic solver, Ansoft HFSS simulation software [20]. Designing of radiating patch element includes the estimation of its

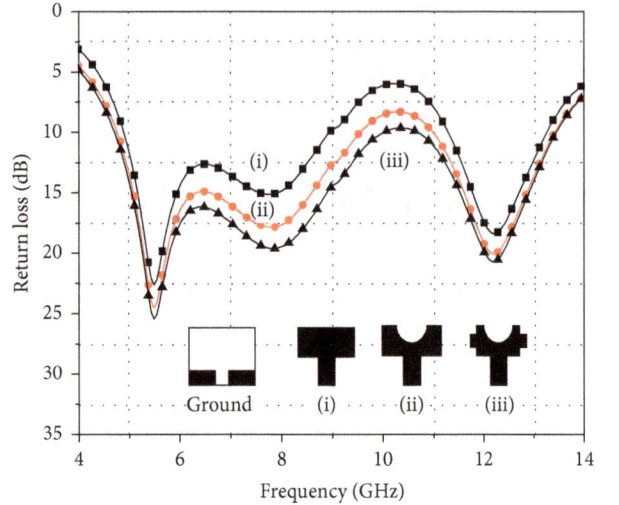

FIGURE 3: Simulated return loss against frequency for the (i) rectangle shape radiator, (ii) arc cut rectangle shape radiator, and (iii) proposed radiator of defected substrate wideband antenna.

dimensions. The patch width (W) has small effect on the resonance and it has been obtained by using the mathematical modelling as shown below [21].

$$W = \frac{v_0}{2f_r}\sqrt{\frac{2}{\epsilon_r + 1}}, \tag{1}$$

where v_0 is the speed of light in free space and ϵ_r is the relative permittivity of the substrate material of the proposed antenna. The microstrip patch is on the top of dielectric material; therefore the electromagnetic wave has an effective permittivity (ϵ_{eff}) which is given by [21]

$$\epsilon_{\text{eff}} = \frac{\epsilon_r + 1}{2} + \frac{\epsilon_r - 1}{2}\left[1 + \frac{10h}{W}\right]^{1/2}. \tag{2}$$

The length of the radiating patch (L) plays a major role in finding the resonant frequency and it is an important parameter in designing of patch antenna due to the inherent narrow bandwidth of the patch. The following value of L can be determined by using the following formula:

$$L = \frac{v_0}{2f_r\sqrt{\epsilon_{\text{eff}}}} - 2\Delta l, \tag{3}$$

where ϵ_{eff} is the effective permittivity of the substrate material of the proposed antenna. The additional line length on ΔL both ends of the patch length, due to the effect of fringing fields, is given by [22]

$$\frac{\Delta L}{h} = 0.412\left[\frac{\epsilon_{\text{eff}} + 0.3}{\epsilon_{\text{eff}} - 0.258}\right]\left[\frac{W/h + 0.264}{W/h + 0.813}\right]. \tag{4}$$

The effective length patch length L_e can be written as [21]

$$L_e = l + 2\Delta l. \tag{5}$$

The further variations in the radiator are shown in Figure 3, having dimensions listed as Table 1. The base of the microstrip

TABLE 1: Design parameters of the proposed defected substrate CPW-fed wideband antenna.

Parameters	Units (mm)
W_s	42
L_s	36
W_f	3
L_{s1}	12
L_{s2}	12
W_g	19.1
R_p	3
L_g	10
L_{pg}	0.8
L_{p1}	7
L_{p2}	3
W_{p1}	12
W_{p2}	2
W_{p3}	16

antenna is rectangular in shape which is shown in trace (i) of Figure 3. Initially the shape of the radiating patch antenna is taken as rectangular slot $L_{p1} \times W_{p3}$. The simulated result shows that this structure excites at resonating frequencies at 5.48, 7.86, and 12.17 GHz but does not cover the entire operating band from 4.5 to 13.5 GHz. After that a semicircular ring is etched in the middle of the radiating patch with radius R_p which is shown in trace (ii) of Figure 3; the simulated result shows that this structure improves the return loss condition but still does not cover the entire operating band. Finally, the edges having dimensions $(L_{p1} - L_{p2}) \times W_{p2}$ are etched on both sides of the radiating patch which is shown in trace (iii) of Figure 3; the simulated result shows that the proposed radiating patch obtains good return loss value but still the return loss is above 10 dB from 9.96 to 10.68 GHz. The detailed dimensions of the proposed defected substrate CPW-fed wideband antenna are listed in Table 1.

The ground plane is on the same plane as radiator with two rectangular slits having dimensions $W_g \times L_g$. The length of gap between the radiating patch and the ground plane is taken as L_{pg}. The width of CPW-fed line is fixed at W_f to achieve 50 ohm characteristics impedance. The gap between the feed and ground plane is taken as 0.4 mm. The radiator is surrounded by a ground plane with etched substrate that helps to reduce the area. The small gap between the radiator and the ground is a foremost factor to provide strong capacitive coupling.

The simulated surface current distributions on resonant frequencies of 5.48 GHz, 8.03 GHz, and 12.01 GHz are shown in Figure 4. When the microstrip patch antenna is provided with power, a charge distribution appears on the upper and lower part of the patch, as well as on the ground plane. Due to this charge distribution the current will flow at the top and bottom surface of the patch. From this closed analysis of the surface current distribution of the proposed antenna it is found out that at 5.48 GHz, 8.03 GHz, and 12.01 GHz resonant frequency the proposed antenna resonates in TM11 mode, TM21 mode, TM12 mode, respectively. The above mentioned

mode can be explained on the basis of surface current distribution of the proposed antenna; from Figure 4(a) it can be observed that the direction of surface current is aligned by the side of circumference and terminated at one point on the circumference of patch; that is, at this location of patch negative node are located and on the opposite side of the patch positive node are located. Thus only one positive node and one negative node are located, which is the case of TM11 mode [23, 24]. Similarly, from Figure 4(b) it can be observed that the current is aligned and terminated at two points on the circumference of patch. Thus there are two positive and two negative nodes are located, which is the case of TM21 mode. While in Figure 4(c) it can be observed that the current is aligned towards tangential of circumference at two points, which is the case of TM12 mode.

The substrate is etched in the form of crown shape. The variation in substrate is shown in Figure 5, having radiator and ground plane dimensions as listed in Table 1. Initially, a rectangular substrate is taken as shown in trace (i). The simulated result shows that this structure excites at resonating frequencies at 5.48, 7.86, and 12.17 GHz but does not cover the entire operating band from 4.5 to 13.5 GHz as the return loss is above 10 dB from 9.96 to 10.68 GHz. However, when the substrate is taken in the form of triangle shape (trace (ii)), the simulated result shows better return loss but does not cover the entire operating band as return loss is above 10 dB from 10.13 to 10.46 GHz. At last, when the substrate is defected into crown shape, as shown in trace (iii), the simulated result covers entire operating bandwidth (4.5 to 13.5 GHz) with three resonating bands at 5.48, 8.03, and 12.01 GHz. Therefore, it is decided to take defected substrate (crown shaped) antenna for further investigations as it is smaller in size and improves the impedance matching conditions for the entire band.

3. Parametric Study of the Proposed Antenna

In this section, the influence of the different design parameters on proposed antenna performance is presented and discussed. At a time, variation in single parameter is done while others are kept constant. The optimization of parameters is helpful for the fabrication of the proposed defected substrate antenna. The effect of change in radiating patch length (L_{p1}, L_{p2}), width (W_{p1}, W_{p2}), inner circle radius (R_p), microstrip feed line (W_f), and length between radiating patch and ground plane (L_{pg}) is considered for parametric study.

Figure 6, shows the variation in length of radiating patch (L_{p1}) of the proposed antenna from 6 to 8 mm. When $L_{p1} = 6$ mm, the return loss remains lower than 10 dB but bandwidth is reduced (4.8 to 13.28 GHz). With further increase in $L_{p1} = 7$ mm, the return loss remains lower than 10 dB with improved impedance bandwidth from 4.5 to 13.5 GHz (9.0 GHz). However, as L_{p1} increases to 8 mm, the bandwidth for the return loss does not remain lower than 10 dB for the entire band. Therefore, it is decided to take $L_{p1} = 7$ mm as the optimum value from 4.5 to 13.5 GHz, covering the entire wideband.

The simulated results of the proposed antenna for patch length L_{p2}, from 1 to 4 mm are depicted in Figure 7. When $L_{p2} = 1$ mm, the bandwidth for the return loss does not

(a)

(b)

(c)

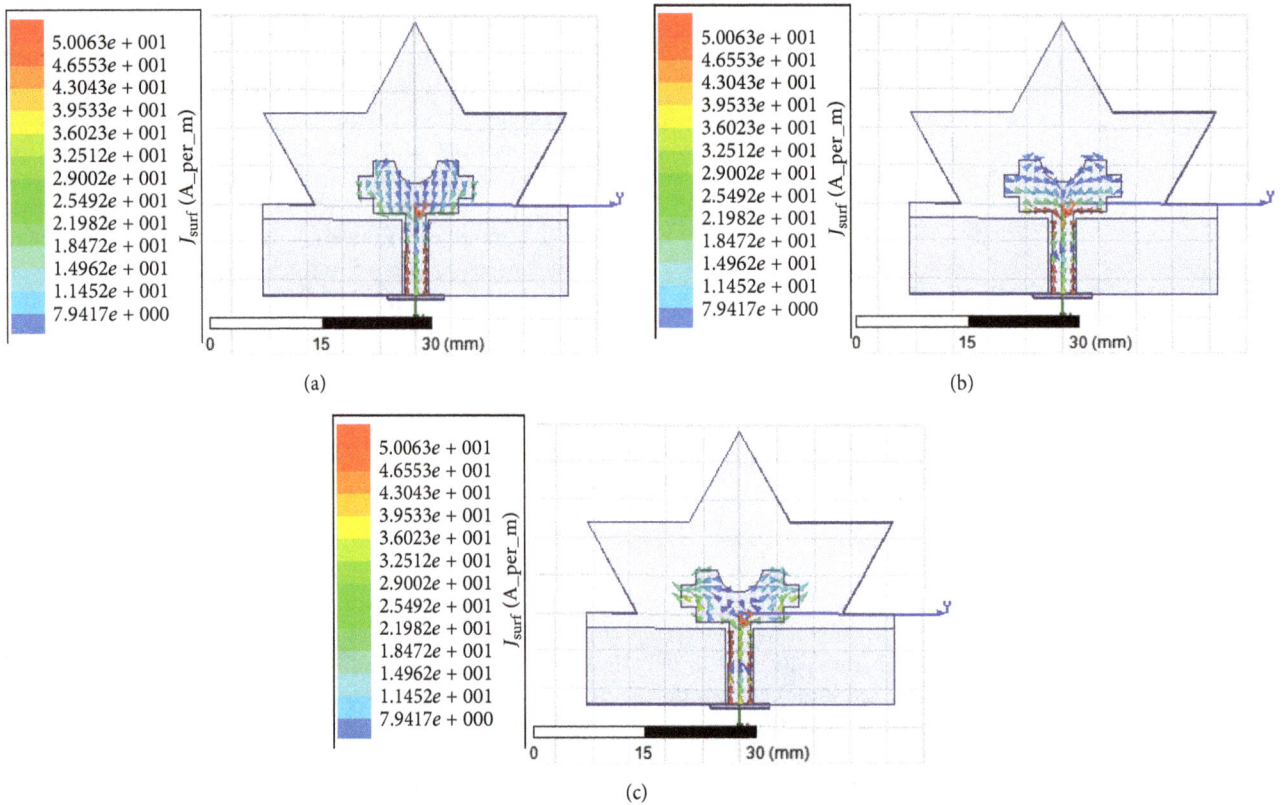

FIGURE 4: The surface current distribution on the proposed defected substrate wideband antenna at (a) 5.48 GHz, (b) 8.03 GHz, and (c) 12.01 GHz.

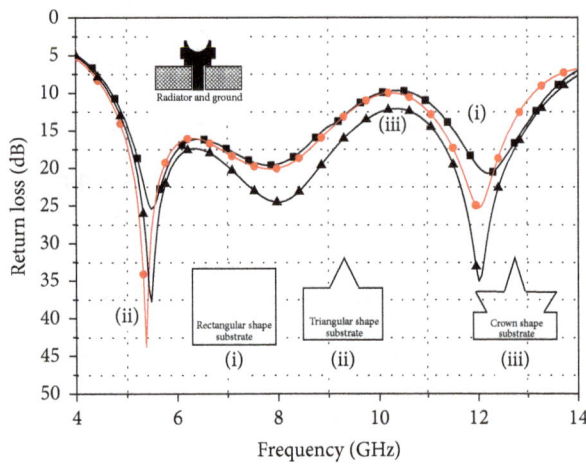

FIGURE 5: Simulated return loss against frequency for the proposed defected substrate (crown shaped) CPW-fed wideband antenna, defected substrate (triangle shape) CPW-fed antenna, and rectangular substrate CPW-fed antenna.

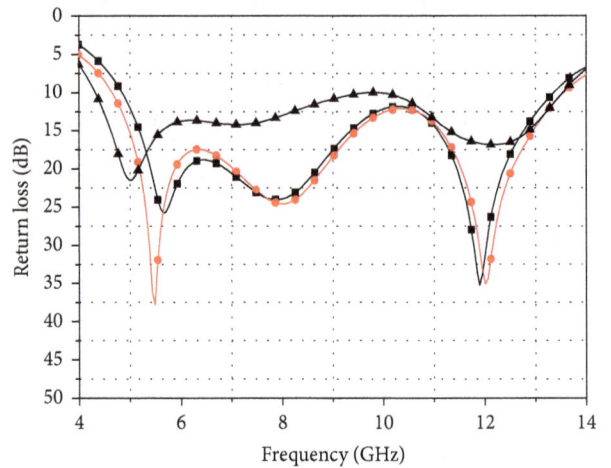

$L_{p1} = 6$ mm
$L_{p1} = 7$ mm
$L_{p1} = 8$ mm

FIGURE 6: Simulated return loss against frequency for the proposed defected substrate wideband antenna with various L_{p1}; other parameters are the same as listed in Table 1.

remain lower than 10 dB for the entire band as it is above 10 dB between frequencies 9.41 and 10.90 GHz. When $L_{p2} =$ 2 mm the return loss does not remain lower than 10 dB for frequencies between 10.24 and 10.60 GHz. For $L_{p2} =$ 3 mm, the bandwidth improves significantly covering the entire band with improved impedance matching. On further

increase in $L_{p2} = 4$ mm, the operational bandwidth decreases and a worse matching condition appears over the frequency band. Therefore, it is decided to take $L_{p1} = 3$ mm as the optimum value covering the band from 4.5 to 13.5 GHz, covering the entire wideband.

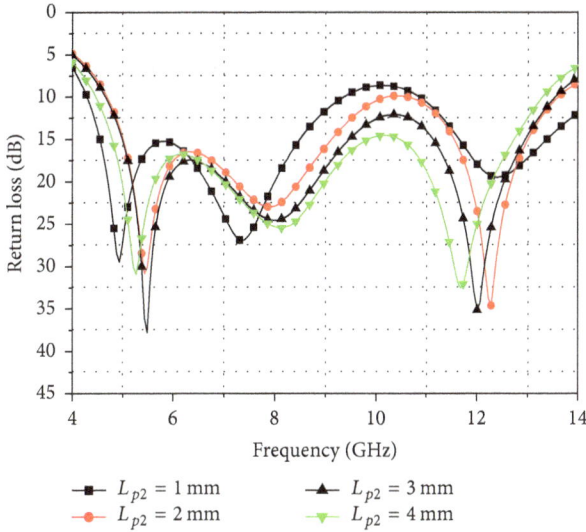

FIGURE 7: Simulated return loss against frequency for the proposed defected substrate wideband antenna with various L_{p2}; other parameters are the same as listed in Table 1.

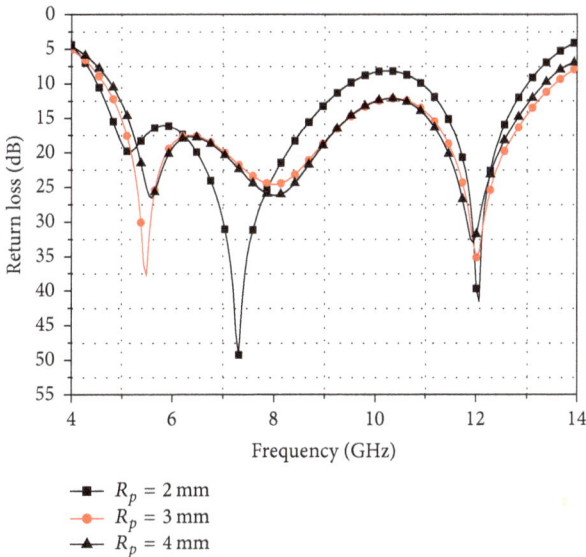

FIGURE 8: Simulated return loss against frequency for the proposed defected substrate wideband antenna with various R_p; other parameters are the same as listed in Table 1.

Figure 8 shows variation of radius of inner circle of radiating patch (R_p) of the proposed antenna from 2 to 4 mm. As R_p = 2 mm, it is observed that the bandwidth of antenna is above 10 dB from 9.52 to 11.01 GHz and does not cover the entire operating band. However, as the value of R_p increases to 3 mm, the impedance matching of the radiating patch and the input impedance improve and cover the entire bandwidth. On further increase of R_p = 4 mm, the impedance matching and input impedance of radiating patch deteriorate. Therefore, it is decided to take R_p = 3 mm as the optimum value with the bandwidth from 4.5 to 13.5 GHz.

The simulated results of the proposed antenna, with microstrip feed (W_f), varying from 2.9 to 3.1 mm are illustrated in Figure 9. It can be seen that the bandwidth of return

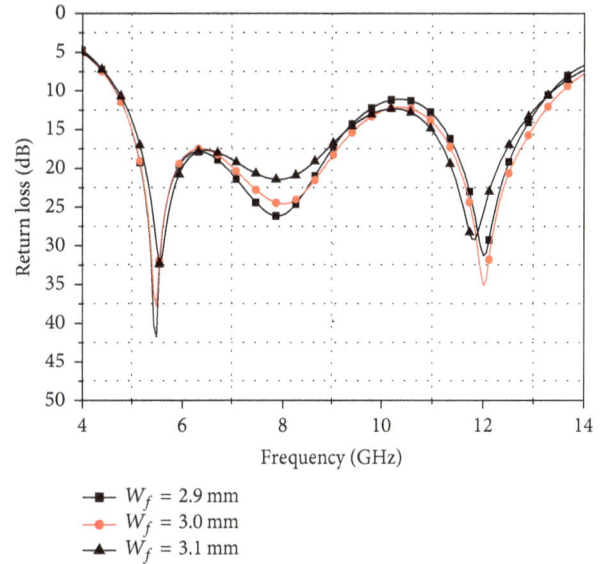

FIGURE 9: Simulated return loss against frequency for the proposed defected substrate wideband antenna with various W_f; other parameters are the same as listed in Table 1.

loss less than 10 dB of the antenna remains constant as W_f increases from 2.9 to 3.1 mm. However, as W_f increases from 2.9 to 3.0 mm, the impedance matching of the radiating patch and the input impedance of the frequency band from 11.20 to 12.32 GHz are improved significantly. With further enhancement in microstrip feed (W_f), it deteriorates. Therefore, it is decided to take W_f = 3.0 mm as the optimum value, with minimum mismatch at higher frequency range.

The variation in patch width (W_{p1}) from 10 to 13 mm is shown in Figure 10. For W_{p1} = 10 mm the bandwidth is above 10 dB for frequencies from 10.68 to 11.40 GHz and does not cover the entire operating band. For W_{p1} = 11 mm, the bandwidth of return loss remains less than 10 dB, but the total band is reduced from 4.5 to 11.40 GHz (6.9 GHz). However, as width W_{p1} increases to 12 mm, the impedance bandwidth significantly improves covering the entire band from 4.5 to 13.5 GHz. With further increase in W_{p1} = 13 mm, the bandwidth of return loss is higher than 10 dB from frequency 10.02 to 10.68 GHz. Therefore, it is decided to take W_{p1} = 12 mm as the optimum value for the bandwidth from 4.5 to 13.5 GHz, covering the entire bandwidth.

Figure 11, shows the variation of radiator patch width (W_{p2}) from 1.0 to 3.0 mm. As W_{p2} = 1.0 mm, the bandwidth of return loss remains less than 10 dB for the entire band from 4.6 to 13.8 GHz, but the impedance matching of the radiating patch with input impedance deteriorates at higher frequencies. When W_{p2} = 1.5 mm, the return loss does not remain less than 10 dB for the entire band as it is above than 10 dB from frequencies 9.7 to 10.8 GHz. As W_{p2} = 2.0 mm, the impedance matching of the radiating patch and the input impedance at the frequency improve covering the entire bandwidth from 4.5 to 13.5 GHz. With further increase in W_{p2} the bandwidth shrinks and remains between 4.3 and 12.26 GHz for W_{p2} = 2.5 mm; for W_{p2} = 3.0 mm the bandwidth remains 4.7 to 12.45 GHz and does not cover the

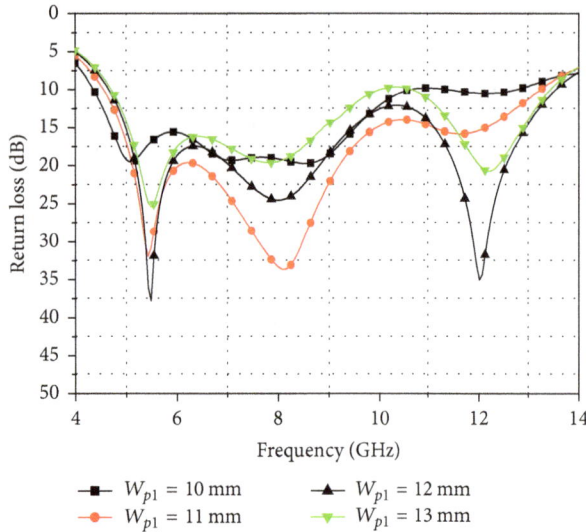

FIGURE 10: Simulated return loss against frequency for the proposed defected substrate wideband antenna with various W_{p1}; other parameters are the same as listed in Table 1.

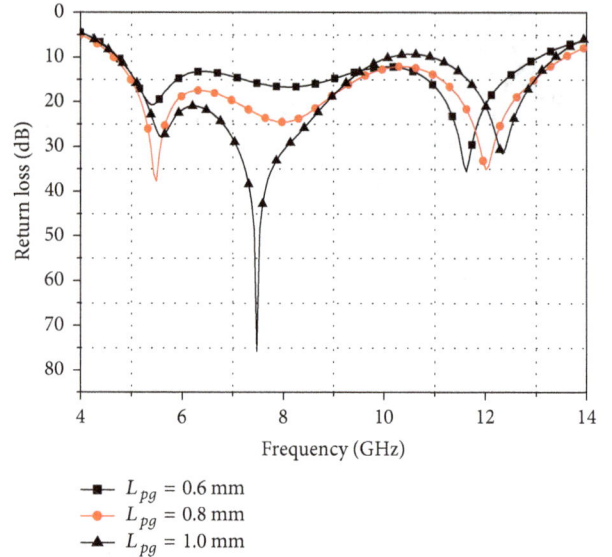

FIGURE 12: Simulated return loss against frequency for the proposed defected substrate wideband antenna with various L_{pg}; other parameters are the same as listed in Table 1.

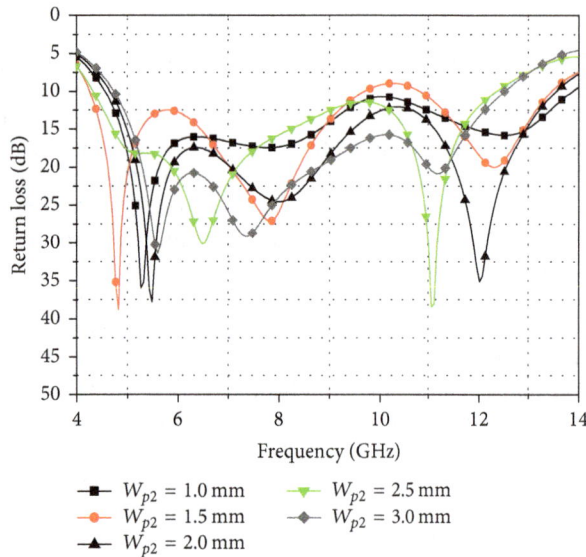

FIGURE 11: Simulated return loss against frequency for the proposed defected substrate wideband antenna with various W_{p2}; other parameters are the same as listed in Table 1.

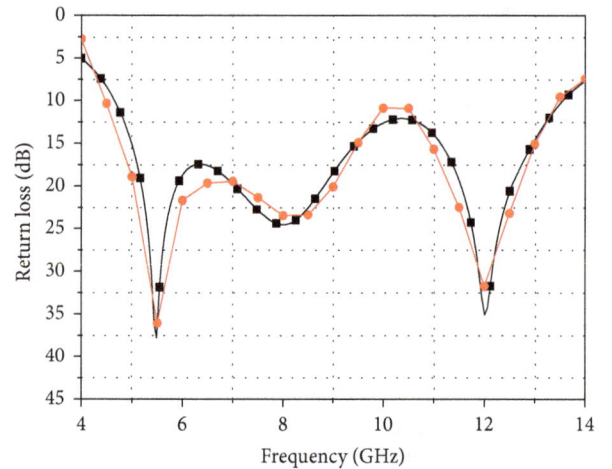

FIGURE 13: Simulated and measured return loss for the proposed defected substrate wideband antenna.

entire operating band. Therefore, it is decided to take W_{p2} = 2.0 mm as the optimum value covering the entire bandwidth from 4.5 to 13.5 GHz.

The variation in patch width (L_{pg}) from 0.6 to 1.0 mm is shown in Figure 12. When L_{pg} = 0.6 mm, the return loss is less than 10 dB but the bandwidth is 4.7 to 13.0 GHz which is less than the desired bandwidth. When L_{pg} = 0.8 mm the bandwidth of return loss remains less than 10 dB with good impedance matching between radiating patch and input impedance. On further increase of L_{pg} the bandwidth of the return loss is not less than 10 dB and does not cover the entire bandwidth. Therefore, it is decided to take L_{pg} = 0.8 mm as

the optimum value with the bandwidth from 4.5 to 13.5 GHz, covering the entire wideband.

4. Experimental Results and Discussion

The performances of the proposed antenna such as return loss and radiation pattern are measured using Agilent 8757E scalar network vector analyzer. There is a good agreement between simulated and measured results of the proposed defected substrate CPW-fed wideband antenna as shown in Figure 13. The minute variation between measured and simulated result is due to the effect of SMA (subminiature version A) connector soldering and fabrication tolerance.

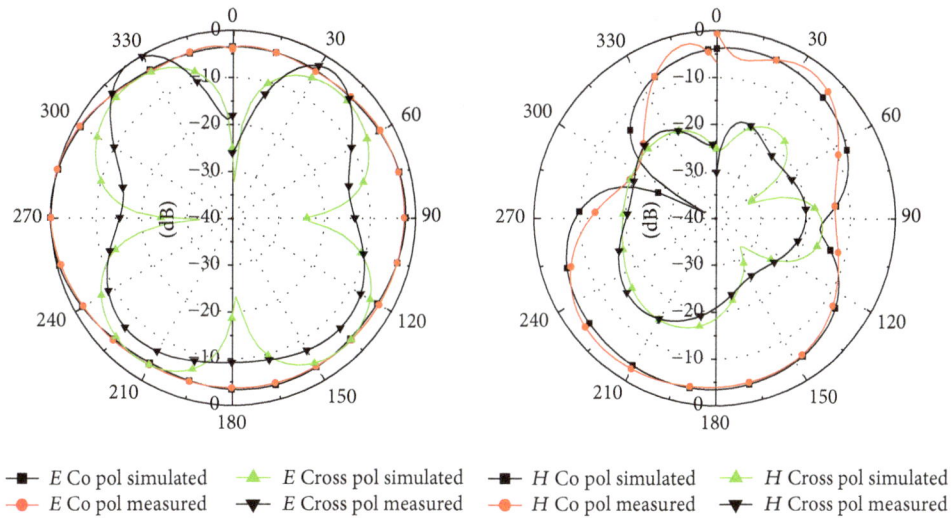

(a)

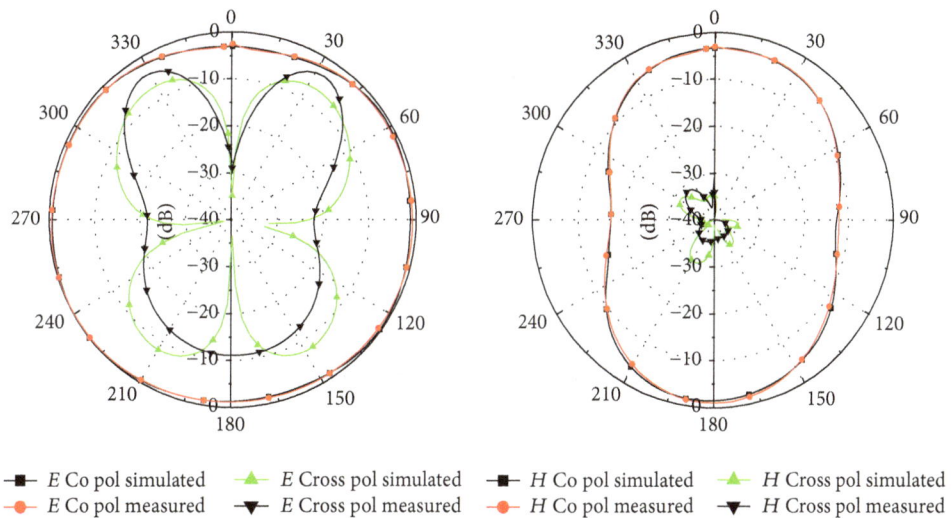

(b)

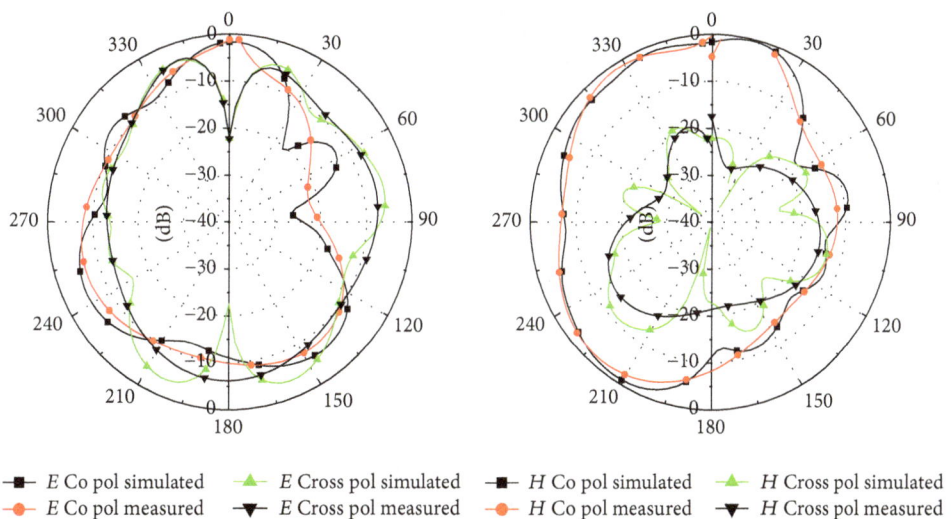

(c)

FIGURE 14: Radiation patterns at various frequencies of proposed defected substrate wideband antenna: (a) 5.48 GHz, (b) 8.03 GHz, and (c) 12.01 GHz.

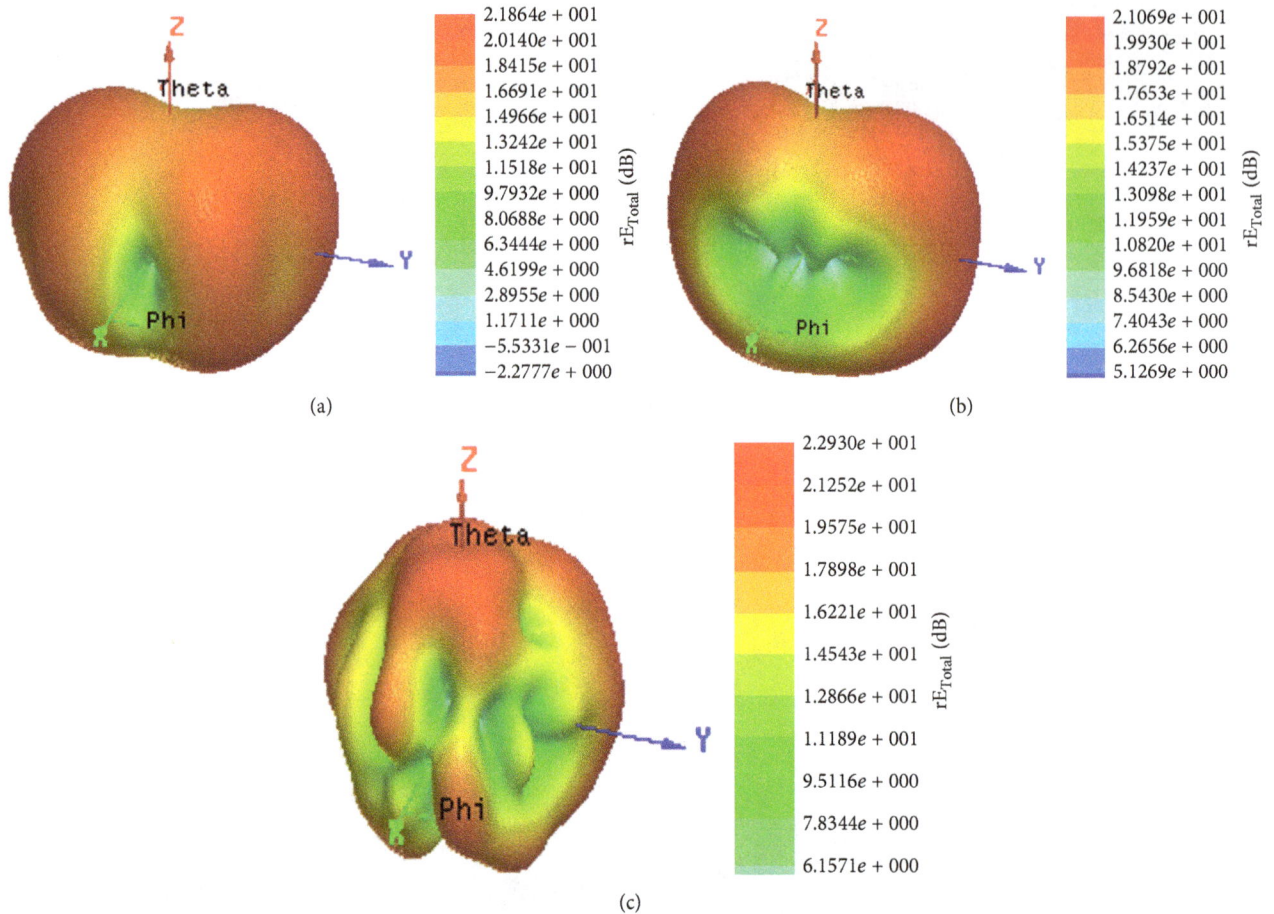

FIGURE 15: 3D polar plot at various frequencies of proposed defected substrate wideband antenna: (a) 5.48 GHz, (b) 8.03 GHz, and (c) 12.01 GHz.

The designed antenna offers a bandwidth of 9.0 GHz (4.5 to 13.5 GHz) that meets the bandwidth requirements for C and X band applications.

The proposed antenna illustrates good radiation pattern characteristics as shown in Figures 14(a)–14(c). The radiation patterns in H and E planes are at sampling frequencies of 5.48, 8.03, and 12.01 GHz, respectively. Patterns are distorted because the ground plane is a part of loop path; the surface current on the radiating plane changes the effective current distribution of the loop and results in distortion. These patterns are suited for application in numerous wireless communication systems, as expected. Measured and simulated results of radiation patterns show good agreement.

Figure 15 shows good results of 3D polar plot of proposed defected substrate wideband antenna at three resonant frequencies 5.48 GHz, 8.03 GHz, and 12.01 GHz. The 3D polar plot gives and additional view for the distribution of the power radiated in space.

Figure 16 illustrates the group delay of the proposed antenna. Group delay is a significant factor in the designing of wideband antenna as it tells about the distortion of the transmitted pulses in the wireless communication. It is observed that the group delay for the proposed antenna is stable and less than 1 ns for entire operating bandwidth 4.5 to

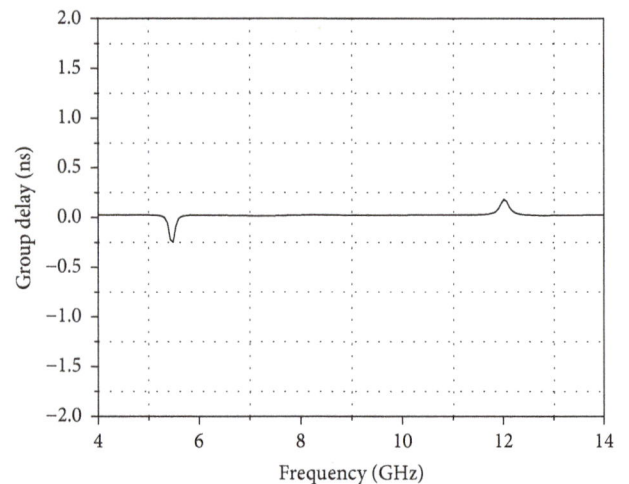

FIGURE 16: Group delay for the proposed defected substrate wideband antenna.

13.5 GHz. For distortion less transmission, group delay should be less than 1 ns in the wideband antenna.

Gain is an important parameter in the designing of wideband microstrip patch antennas. Figure 17 demonstrates

TABLE 2: Comparison of reference antennas with proposed antenna.

S. number	Reference number	Antenna type	Overall size (mm^3)	Operating frequency band (GHz)	Relative dielectric constant (ϵ_r)	Applications
1	[1]	Wideband	32 × 30 × 1.58	3.32–6.5	4.4	WLAN/WiMAX
2	[2]	Wideband	25 × 25 × 1.6	2.96–7.95	4.4	S and C band
3	[4]	Wideband	38 × 25 × 1.6	2.4–6.0	4.4	WLAN/WiMAX
4	[13]	UWB	25 × 25 × 1.6	2.9–11.5	4.4	UWB
5	[19]	UWB	18 × 25 × 1.25	2.0–10.6	4.4	UWB
6	Proposed antenna	Wideband	42 × 36 × 1.6	4.5–13.5	4.4	C and X band

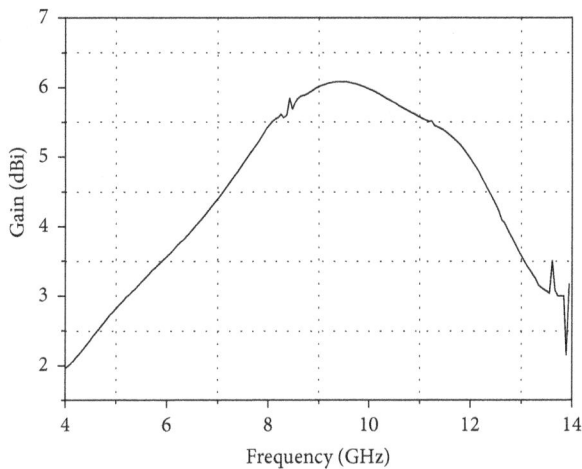

FIGURE 17: Gain for the proposed defected substrate wideband antenna.

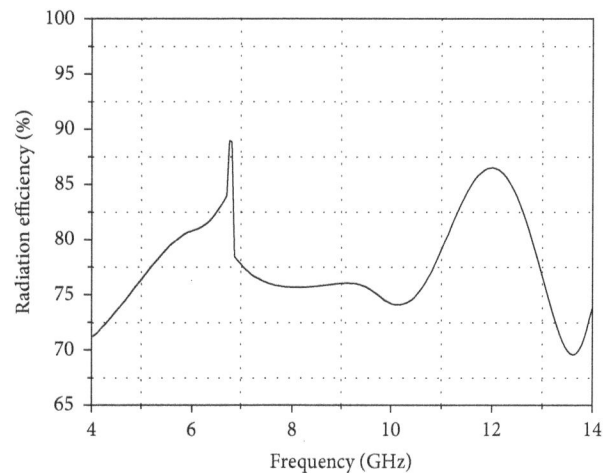

FIGURE 18: Radiation efficiency for the proposed defected substrate wideband antenna.

the gain of the proposed antenna. It was found out that the gain of the antenna varies within 1.9 to 6.08 dBi against the frequency band of 4.5 to 13.5 GHz.

Antenna radiation efficiency is defined as the ratio of power radiated to the input power. It is related to the gain and directivity of the patch antenna. Radiation efficiency also considers the conduction and dielectric losses. The radiation efficiency of the proposed antenna is shown in Figure 18. It exhibits 79.35%, 75.66%, and 86.44% radiation efficiency at three resonant frequencies 5.48 GHz, 8.03 GHz, and 12.01 GHz.

Table 2, illustrates the comparison between the proposed (defected substrate) antenna and some other existing antennas in terms of the antenna type, overall size, operating frequency band, dielectric constant and applications. The comparative chart shows that the proposed antenna has defected substrate and wideband applications with respect to other antenna of different dimension and shapes.

5. Conclusion

A novel CPW-fed defected substrate patch antenna is fabricated and proposed for wideband applications. The proposed antenna achieves good return loss, constant group delay, and good radiation patterns over the entire operating bandwidth

from 4.5 to 13.5 GHz (9.0 GHz) with 100% impedance bandwidth. The gain of the proposed antenna reaches a peak value of 6.08 dBi, while the radiation efficiency of proposed antenna reaches a maximum value of 88%. The simulated and measured results of the projected antenna show balanced agreement. The proposed antenna can be used in numerous wireless applications.

Competing Interests

The authors declare that there are no competing interests regarding publication of this paper.

References

[1] A. Singh and S. Singh, "A novel CPW-fed wideband printed monopole antenna with DGS," *AEU—International Journal of Electronics and Communications*, vol. 69, no. 1, pp. 299–306, 2014.

[2] P. Khanna, A. Sharma, K. Shinghal, and A. Kumar, "A defected structure shaped CPW-fed wideband microstrip antenna for wireless applications," *Journal of Engineering*, vol. 2016, Article ID 2863508, 7 pages, 2016.

[3] A. Nouri and G. R. Dadashzadeh, "A compact UWB band-notched printed monopole antenna with defected ground

structure," *IEEE Antennas and Wireless Propagation Letters*, vol. 10, pp. 1178–1181, 2011.

[4] A. K. Gautam, A. Bisht, and B. K. Kanaujia, "A wideband antenna with defected ground plane for WLAN/WiMAX applications," *AEU—International Journal of Electronics and Communications*, vol. 70, pp. 354–358, 2016.

[5] J. Pei, A.-G. Wang, S. Gao, and W. Leng, "Miniaturized triple-band antenna with a defected ground plane for WLAN/WiMAX applications," *IEEE Antennas and Wireless Propagation Letters*, vol. 10, pp. 298–301, 2011.

[6] M. A. Antoniades and G. V. Eleftheriades, "A compact multi-band monopole antenna with a defected ground plane," *IEEE Antennas and Wireless Propagation Letters*, vol. 7, pp. 652–655, 2008.

[7] K. H. Chiang and K. W. Tam, "Microstrip monopole antenna with enhanced bandwidth using defected ground structure," *IEEE Antennas and Wireless Propagation Letters*, vol. 7, pp. 532–535, 2008.

[8] M. K. Khandelwal, B. K. Kanaujia, S. Dwari, S. Kumar, and A. K. Gautam, "Analysis and design of wide band Microstrip-line-fed antenna with defected ground structure for Ku band applications," *AEU—International Journal of Electronics and Communications*, vol. 68, no. 10, pp. 951–957, 2014.

[9] S. S. Kumar, G. S. Rao, and R. Pillalamarri, "Rectangular slotted microstrip line fed compact printed antenna with etched ground plane for UWB communications," *Microsystem Technologies*, vol. 21, no. 10, pp. 2077–2081, 2014.

[10] M. Bitchikh, R. Aksas, A. Azrar, and H. Kimouche, "A 2.3-14 GHz UWB planar octagonal antenna with modified ground plane," *Microwave and Optical Technology Letters*, vol. 55, no. 3, pp. 479–482, 2013.

[11] A. A. Adam, S. K. A. Rahim, K. G. Tan, and A. W. Reza, "Design of 3.1–12 GHz printed elliptical disc monopole antenna with half circular modified ground plane for UWB application," *Wireless Personal Communications*, vol. 69, no. 2, pp. 535–549, 2013.

[12] A. C. Shagar and R. S. D. Wahidabanu, "New design of CPW-fed rectangular slot antenna forultra wideband applications," *International Journal of Electronics Engineering*, vol. 2, pp. 69–73, 2010.

[13] M. K. Shrivastava, A. K. Gautam, and B. K. Kanaujia, "A novel a-shaped monopole-like slot antenna for ultrawideband applications," *Microwave and Optical Technology Letters*, vol. 56, no. 8, pp. 1826–1829, 2014.

[14] B. R. S. Reddy and D. Vakula, "Compact zigzag-shaped-slit microstrip antenna with circular defected ground structure for wireless applications," *IEEE Antennas and Wireless Propagation Letters*, vol. 14, pp. 678–681, 2015.

[15] Km. Kamakshi, A. Singh, M. Aneesh, and J. A. Ansari, "Novel design of microstrip antenna with improved bandwidth," *International Journal of Microwave Science and Technology*, vol. 2014, Article ID 659592, 7 pages, 2014.

[16] S. M. Naveen, R. M. Vani, and P. V. Hunagund, "Compact wide-band rectangular monopole antenna for wireless applications," *Wireless Engineering and Technology*, vol. 3, no. 4, pp. 240–243, 2012.

[17] N. Prombutr, P. Kirawanich, and P. Akkaraekthalin, "Bandwidth enhancement of UWB microstrip antenna with a modified ground plane," *International Journal of Microwave Science and Technology*, vol. 2009, Article ID 821515, 7 pages, 2009.

[18] R. Kumar, P. Naidu V, and V. Kamble, "A compact asymmetric slot dual band antenna fed by CPW for PCS and UWB applications," *International Journal of RF and Microwave Computer-Aided Engineering*, vol. 25, no. 3, pp. 243–254, 2015.

[19] K. Q. Da Costa and V. Dmitriev, "Planar monopole UWB antenna with cuts at the edges and two parasitic loops," *Journal of Microwaves and Optoelectronics*, vol. 8, no. 1, pp. 92–100, 2009.

[20] Ansoft Corporation, Ansoft High-Frequency Structure Simulator (HFSS), Version 14.0, Ansoft Corporation, Pittsburgh, Pa, USA.

[21] C. A. Balanis, *Antenna Theory: Analysis and Design*, Wiley-Interscience, 3rd edition, 2012.

[22] M. Habib Ullah, M. T. Islam, J. S. Mandeep, N. Misran, and N. Nikabdullah, "A compact wideband antenna on dielectric material substrate for K band," *Electronics and Electrical Engineering*, vol. 123, no. 7, pp. 75–78, 2012.

[23] Y. S. Wu and F. J. Rosenbaum, "Mode chart for microstrip ring resonators (Short Papers)," *IEEE Transactions on Microwave Theory and Techniques*, vol. 21, no. 7, pp. 487–489, 1973.

[24] P. Mythili and A. Das, "Theoretical investigations of an annular elliptical ring microstrip antenna using Green's function technique," *IEEE Proceedings Microwave, Antenna Propagation*, vol. 146, no. 6, pp. 379–384, 1999.

8

PAPR Reduction using Fireworks Search Optimization Algorithm in MIMO-OFDM Systems

Lahcen Amhaimar ⓘ,[1] **Saida Ahyoud**,[2] **Ali Elyaakoubi** ⓘ,[1] **Abdelmoumen Kaabal** ⓘ,[1] **Kamal Attari** ⓘ,[1] **and Adel Asselman**[1]

[1]*Optics & Photonics Team, Faculty of Sciences, Abdelmalek Essaadi University, Tétouan, Morocco*
[2]*Information Technology and Systems Modeling Team, Faculty of Sciences, Abdelmalek Essaadi University, Tétouan, Morocco*

Correspondence should be addressed to Lahcen Amhaimar; lamhaimar@uae.ac.ma

Academic Editor: Tho Le-Ngoc

The transceiver combination technology, of orthogonal frequency division multiplexing (OFDM) with multiple-input multiple-output (MIMO), provides a viable alternative to enhance the quality of service and simultaneously to achieve high spectral efficiency and data rate for wireless mobile communication systems. However, the high peak-to-average power ratio (PAPR) is the main concern that should be taken into consideration in the MIMO-OFDM system. Partial transmit sequences (PTSs) is a promising scheme and straightforward method, able to achieve an effective PAPR reduction performance, but it requires an exhaustive search to find the optimum phase factors, which causes high computational complexity increased with the number of subblocks. In this paper, a reduced computational complexity PTS scheme is proposed, based on a novel swarm intelligence algorithm, called fireworks algorithm (FWA). Simulation results confirmed the adequacy and the effectiveness of the proposed method which can effectively reduce the computation complexity while keeping good PAPR reduction. Moreover, it turns out from the results that the proposed PTS scheme-based FWA clearly outperforms the hottest and most important evolutionary algorithm in the literature like simulated annealing (SA), particle swarm optimization (PSO), and genetic algorithm (GA).

1. Introduction

The transmission data through the use of multicarrier modulation techniques like orthogonal frequency division multiplexing (OFDM) has been considered as a good choice for data transmission over single carrier systems. The OFDM technique has various advantages and now being used in a number of wireless communication systems. Meanwhile, the multiple-input multiple-output (MIMO) with OFDM system, has recently attracted a great deal of attention due to its various advantages such as high data rate, spectral efficiency, diversity in a fading environment, and robustness to channel fading [1]. Hence, a system with OFDM modulation and multiple transmit and multiple receive antennas (MIMO-OFDM) is now becoming adopted by several applications such as digital audio broadcasting (DAB), Worldwide Interoperability for Microwave Access (WiMAX), the fourth generation of telecommunication systems (4G), digital video

broadcasting (DVB), high speed WLAN standards, and many others application areas of MIMO-OFDM [2]. But besides these useful advantages, it still suffers from the high envelope fluctuations of the transmitted signal called the peak-to-average power ratio (PAPR), which decreases the efficiency of high power amplifiers (HPA), improves the complexity of nonlinear elements, and causes out-of-band radiation with degradation of bit error rate (BER).

To deal with this problem, several solutions have been proposed to mitigate the high PAPR of OFDM [3, 4] and MIMO-OFDM signals [5, 6], as clipping [7], clipping and filtering [8], coding [9], tone injection [10], peak windowing [11], selected mapping [12], and partial transmit sequence (PTS) [13–15]. All these methods have their own advantages and disadvantages, but the PTS technique is still the most attractive one due to its efficiency in PAPR reduction. However, the exhaustive search complexity of finding the optimal phase combination for PTS increases exponentially with number of subblocks and to reduce the computational complexity, many evolutionary algorithms for optimization-based PTS schemes have been proposed such as genetic algorithm (GA) [16, 17], ant colony optimization (ACO) [18], simulated annealing algorithm (SA) [19], and the most well-known algorithm of particle swarm optimization (PSO) [20, 21]. In this paper, we propose a novel swarm intelligence algorithm called fireworks algorithm (FWA) [22], to reduce the PAPR with less complexity and more easy implementation, and this developed search optimization algorithm is based on the explosion process simulation of fireworks. It turns out from the results that the proposed method FWA-PTS effectively reduces PAPR of MIMO-OFDM signal and clearly outperforms the old algorithms in both global high precision of calculated solution and convergence speed.

This manuscript is organized as follows. In Section 2, MIMO-OFDM system model and the PAPR problem is formulated, and then the principles of PTS techniques are introduced. Section 3 describes the framework of the FWA-based PTS and mechanisms of the algorithm with some improved version, while Sections 4 and 5 are devoted to the analysis of simulation results and conclusions successively.

2. MIMO-OFDM System and PTS Approach

2.1. PAPR of the MIMO-OFDM Signal. The transmission of signal through the use of transceiver based on orthogonal frequency division multiplexing (OFDM) system is a typical technique which divides the effective spectrum channel to a number of orthogonal subchannels, and with equal bandwidth, each subchannel handles independently with its own data using individual subcarrier, and the OFDM signal is the sum of all independent subcarriers. In transmission systems with multicarrier signal, the input data of binary sequences are mapped into symbols by a modulator (PSK, QPSK, QAM, etc.). Then, the N symbols $X = [X_0, X_1, \ldots, X_{N-1}]^T$ are inserted into the IFFT block to modulate each subcarrier independently and to obtain the OFDM signal in time domain $x = [x_0, x_1, \ldots, x_{N-1}]^T$.

The complex envelope of OFDM signal in the discrete time domain with oversampled factor L (usually used $L = 4$) can be mathematically written as

$$x[n] = \frac{1}{\sqrt{N}} \sum_{k=0}^{N-1} X_k e^{(j2\pi nk/LN)}, \quad 0 \leq n \leq LN - 1, \quad (1)$$

where N is the number of subcarriers and X_k is the nth complex symbol carried and transmitted by the kth subcarrier.

From equation (1), the signal in time domain generated by IFFT operation consists N number of independently modulated and orthogonal subcarriers with large peak values (PAPR) when added up at the output of IDFT block. The PAPR of the OFDM signal in discrete time is defined as the ratio between the maximum power and the average power of the complex OFDM signal, and it can be defined as

$$\text{PAPR}\{x[n]\} = \frac{\max\{|x[n]|^2\}}{E\{|x[n]|^2\}}, \quad 0 \leq n \leq LN - 1, \quad (2)$$

where $x[n]$ is given by (1) and $E\{\cdot\}$ denotes the expected value (average power).

MIMO-OFDM is a generalized case of OFDM systems based on space time block code (STBC) [23–25] for two, three, and four antennas. The encoder signal with two transmitting antennas, using Alamouti code and an input signal $X = [X(0), X(1), \ldots, X(N-1)]$ is written as

$$X_1 = [X(0), -X^*(1), \ldots, X(N-2), -X^*(N-1)]^T,$$
$$X_2 = [X(1), X^*(0), \ldots, X(N-1), X^*(N-2)]^T. \quad (3)$$

The signals X_1 and X_2 are transmitted by antennas 1 and 2, respectively.

At each antenna of MIMO-OFDM system, the peak-to-average power ratio is defined as

$$\text{PAPR}(x_i) = \frac{\max\{|x_i[n]|^2\}}{E\{|x_i[n]|^2\}}, \quad 0 \leq n \leq NL - 1, \quad (4)$$

where $i = 1, 2, \ldots, N_T$ number of transmitted antennas. The time domain signal at each transmit antennas can be presented as

$$x_i[n] = \frac{1}{\sqrt{N}} \sum_{k=0}^{N-1} X_k^i e^{(j2\pi nk/LN)}. \quad (5)$$

The expression characterized the peak power variation of MIMO-OFDM systems is defined as

$$\text{PAPR}_{\text{MIMO-OFDM}} = \max\{\text{PAPR}(x_i)\}, \quad i = 1, \ldots, N_T, \quad (6)$$

2.2. PAPR Reduction by Partial Transmit Sequence. The probabilistic distortionless technique of partial transmit sequence presented in Figure 1, divided an input data block $X = [X_1, X_2, \ldots, X_N]^T$ into V disjoint subblocks, represented by the vector X_v, $v = 1, 2, \ldots, V$ such that $X = \sum_{v=1}^{V} X_v$. The partition of subblocks was performed with a simple method, where all the used subcarriers by another block must be zero so that the sum of all the disjoint subblocks constitutes the original signal. Then, the subblocks are oversampled and transformed to time domain by LN-point IFFT (inverse fast Fourier transform) and using an optimization algorithm or conventional searching method, each subblock is multiplied by a phase factor as follows:

$$x = \text{IFFT}\left\{\sum_{v=1}^{V} b_v \cdot X_v\right\} = \sum_{v=1}^{V} b_v \cdot \text{IFFT}\{X_v\} = \sum_{v=1}^{V} b_v \cdot x_v, \quad (7)$$

where x_v is the time domain signal of each subblock v. The complex phase factor $b_v = e^{j\varphi_v}$, $v = 1, 2, \ldots, V$, rotates the sequences independently to obtain the OFDM signals with the lowest PAPR possible. The phase vector is chosen within $[0, 2\pi]$ interval, and the optimum one can be presented as

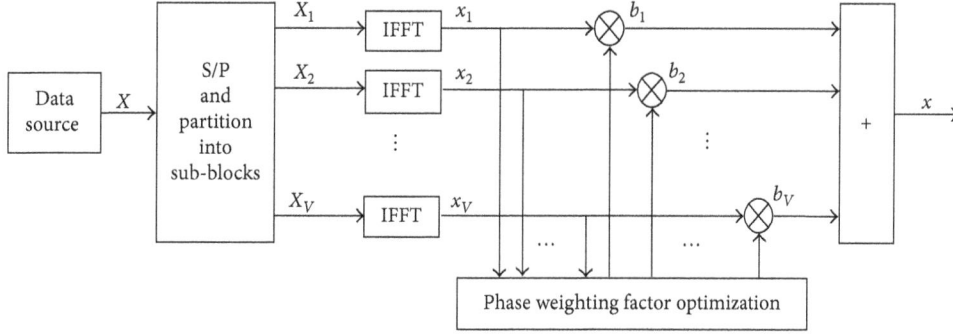

FIGURE 1: Block diagram of PTS technique.

$$[b_1, \ldots, b_V] = \arg \min_{[b_1, \ldots, b_V]}$$
$$\cdot \left(\max_{n=0,1,\ldots,N-1} \left| \sum_{v=1}^{V} b_v \cdot x_v[n] \right| \right). \tag{8}$$

In the practical application of wireless communication systems using the PTS approach, the PAPR performance will be improved as the number of subblocks V is increased and to match the optimal phase weighting sequence for each input data sequence, W^V possible combinations should be checked (W number of phase factors), which needs a huge quantity of computations to analyze all possible applicant rotation phase vectors. Therefore, the computational complexity increases.

3. FWA Optimization-Based PTS

3.1. FWA. In the PTS technique, finding the optimal combination of phase factors requires exhaustive searching by full enumeration of all W^V possible combinations of phase vectors. In recent years, swarm intelligence (SI) has become popular among researchers working on optimization problems. SI algorithms have advantages in solving many optimization problems. The firework algorithm (FWA) is the most recently discovered swarm intelligence [22, 26], which seems effective at problem solving and finding a good enough solution for global optimization of complex functions. In this part, we propose a novel approach based on the FWA to find the optimal phase vector combination to reduce the PAPR with less complexity compared with old algorithms. The objective function f_{Obj} is to minimize the PAPR of the transmitted OFDM signals as follows:

$$\text{minimize} \quad f_{\text{Obj}}(b, x) = \sum_{v=1}^{V} b_v \cdot x_v,$$
$$\text{subject to} \quad b_v = \{e^{j\varphi_v}\}, \quad v = 1, 2, \ldots, V, \ 0 \le \varphi_v \le 2\pi. \tag{9}$$

And the bounds of the potential space are defined by $0 \le \varphi_v \le 2\pi$.

As any probabilistic technique without downward compatibility and in order for the receiver to be able to recover the original data block, the information about the selected phase sequence should be transmitted as a side information. In conventional PTS schemes or the PTS with metaheuristic algorithm, side information about this choice needs to be explicitly transmitted along with the chosen candidate signal. If the side information (phase vector sequence) about the transmitter's choice is received in error, then the information in the transmitted candidate signal cannot be recovered in the receiver and is completely lost. Therefore, the side information needs a particularly strong protection against transmission errors.

In this proposed technique PTS-FWA, we proposed the transmission of particularly protected side information with a simple way. Our approach is based on the use of null subcarriers of each symbol OFDM to transmit the selected phase vectors in order to perform data detection in the receiver, which can easily detect the position of the transmitted phase sequence after deleting the guard interval of each symbol.

FWA is an iterative algorithm which starts to run until the termination criteria are reached, and it is made up of four key parameters: explosive operator, mutation operation, mapping rule, and selection strategy.

3.1.1. Explosion Operator. Based on these four steps, the realization of FWA starts by randomly generating N fireworks in the feasible space $[0, 2\pi]$ as the initial swarm, then the explosion operator generate sparks around N fireworks, and it can be divided further into explosion amplitude, explosion strength, and displacement operation [22]. Let us assume N fireworks with a number of dimensions d, and the number of sparks generated by each firework b_i and the explosion amplitude A_i is defined as follows:

$$s_i = m \cdot \frac{y_{\max} - f_{\text{Obj}}(b_i) + \varepsilon}{\sum_{i=1}^{n} (y_{\max} - f_{\text{Obj}}(b_i)) + \varepsilon}, \tag{10}$$

$$A_i = \widehat{A} \cdot \frac{f_{\text{Obj}}(b_i) - y_{\min} + \varepsilon}{\sum_{i=1}^{n} (f_{\text{Obj}}(b_i) - y_{\min}) + \varepsilon}, \tag{11}$$

where m and $\widehat{A}$ are two constants to control the total number of sparks and the explosion amplitude, respectively, $y_{\max} = \max(f_{\text{Obj}}(b_i))$, $y_{\min} = \min(f_{\text{Obj}}(b_i))$, (maximum and minimum PAPR), and ε is the machine constant.

In the process of FA, the number of sparks s_i, is bounded in order to avoid the overwhelming effects of splendid fireworks as follows:

$$s_i = \begin{cases} \text{round}\,(a \cdot m), & \text{if } s_i < am, \\ \text{round}\,(b \cdot m), & \text{if } s_i > am, \quad a < b < 1 \\ \text{round}\,(s_i), & \text{otherwise,} \end{cases} \quad (12)$$

where a and b are constant parameters.

3.1.2. Gaussian Mutation Operator.

In order to keep the diversity of a population after the explosion, the Gaussian mutation process is performed to generate Gaussian sparks. We select arbitrarily a firework from the present population with arbitrarily dimension Z. If we consider that b_j^d is the current individual with $d \in Z$, the Gaussian mutation applied to the firework to generate the second type of sparks is calculated by $b_j^d = b_j^d * \text{Gaussian}(1,1)$, where the function $\text{Gaussian}(1,1)$ denotes a Gaussian distribution with mean 1 and variance 1. Although the Gaussian mutation operator improves the diversity by conducting search in a local Gaussian space around a firework, it may produce some sparks that exceed the workable space $[0, 2\pi]$, for that the mapping rule will be done to guarantee all the individual sparks remain in the feasible space.

3.1.3. Mapping Rule.

During the explosion generation of Gaussian sparks, some obtained location of generated individuals may lie out of the feasible space. Therefore, it needs to be mapped back into the potential space using the mapping rule to ensure that all generated sparks stay in the feasible space and is stated as follows:

$$b_j^d = B_{\min,d} + b_j^d \% \left(B_{\max,d} + B_{\min,d} \right). \quad (13)$$

where b_j^d is the individual spark's position that is found to fall out of maximum and minimum boundaries $B_{\max,d}$ and $B_{\min,d}$, respectively. And % represents the mathematics modular arithmetic (remainder of division).

3.1.4. Selection Strategy.

At the end of each iteration, after applying the first three steps, a new population of fireworks is selected to be a part of the next generation based on selection strategy, to keep information exchange. In the process of this step, the best spark having the best location is constantly preserved for the next iterations and the remaining $N - 1$ individuals are chosen in light of their distance to other individuals so as to keep diversity of the population.

In selection operation, the distance between two individuals b_i and b_j is defined as

$$R(b_i) = \sum_{j \in K} d(b_i, b_j) = \sum_{j \in K} \left\| b_i - b_j \right\|, \quad (14)$$

where K is the set of all current locations combining fireworks and both types of sparks.

Then, the selection probability $p(b_i)$ to choose individuals for next generation is calculated by

$$p(b_i) = \frac{R(b_i)}{\sum_{j \in K} R(b_i)}. \quad (15)$$

As a result, the fireworks or sparks with larger distances have greater chance to be selected for next generation.

3.1.5. Firework Algorithm-Based PTS.

The following algorithm 1, summarizes in detail the steps employed by FWA to search the optimal phase vector for PTS.

3.2. Comparison of FWA with Other SI Algorithms.

One of the most important categories in computational intelligence community is the swarm intelligence algorithm which is an active branch of evolutionary computation. The fireworks algorithm (FWA), particle swarm optimization (PSO), and genetic algorithms (GA) are the most respected and popular SI algorithms, which are inspired by the intelligent behavior existing in nature and widely used to solve combinatorial optimization problems and real parameter optimization. The FWA algorithm has several characteristics and advantages, such as explosiveness, instantaneity, parallelism (i.e., there is no central control mechanism among fireworks), diversity, robustness (i.e., fireworks are highly independent), and flexibility (i.e., the problem does not need to have an explicit expression to be optimized by FWA), and with these advantages FWA surpasses other SI algorithms.

3.2.1. Comparison between GA and FWA.

The comparison between FWA and GA leads to see that the FWA and the genetic algorithms have a lot in common, for example, both algorithms initialize a population randomly, evaluate the individuals of each iteration according to their fitness values (objective function), and perform a certain random search. In addition, FWA and GA are not guaranteed to find the optimal values. Comparing both algorithms, the mechanisms of sharing the information are quite different. In the fireworks algorithm, a distributed information sharing mechanism is used, while the number of sparks and the explosion amplitudes are determined by the fitness values of fireworks which are located in different areas. However, in the GA, chromosomes share information with each other so the entire population moves relatively homogeneously in the feasible space.

The strategy of selection in FWA sounds like the selection operator in GA, but they are different. A spark which has less similar sparks is selected with a higher probability, on the contrary, a spark which has more similar sparks is chosen with a lower chance. Hence, the sparks with lower fitness values have the chance to be selected such that the diversity of the sparks is ensured. Comparing this strategy with the GA, the selection operator is based on the roulette to determine which individual is to be selected by the fitness values of the individuals. But the diversity of the population is not guaranteed compared with FWA. In addition, the fireworks are always selected from different areas and can hardly stay together due to the immune-based selection. Yet, the FWA has more mechanisms to avoid premature compared to genetic algorithm. Finally, There is no crossover operator in the FWA, and the mutation operator in the GA is totally different from that in the fireworks algorithm.

3.2.2. Comparison between PSO and FWA.

The particle swarm optimization is one of the popular SI algorithms which is inspired by the social behavior of bird flocking, and

---

(1) Set general input parameters, stopping criterion, phase factors, etc;
(2) select N locations for fireworks randomly;
(3) computing the fitness of each seeds (PAPR);
(4) **while** stop criteria is not met **do**
(5)   Set off n fireworks respectively at the n locations:
(6)   **for** each firework b_i **do**
(7)   Calculate the number of sons that each firework should generate $\hat{s}_i$ (10);
(8)   Calculate the amplitude of each individual (sparks) A_i (11);
(9)   **end for**
(10)   **for** $k = 1$ to $\hat{m}$ (number of generated sparks by Gaussian Mutation Operator) **do**
(11)   Generate a specific spark for a randomly selected firework b_j;
(12)   **end for**
(13)   selection strategy to choose the best sparks for next generation using the selection probability given in (15);
(14) **end while**

ALGORITHM 1: Employ FWA based PTS to reduce the PAPR.

---

compared with the FWA, we can conclude that both algorithms have much in common. They adopt random initial populations, evaluate the objective functions, and perform the search based on the fitness values. Also, all the algorithms are not guaranteed to find the optimal solution. In PSO, there is no mutation operation, while there is both displacement operation and Gaussian mutation in FWA. Hence, the diversity of the population is not guaranteed in PSO compared with FWA. Besides, FWA utilizes the idea of immune concentration to keep the diversity of the population, whereas the idea is not contained in PSO.

Furthermore, the mechanism of sharing the information is different. FWA uses a distributed information sharing mechanism, so as to determine the number of sparks and explosion amplitude by the fitness values of each spark in different regions. It also needs to maintain the best firework throughout the iterative process. But on the other hand, with PSO, only gbest gives information to other particles, which is one way of information delivery, as the search process follows the information about the best particle which limits optimization process.

3.3. Enhanced Fireworks Algorithm.
Enhanced fireworks algorithm (EFWA) [27], is an improved extension version of the FWA, and it was proposed to tackle some of the limitations inherent in the original algorithm. The results of FWA decline when being applied on some functions which have a distance between their optimum and origin of the search space (shifted functions). The main factors causing this behavior are the Gaussian sparks and mapping operator. Moreover, the fireworks algorithm suffers from runtime disadvantage, and compared with other optimization algorithms, the FWA has a significant computational cost per iteration, leading to high power consumption. Comparing the two algorithms, EFWA outperforms FWA in terms of results stability and time of convergence. EFWA based on five major improvements is summarized as follows [27].

3.3.1. New Minimal Explosion Amplitude Operator.
The explosion amplitude A_i (11) in FWA shows that the fireworks

having the best fitness value will have a very small explosion amplitude (close to 0). If this operator is close to 0, the explosion sparks will be located at (almost) the same location as the firework itself. As a result, the location of these sparks cannot be improved. In order to overcome this situation, EFWA binds the explosion amplitude of each firework using a minimal explosion amplitude strategy as follows:

$$A_i^k = \begin{cases} A_{\min}^k, & \text{if } A_i^k < A_{\min}^k, \\ A_i^k, & \text{otherwise,} \end{cases} \tag{16}$$

where $A_{\min}$, the minimal explosion amplitude, is calculated in each iteration according to [27].

3.3.2. A New Explosion Sparks Operator.
In the explosion sparks of firework algorithm, the offset displacement is similar, and it is just computed once for all selected dimensions and in each selection step, the same value is added to the location of selected dimensions. However, this operation leads to bad local search ability. To avoid this bad behavior of generating explosion sparks, EFWA proposed a new process to compute a new and different offset displacement for each explosion sparks s_i [27].

3.3.3. A New Mapping Strategy.
To avoid the problem of mapping operator in the traditional algorithm of firework, a new uniform random mapping strategy is proposed in EFWA for the out-of-band sparks which exceed the search space $[0, 2\pi]$. The following operator maps each spark s_i to another location with uniform distribution within the search space:

$$B_i^k = B_{\min}^k + \text{rand} * \left(B_{\max}^k - B_{\min}^k \right), \tag{17}$$

where rand is a function of uniformly distributed random numbers.

3.3.4. Gaussian Sparks Operator.
After the explosion, the Gaussian mutation operator is employed to generate another type of spark using an improved operator for generating

Gaussian sparks in order to avoid FWA mutation disadvantages. The operator is computed by [26, 27]

$$B_i^k = B_i^k + \left(B_b^k - B_i^k\right) * e,$$
$$e = \text{Gaussian}\,(0, 1),$$

(18)

where B_b is the best firework (phase factor) position found so far for each dimension and Gaussian$(0, 1)$ is a random selection from a Gaussian distribution with mean 0 and variance 1.

3.3.5. Selection Operator.

The selection operator is among the most critical key which aims to keep diversity for the next explosion generation. In fireworks algorithm, the selection strategy is processed based on the distance between two individuals, which favors the selection of fireworks or sparks in less crowded regions of the feasible space ((14) and (15)). However, it has the downside of being responsible of most of the runtime within optimization process. To speed the selection process, EFWA applies a simple selection method, which is referred to as the elitism-random selection method [28]. In this selection procedure, the optima of the set will be chosen first. Then, the other individuals are selected randomly. The comparison between two algorithms FWA and EFWA indicated that the new operator diminishes the runtime largely in EFWA.

3.4. Dynamic Search in FWA.

One of the developed versions of enhanced fireworks algorithm (EFWA) is a fireworks algorithm-based dynamic search mechanism which enhances the results of the optimization research and also deals some limitations of EFWA. The explosion amplitude is the key operator which is used to control locality and globality of search based on the quality of the fireworks and also the present number of iterations. As a result, a firework with smaller fitness value (i.e., small explosion amplitude) will perform local search while the global search produces a smaller population (i.e., higher explosion amplitude) performed by fireworks with a fitness value significantly higher. Therefore, in [29], the authors propose a dynamic search fireworks algorithm (dynFWA) to tackle EFWA limitations, using an appropriate strategy applied on the firework at the present best position with changing the explosion amplitude dynamically. Moreover, the authors improve the efficiency of dynFWA by removing the Gaussian sparks operator compared with EFWA without a loss in accuracy.

3.5. Adaptive Fireworks Algorithm.

Several versions of the fireworks algorithm are developed to ensure the quality of optimization. EFWA and dyFWA are two examples of this improvement. However, these algorithms remain to be improved in many aspects. The adaptive fireworks algorithm (AFWA) [30], is an improved version of the recently developed EFWA where the amplitude operator is replaced by new adaptive amplitude so as to enhance the computational mechanism of explosion amplitude. The amplitude is a critical factor influencing the performance of EFWA, and for that reason the AFWA tries to control it precisely, in order to compute the explosion amplitude operator of the best firework using the old sparks with an adaptive method. And in the next generation, we use the retain information from the previous generation to calculate the amplitude of the best firework according to the algorithm process described in [26, 30].

4. Simulation Results

4.1. Parameters.

In this section, numerous computer simulations have been examined based on IEEE 802.11a (wireless LAN) and IEEE 802.16e (WiMAX) standards to verify the performance of PTS-OFDM with FWA searching the optimal combination of phase factors for PAPR reduction. The IEEE 802.16e standard form includes 256 subcarriers (IFFT size), in which the data are carried by 192 subcarriers, 8 subcarriers are reserved as a pilot for channel estimation by the receiver, and the rest are used as guard band subcarriers.

Similarly, the WLAN design has 52 subcarriers, which includes 48 data, 4 pilots, and 12 null subcarriers. For simulation, 10^4 random OFDM symbols have been generated and QPSK modulation is employed, and $L = 4$ is the oversampling factor chosen and the communication channel was AWGN models. The phase weighting factors $W = [0, \ 2\pi]$ have been used, and subblocks $V = 2$; 4; 6; 8 are chosen. For the FWA, the parameters were chosen by some preliminary experiments, and as described in [22], the FWA worked quite well at the setting: $\hat{m} = 5$, $\hat{A} = 40$, $N = 5$, $m = 50$, $a = 0.04$, and $b = 0.8$. These parameters are applied in all the comparison. Firstly, the simulation of MIMO-OFDM system is carried out using two transmit antennas 2Tx and one receiver antenna (2×1), and then the simulation generalized to the case of four transmit and one receiver antenna (4×1). Table 1 documented all the parameters used in this simulation.

4.2. PAPR Reduction Performance.

In this section, we compare the PAPR performance of the PTS-based FWA with the most popular algorithms used to search the phase factor as like standard particle swarm optimization (SPSO) [21], genetic algorithm (GA), simulated annealing (SA), conventional PTS, and SLM methods in terms of both optimization accuracy for PAPR reduction and convergence speed.

The CCDF curves of the simulations results are given in Figure 2. The OFDM system was simulated with 64 subcarriers, in which 4 subblocks are employed, and the PTS, SLM, and GA used $W = 4 \{1 \ {-}1 \ j \ {-}j\}$ phase weighting factors to optimize the PAPR of the modulated OFDM symbol, while other algorithms chose randomly the four phases within the interval $W = [0, \ 2\pi)$. It can be seen that the proposed scheme (PTS-FWA) yields the best performance while the SLM and GA-PTS yields the worst performance for the PAPR reduction of OFDM systems. For example, for the probability of 10^{-3}, the PAPRs are 4 dB, 4.421 dB, 4.948 dB, 5.226 dB, 5.879 dB, 7.034 dB, and 10.66 dB for the FWA, SPSO, SA, PTS, GA, SLM scheme, and original OFDM signals, respectively.

TABLE 1: Simulation parameters.

Parameter	802.11a	802.16e
FFT size	64	256
User carriers	52	200
Pilot carriers	4	8
Number of null/guard band subcarriers	12	56
Cyclic prefix or guard time	1/4, 1/8, 1/16, 1/32	
Modulation	QPSK, 3/4	
Oversampling factor	$L = 4$	
Number of subblocks	$V = 2, 4, 6, 8$	
Phase factors	$W = [0, \ 2\pi]$	
Channel type	AWGN	
Population size N	5	
Gaussian mutation	5	
Number of sparks (m)	50	
Parameters a and b	0.04 and 0.8	
$\widehat{m}$ and $\widehat{A}$	5 and 40	
Channel bandwidth (MHz)	3.5	

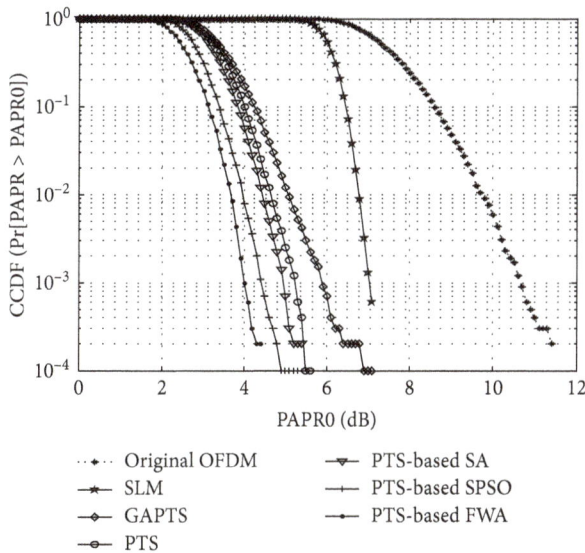

FIGURE 3: PAPR reduction comparison-based improved version of FWA.

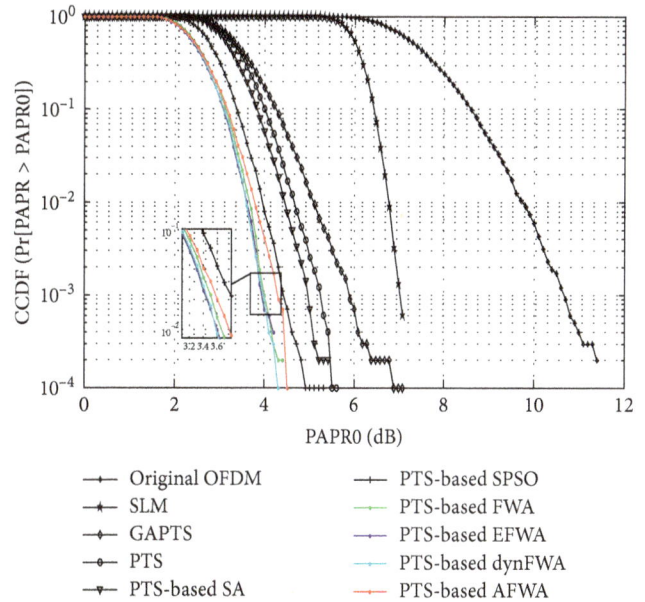

FIGURE 2: CCDFs comparison with different searching algorithms and conventional PTS.

Figure 3 shows the simulated results of the fireworks algorithm-assisted PTS technique and the recently improved version of FWA, in comparison against normal OFDM and different PAPR reduction approaches. The PTS scheme based on the EFWA, dynFWA, and AFWA algorithms are given to compare their PAPR reduction performances in OFDM systems. It is clear from the figure that the FWA and all developed version here can all effectively reduce the PAPR in OFDM systems. However, their PAPR reduction performances and convergence speed are different. In general, EFWA and dynFWA show a few improvements over conventional FWA. For CCDF = 10^{-3}, we have 3.942 dB and 3.979 dB for EFWA and dynFWA, respectively, while AFWA has 4.283 dB compared with FWA (4 dB). In addition, the PAPR reduction based-new version of FWA shows a good improvement in terms of computational cost.

Besides the PAPR reduction performances of the novel swarm intelligence algorithm, convergence speed is quite

essential parameter for any optimization algorithm. Figure 4 depicts the convergence curves of the proposed schemes in comparison with GA and SPSO in order to validate the convergence speed of the FWA. The simulation is carried out for random OFDM symbol with 10 independent runs and 3000 iterations. From these results, we can conclude that the four methods of FWA have a much faster speed convergence than the SPSO and the GA, while dynFWA and AFWA were a very promising schemes due to their efficiency and simplicity.

Figure 5 shows the influence of different numbers of subblocks V on PAPR performance for a QPSK-OFDM system and 10^4 randomly OFDM symbols using the dynamic search fireworks algorithm, which is characterized by a good simulation runtime without losing optimization accuracy. It is seen in Figure 5 that the PAPR performance in OFDM systems-based dynFWA enhances as the number of subblocks increases with $V = 2, 4, 6,$ and 8. Therefore, the system with the bigger number of subblocks performs the better PAPR.

Finally, Figure 6 depicts the PAPR reduction performance for WiMAX OFDM systems with AFWA. For randomly generated QPSK symbols, both CCDF graphs of original and optimized signals have been illustrated, and space time block coding (STBC) is used in order to diminish the computational complexity at the receiver side and achieve more diversity gain at the transmit side, where the test starts with two transmit antennas-based Alamouti scheme [23] (2×1) and finished by using four transmit antennas (4×1) [24, 25]. With a half-rate orthogonal STBC encoding and 10^4 OFDM symbol, it can be observed that PAPR of MIMO OFDM with AFWA was ≈ 5.5 dB and ≈ 5.812 dB at CCDF = 10^{-3} compared with original PAPR ≈ 11.25 dB and ≈ 11.76 dB for WiMAX with two and four transmit antennas, respectively.

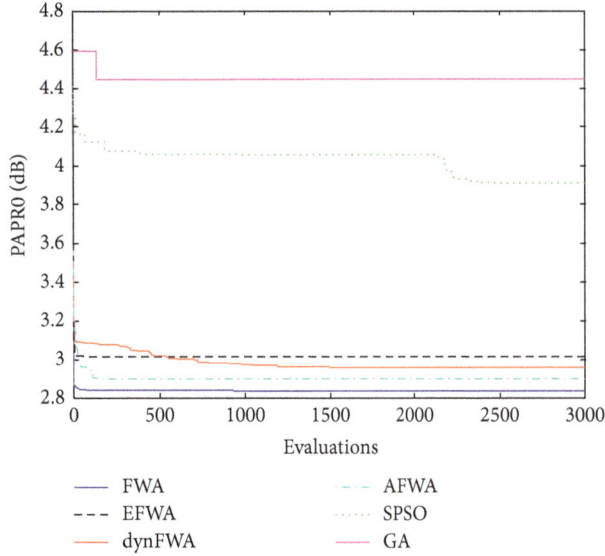

FIGURE 4: Convergence curves of the FWA, EFWA, dynFWA, AFWA, SPSO, and GA algorithms-based PTS for one symbol OFDM.

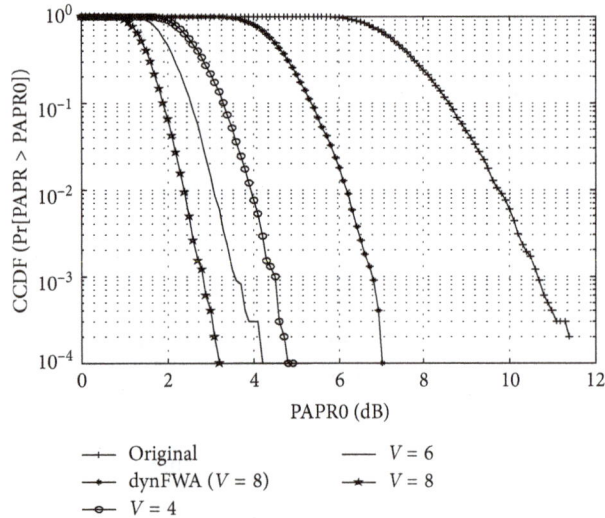

FIGURE 5: PAPR performance of a QPSK/OFDM system with dynFWA-based PTS technique when the number of subblocks varies.

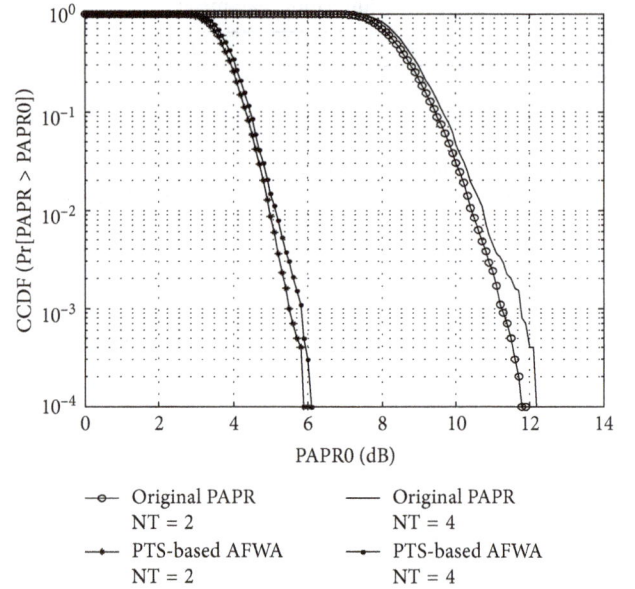

FIGURE 6: PAPR reduction performances in MIMO-OFDM-based WiMAX standard.

TABLE 2: Computational complexity of the different PTS schemes for CCDF $= 10^{-3}$.

Methods	Number of searches	PAPR (dB)
Original	0	10.66
PTS	$W^V = 4^4 = 256$	5.226
PTS-GA	$P \times G = 10 \times 10 = 100$	5.879
PTS-SPSO	$S \times \text{Iter} = 100 \times 2 = 200$	4.421
PTS-FWA	$\text{Iter} \times (N + m + \hat{m}) = 2 \times (5 + 50 + 5) = 120$	4

TABLE 3: Performance evaluation by the FWA, EFWA, dynFWA, AFWA, SPSO, and GA on one symbol OFDM over 10 independent runs.

Methods	Function evaluations	PAPR (dB)
Original	—	10.66
SPSO	2000	4.055
GA	2000	4.447
FWA	500	2.839
EFWA	500	3.015
dynFWA	500	3.021
AFWA	500	2.9

4.3. Complexity Comparison.

Table 2 shows the computational complexity to find the phase factors and the reduced PAPR values of the various PTS schemes at CCDF $= 10^{-3}$. These proposed techniques are diversified in terms of reducing the PAPR as well as computational complexity. It must be kept in mind that the complexity of a PAPR reduction system increases linearly with the number of iterations of such an algorithm. There will therefore be a trade-off between reducing the PAPR and the complexity of the system.

In the original OFDM signal, the PAPR value is calculated before the PTS optimization and therefore its search number is 0. Optimum PTS considers all the phase factors and thus it needs $W^V = 256$ iterations. For GA-PTS, the search complexity is proportional to $P \times G = 100$, where P is the maximum size of the population and G is the generation. In PTS-PSO, the search number is equal to $S \times \text{Iter} = 200$, where Iter is the maximum iteration number and S is the size of particle swarm. Finally, in the FWA-PTS, approximate $(n + m + m)$ function evaluations are done in each generation. Suppose the optimum of a function can be found in Iter generations, then we can deduce that the complexity of the FWA is O $(\text{Iter} \times (n + m + m)) = 120$.

Table 3 depicts the optimization accuracy of the proposed algorithms on one symbol OFDM in comparison with SPSO and GA. It can be seen that the proposed schemes

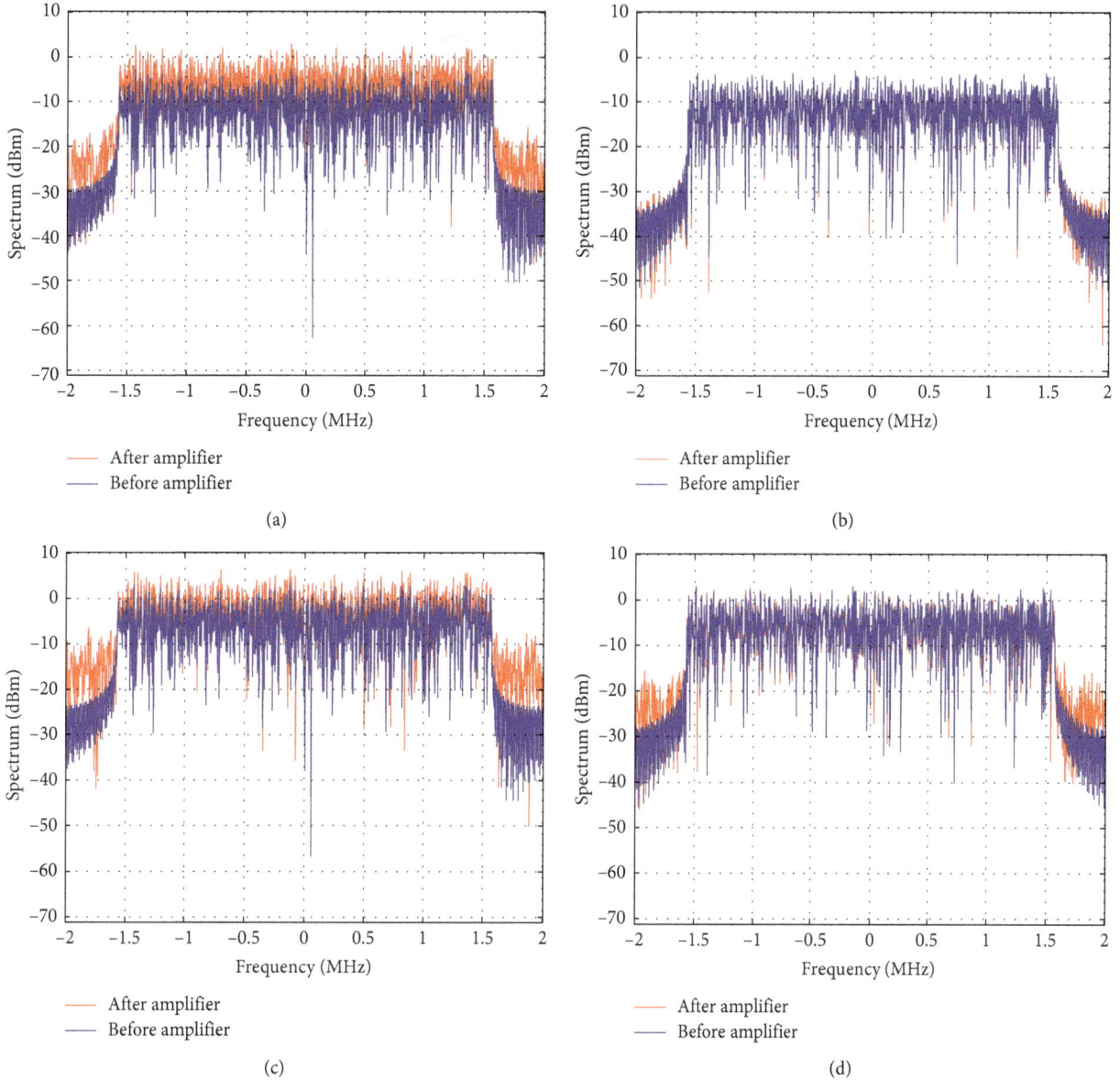

FIGURE 7: Power Spectral Density comparison performance of an OFDM signal without and with the proposed method. (a, b) 7 dB from saturation point (moderate nonlinearity) and (c, d) 1 dB from saturation point (severe nonlinearity). (a) Without PTS-dynFWA-DPD, (b) with PTS-dynFWA-DPD, (c) without PTS-dynFWA-DPD, and (d) with PTS-dynFWA-DPD.

clearly outperform both the SPSO and GA. In addition, the fireworks algorithm and its improved versions can find optimal solutions in less than 500 function evaluations.

4.4. Spectrum Performance.
One of the major problems for wireless communications is the RF sidelobe generation [31]. Out-of-band radiations have decisive effects on adjacent channel interference (ACI) performances and reduce the spectrum efficiency, which are of great importance especially for multiuser and RF wireless communications systems. The transmission under ideal conditions leads to no spectral spread, while the transmission of original OFDM with height PAPR via a nonlinear device produces considerable out-of-band power when approaching the saturation zone

(6, 7, 8, and 9 dB backoff from saturation point). Our approach consists of finding an alternative representation of the OFDM signal with minimum PAPR based on fireworks algorithm with dynamic search mechanism (dynFWA), and then the output signal will be amplified after digital predistortion (DPD) process. This efficient approach suggested to reduce the PAPR and to compensate the high power amplifiers (HPA), by making a combination between the proposed scheme (PTS-dynFWA) of PAPR reduction and digital predistortion (DPD).

Figure 7 shows the signal of spectrum performance before and after passing through the nonlinear power amplifier model for conventional OFDM system and OFDM system with the proposed approach (PAPR reduction method PTS-dynFWA and DPD). The results are obtained for two values of the HPA backoff, 7 dB and 1 dB from saturation point.

Figure 7(b) shows the frequency spectra with 7 db backoff below the input power that causes amplifier saturation. It is clearly seen that the proposed method (PTS-dynFWA with DPD) has the ability to suppress the out-of-band distortion completely. Compared to the case (a) where no PAPR reduction and no predistortion are used, Figures 7(c) and 7(d) show the spectrum with severe nonlinearity (1 db from saturation point). Comparing the two spectra (Figure 7(c)), we can view that the transmission of original OFDM signal via a HPA device produces considerable out-of-band power. However, the proposed method was able to reduce the spectral regrowth around each of the two bands.

5. Conclusion

In this paper, we propose a FWA-based PTS technique to reduce the PAPR of MIMO-OFDM system with low computational complexity. The FWA-PTS is compared with GA-PTS, SA-PTS, SPSO-PTS, SLM, and conventional PTS, in terms of CCDF. The simulation results show that the proposed method has a promising performance in both optimization accuracy of PAPR and convergence speed over conventional schemes and algorithms. Moreover, in order to eliminate the drawbacks of conventional FWA, we have presented in this paper some improved version of FWA, like EFWA, dynFWA, and AFWA. Simulation results showed that the improved versions significantly improve the results in term of PAPR reduction and computation cost over conventional FWA, and clearly outperform SA, GA, and SPSO without loss in optimization accuracy. Besides that, the results analyzed of each method alone come to the conclusion that dynFWA-based PTS and AFWA-based PTS were very successful and have good prospects.

Conflicts of Interest

The authors declare that they have no conflicts of interest.

References

[1] Y. Wu and W. Y. Zou, "Orthogonal frequency division multiplexing: a multi-carrier modulation scheme," *IEEE Transactions on Consumer Electronics*, vol. 41, no. 3, pp. 392–399, 1995.

[2] Y. S. Cho, J. Kim, W. Y. Yang, and C. G. Kang, *MIMO-OFDM Wireless Communications with MATLAB*, John Wiley & Sons, Hoboken, NJ, USA, 2010.

[3] S. P. DelMarco, "A constrained optimization approach to compander design for OFDM PAPR reduction," *IEEE Transactions on Broadcasting*, vol. 64, no. 2, pp. 307–318, 2018.

[4] S. H. Han and J. H. Lee, "An overview of peak-to-average power ratio reduction techniques for multicarrier transmission," *IEEE Wireless Communications*, vol. 12, no. 2, pp. 56–65, 2005.

[5] L. Amhaimar, S. Ahyoud, and A. Asselman, "An efficient combined scheme of proposed PAPR reduction approach and digital predistortion in MIMO-OFDM systems," *International Journal on Communications Antenna and Propagation*, vol. 7, no. 5, 2017.

[6] P. Ramakrishnan, P. T. Sivagurunathan, and N. S. Kumar, "An optimization technique for reducing the PAPR in MIMO-OFDM technique by using FPPC scheme," in *Proceedings of IEEE International Conference on Electrical, Instrumentation and Communication Engineering (ICEICE)*, pp. 1–8, Karur, India, April 2017.

[7] J. Wang, Y. Guo, and X. Zhou, "PTS-clipping method to reduce the PAPR in ROF-OFDM system," *IEEE Transactions on Consumer Electronics*, vol. 55, no. 2, pp. 356–359, 2009.

[8] K. Anoh, C. Tanriover, B. Adebisi, and M. Hammoudeh, "A new approach to iterative clipping and filtering PAPR reduction scheme for OFDM systems," *IEEE Access*, vol. 6, pp. 17533–17544, 2018.

[9] A. E. Jones, T. A. Wilkinson, and S. K. Barton, "Block coding scheme for reduction of peak to mean envelope power ratio of multicarrier transmission schemes," *Electronics Letters*, vol. 30, no. 25, pp. 2098-2099, 1994.

[10] J.-C. Chen and C.-K. Wen, "PAPR reduction of OFDM signals using cross-entropy-based tone injection schemes," *IEEE Signal Processing Letters*, vol. 17, no. 8, pp. 727–730, 2010.

[11] G. Chen, R. Ansari, and Y. Yao, "Improved peak windowing for PAPR reduction in OFDM," in *Proceedings of VTC Spring 2009-IEEE 69th Vehicular Technology Conference*, pp. 1–5, Barcelona, Spain, April 2009.

[12] X. Cheng, D. Liu, S. Feng, H. Fang, and D. Liu, "An artificial bee colony-based SLM scheme for PAPR reduction in OFDM systems," in *Proceedings of 2017 2nd IEEE International Conference on Computational Intelligence and Applications (ICCIA)*, pp. 449–453, Beijing, China, September 2017.

[13] H.-S. Joo, K.-H. Kim, J.-S. No, and D.-J. Shin, "New PTS schemes for PAPR reduction of OFDM signals without side information," *IEEE Transactions on Broadcasting*, vol. 63, no. 3, pp. 562–570, 2017.

[14] L. J. Cimini and N. R. Sollenberger, "Peak-to-average power ratio reduction of an OFDM signal using partial transmit sequences," *IEEE Communications Letters*, vol. 4, no. 3, pp. 86–88, 2000.

[15] A. Joshi and D. S. Saini, "Peak-to-average power ratio reduction of OFDM signals using improved PTS scheme with low computational complexity," *WSEAS Transactions on Communications*, vol. 12, pp. 630–640, 2013.

[16] D. E. Goldberg, *Genetic Algorithms in Search, Optimization, and Machine Learning*, Addison-Wesley, Reading, PA, USA, 1989.

[17] C. Ergun and K. Hacioglu, "Multiuser detection using a genetic algorithm in CDMA communications systems," *IEEE Transactions on Communications*, vol. 48, no. 8, pp. 1374–1383, 2000.

[18] J.-K. Lain and J.-J. Lai, "Ant colony optimisation-based multiuser detection for direct-sequence CDMA systems with diversity reception," *IET Communications*, vol. 1, no. 4, pp. 556–561, 2007.

[19] T. Jiang, W. Xiang, P. C. Richardson, J. Guo, and G. Zhu, "PAPR reduction of OFDM signals using partial transmit sequences with low computational complexity," *IEEE Transactions on Broadcasting*, vol. 53, no. 3, pp. 719–724, 2007.

[20] J. Robinson and Y. Rahmat-Samii, "Particle swarm optimization in electromagnetics," *IEEE Transactions on Antennas and Propagation*, vol. 52, no. 2, pp. 397–407, 2004.

[21] D. Bratton and J. Kennedy, "Defining a standard for particle swarm optimization," in *Proceedings of IEEE Swarm Intelligence Symposium (SIS)*, pp. 120–127, Honolulu, HI, USA, April 2007.

[22] Y. Tan and Y. Zhu, "Fireworks algorithm for optimization," in *Advances in Swarm Intelligence*, pp. 355–364, Springer, Berlin, Germany, 2010.

[23] S. Alamouti and S. Patole, "A simple transmit diversity schemes for wireless communications," *IEEE Journal on Selected Areas in Communications*, vol. 16, no. 8, pp. 1451–1458, 1998.

[24] V. Tarokh, H. Jafarkhani, and A. R. Calderbank, "Space-time block codes from orthogonal designs," *IEEE Transactions on Information Theory*, vol. 45, no. 5, pp. 1456–1467, 1999.

[25] V. Tarokh, H. Jafarkhani, and A. R. Calderbank, "Space-time block coding for wireless communications: performance results," *IEEE Journal on Selected Areas in Communications*, vol. 17, no. 3, pp. 451–460, 1999.

[26] Y. Tan, *Firework Algorithm: A Novel Swarm Intelligence Optimization Method*, Springer, Berlin, Heidelberg, Germany, 2015.

[27] S. Zheng, A. Janecek, and Y. Tan, "Enhanced fireworks algorithm," in *Proceedings of 2013 IEEE Congress on Evolutionary Computation (CEC)*, pp. 2069–2077, Cancun, Mexico, June 2013.

[28] A. P. Engelbrecht, *Fundamentals of Computational Swarm Intelligence*, John Wiley & Sons, Hoboken, NJ, USA, 2006.

[29] S. Zheng, A. Janecek, J. Li, and Y. Tan, "Dynamic search in fireworks algorithm," in *Proceedings of IEEE Congress on Evolutionary Computation (CEC)*, pp. 3222–3229, Beijing, China, July 2014.

[30] J. Li, S. Zheng, and Y. Tan, "Adaptive fireworks algorithm," in *Proceedings of IEEE Congress on Evolutionary Computation (CEC)*, pp. 3214–3221, Beijing, China, July 2014.

[31] L. Wang, L. Zhang, R. Xu, and L. Peng, "Precoding for joint spectral sidelobes suppression and PAPR reduction in OFDM system," in *Proceedings of IEEE 9th International Conference on Communication Software and Networks (ICCSN)*, pp. 482–486, Guangzhou, China, May 2017.

9

A Comprehensive Taxonomy and Analysis of IEEE 802.15.4 Attacks

Yasmin M. Amin and Amr T. Abdel-Hamid

Department of Networks Engineering, German University in Cairo, Cairo 11835, Egypt

Correspondence should be addressed to Yasmin M. Amin; yasminmahmoudamin@hotmail.com

Academic Editor: Vinod Sharma

The IEEE 802.15.4 standard has been established as the dominant enabling technology for Wireless Sensor Networks (WSNs). With the proliferation of security-sensitive applications involving WSNs, WSN security has become a topic of great significance. In comparison with traditional wired and wireless networks, WSNs possess additional vulnerabilities which present opportunities for attackers to launch novel and more complicated attacks against such networks. For this reason, a thorough investigation of attacks against WSNs is required. This paper provides a single unified survey that dissects all IEEE 802.15.4 PHY and MAC layer attacks known to date. While the majority of existing references investigate the motive and behavior of each attack separately, this survey classifies the attacks according to clear metrics within the paper and addresses the interrelationships and differences between the attacks following their classification. The authors' opinions and comments regarding the placement of the attacks within the defined classifications are also provided. A comparative analysis between the classified attacks is then performed with respect to a set of defined evaluation criteria. The first half of this paper addresses attacks on the IEEE 802.15.4 PHY layer, whereas the second half of the paper addresses IEEE 802.15.4 MAC layer attacks.

1. Introduction

The IEEE 802.15.4 standard [1] has become the dominant enabling technology for WSNs. WSNs are networks of a large number of tiny sensor devices which target applications in a diverse set of fields, particularly military, healthcare, residential, transport, and industrial fields to name a few. Such applications require very low data rates and relaxed Quality of Service (QoS) requirements over short ranges for wireless devices with very small sizes [2].

Many applications involving WSNs are security-sensitive and possess zero tolerance for error and latency. While error and latency can occur due to network failure and congestion, they can also be triggered by malicious behavior. For instance, eavesdropping on confidential information exchange or injecting false information in battlefield monitoring applications can have severe consequences and can lead to injuries or at worst fatalities. Another example is Home Area Networks (HANs) in smart grids [3], in which adversaries can manipulate data transfer among smart meters

and household electric appliances, for the purpose of manipulating control signals transmitted to targeted appliances or modifying billing and payment information, either in favor of or against the customer to whom the HAN belongs. Wireless Body Area Networks (WBANs) monitor human body functions and the surrounding environment and can lead to dangerous conditions or even death of patients in case of security threats. Also attacks launched against home Intrusion Detection Systems (IDSs) might cause home owners to be informed of the presence of intruders after significant delay at best or may cause intruder alarms to be overlooked altogether at worst.

IEEE 802.15.4 defines the Physical (PHY) and Media Access Control (MAC) layer specifications of Low-Rate Wireless Personal Area Networks (LR-WPANs) [1]. Since the PHY and MAC layers are considered as the base for any WSN [4], they are the main target of most launched attacks and malicious behavior. The primary focus of this paper is on providing a comprehensive classification of the attacks which can be launched on each of the IEEE 802.15.4 PHY and MAC layers. For each layer, the provided classification is used as the basis for performing a comparative analysis between the attacks. *To the best of the authors' knowledge*, this survey is the first to provide a single reference consolidating all IEEE 802.15.4 PHY and MAC layer attacks known to date. The paper is organized as follows. In Section 2, we describe the operation and purpose of each PHY layer attack and its variants. In Section 3, we present our own two novel classifications of PHY layer attacks. In Section 4, we evaluate the differences between the discussed attacks based on a set of defined evaluation criteria. Sections 5, 6, and 7 are reiterations of Sections 2, 3, and 4 respectively, but for IEEE 802.15.4 MAC layer attacks instead. In Section 5, we also identify a new variant of the *GTS Attack*. In Section 6, we present two significant classifications of MAC layer attacks, collectively obtained from external references, along with our own additions to these existing classifications. Section 8 concludes the paper.

2. Attacks on IEEE 802.15.4 PHY Layer

This section explains the purpose and operation of the attacks which can be launched by a malicious adversary against the PHY layer of an IEEE 802.15.4-based network. We refer to the different methods of launching the same attack as the attack's *variants*, and we discuss how an attack's variants can be conducted. We assign names for attacks which have not previously been named in existing literature.

2.1. Radio Jamming. *Radio Jamming* [5] is a PHY layer attack which is launched with the intent of creating a Denial of Service (DoS) against the network. This attack achieves its purpose through the intentional emission of radio signals in order to decrease the Signal to Noise Ratio (SNR) of ongoing radio communications [3], thus disrupting the reception of messages at network nodes.

Sokullu et al. [6] identify four approaches to *Radio Jamming*, which are *Constant*, *Deceptive*, *Random*, and *Reactive Jamming*. Jokar et al. [3] classify Radio Jamming attacks into *Wide-Band Denial*, *Pulse-Band Denial*, and *IEEE 802.15.4-Specific Interruption Denial*. O'Flynn [4] extends the Radio Jamming attacks defined in Jokar et al. [3] by defining two additional attacks, which are *Node-Specific Denial* and *Message-Specific Denial*. Wood et al. [5] divide Radio Jamming into *Interrupt*, *Activity*, *Scan*, and *Pulse Jamming*. Balarengadurai and Saraswalhi [7] distinguish between *Signal*, *Radio*, and *Noise Jamming*. In our opinion, we consider *Signal* and *Noise Jamming* to be the same as *Radio Jamming*, as they possess the same intent and characteristics as *Radio Jamming*. This subsection discusses all of the *Radio Jamming* attacks explained in [3–7].

2.1.1. Wide-Band Denial. This jamming attack transmits radio interference signals with high transmission power over all channels of the related frequency band. This can be achieved either through continuous transmission of a jamming signal over the entire frequency band [3] or by using channel hopping in conjunction with single-channel

Pulse Jamming, also known as *Pulse-Band Denial*, in order to emit pulse signals with short time periods on each and every single 802.15.4 channel before moving on to and repeating the same procedure in the successive channel. *Wide-Band Denial* causes DoS against the network by blocking the whole Radio Frequency (RF) spectrum, which in turn leads to the corruption of all ongoing transmissions [4].

2.1.2. Pulse-Band Denial. In contrast to *Wide-Band Denial*, Pulse-Band Denial targets a single channel rather than all channels within the related frequency band. For this reason, this method is also known as single-channel Pulse Jamming. A malicious adversary can use *Pulse-Band Denial* to prevent legitimate nodes from selecting a particular channel by jamming the target channel during the channel Energy Detection (ED) scan performed by legitimate nodes during the channel selection process. This act of jamming forces the channel to seem unattractive to the nodes, thus coercing the nodes into scanning for other interference-free channels. *Pulse-Band Denial* can be perceived as a misbehavior attack if the adversary not only prevents legitimate nodes from utilizing the target channel, but also selfishly frees up the channel for its own use [4]. We consider the four *Radio Jamming* attacks explained in [6] to be variants of *Pulse-Band Denial*, as shown in Figure 1. This is because the adversary targets a single channel in each of these four variants, as is the case with *Pulse-Band Denial*. These four variants differ with respect to the recurrence of the emitted jamming signal within the target channel.

(a) Constant Jamming. In this variant, a radio signal is continuously transmitted over the target channel [6].

(b) Deceptive Jamming. Similar to *Constant Jamming*, *Deceptive Jamming* involves continuous channel jamming. However, while the former involves the continuous transmission of radio signals over the target channel, an adversary performing the latter emits regular frames over the target channel [6].

(c) Random Jamming. This variant is performed by jamming the target channel at random times, either using a constant radio signal (as in *Constant Jamming*) or using regular frames (as in *Deceptive Jamming*) [6].

(d) Reactive Jamming. A reactive jammer only starts jamming the target channel when it senses ongoing network activity over the channel [6]. There are several ways for a malicious adversary to detect activity within the network. For this reason, we subdivide *Reactive Jamming* into five variants extracted from [4, 5], depending on how an adversary detects activity within its target network, as illustrated in Figure 1. For some of these variants, the adversary initiates jamming only when detected network activity includes frames of a specific nature, destination, or type:

(i) *Interrupt Jamming.* It is also known as *IEEE 802.15.4-Specific Interruption Denial* [4]: this variant triggers channel interference by the adversary only when IEEE 802.15.4 traffic is detected over the target channel.

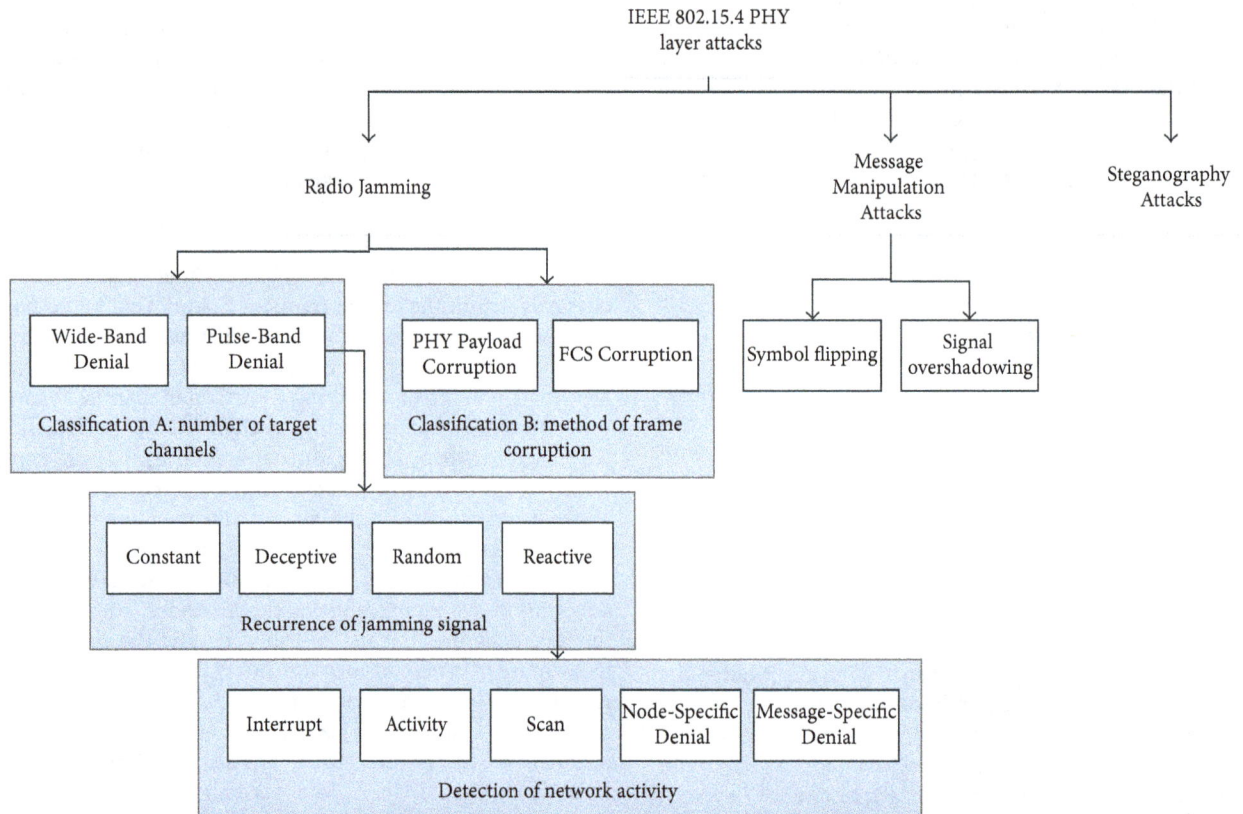

FIGURE 1: Classification of IEEE 802.15.4 PHY layer attacks.

While awaiting frame reception, the adversary's radio constantly scans for a preamble and a Start of Frame Delimiter (SFD), both of which precede the payload of a PHY frame. Upon their detection, the adversary starts transmitting a jamming frame in order to interfere with the ongoing transmission of the legitimate frame over the channel. Wood et al. [5] explain the use of frame masking as a defense against Interrupt Jamming. Frame masking is a mechanism which generates a secret pseudorandom sequence, which is agreed upon by sender and receiver, to be used for the SFD field in each frame. In this way, an adversary's radio will not be able to scan for a fixed SFD because the SFD now differs for each transmitted frame between sender and receiver. As such, launching an interrupt attack will not be possible [5].

(ii) *Activity Jamming.* Unlike *Interrupt Jamming*, an adversary does not need to know the value of each frame's SFD in order to launch an *Activity Jamming* attack. Rather, the adversary periodically samples the Radio Signal Strength Indicator (RSSI) or the radio's Clear Channel Assessment (CCA) output in order to detect the presence of network activity. Network activity is confirmed if the RSSI value is above a predefined threshold. Upon activity detection, the adversary initiates jamming. One defense mechanism

suggested by Wood et al. [5] against *Activity Jamming* is channel hopping. An adversary can only sample RSSI or CCA for the channel on which its radio is listening. If a sender and receiver agree upon a secret pseudorandom channel sequence, they can use that sequence to hop between similar channels, thus leaving the adversary behind [5].

(iii) *Scan Jamming.* This *Reactive Jamming* variant enables an adversary to break the defense of the channel hopping mechanism by allowing the adversary to scan all channels within the frequency band before the termination of a single frame transmission from sender to receiver. Each time the adversary hops over to a new channel, it uses RSSI or CCA sampling, as in *Activity Jamming*, to determine if a frame is being transmitted on the current channel. If frame transmission is detected, the adversary begins jamming. If the channel is determined to be idle, the adversary hops over to the next channel in the frequency band. This attack variant can be overcome by fragmenting each frame and transmitting the different fragments on different channels with different SFDs [5].

(iv) *Node-Specific Denial.* In this *Radio Jamming* variant, an adversary only jams frames which are destined for specific target nodes. This is achieved by examining the addressing information contained in the MAC header of the ongoing data frame [4].

(v) *Message-Specific Denial. Message-Specific Denial* is quite similar to *Node-Specific Denial*, with the difference being that the adversary only jams frames of a specific type rather than a specific destination address. The type of the frame can be known by examining the frame type field of the transmitted frame's MAC header [4].

2.2. Message Manipulation Attacks.

These attacks are used by adversaries to inject false data into the network by transforming a legitimate data frame into a modified frame containing information of the adversary's choice. Wilhelm et al. [8] explain that frame manipulation can be accomplished using one of the following two techniques.

2.2.1. Symbol Flipping. An adversary can emit RF waves whose amplitude and phase are synchronized with those of the original transmitted signal. If these RF waves are combined with the original signal at the correct time, this leads to a new signal containing the falsely injected data. This technique is referred to as *Symbol Flipping* [8].

2.2.2. Signal Overshadowing. This technique is used in conjunction with angular modulation schemes, in which only the stronger of two colliding signals is received. While *Symbol Flipping* is highly sensitive to the physical properties of the original transmitted signal, Signal Overshadowing overcomes this limitation. However, *Signal Overshadowing* is still concerned with tight timing and phase synchronization requirements from a technical, as opposed to a physical, perspective [8].

2.3. Steganography Attacks. Martins and Guyennet [9] explain how attacks can be performed on the PHY and MAC layers of the 802.15.4 protocol using steganography. In a general context, steganography is used to hide the existence of data. This is accomplished by embedding secret data within existing data, known as cover data, thus resulting in a stegomessage. Data can be hidden in 802.15.4 networks by using the PHY header field of PHY frames. The PHY header field is one byte long, with only seven used bits and one reserved bit. The eighth reserved bit within the PHY header field can be used to transmit one bit of an intended stegomessage per PHY frame. A receiver which is aware of the existence of the stegomessage reads every hidden bit of all sent frames until it recovers the entire stegomessage. *Steganography Attacks* create a hidden channel between collaborating adversaries in the network, which opens up endless opportunities for adversaries. For instance, the hidden channel could be used by adversaries to exchange information regarding the execution of new attacks in the network. Moreover, adversaries could use the hidden channel to monitor the network and warn each other when the network detects or senses launched attacks. The latter enables adversaries to stop executing their attacks just in time to remove suspicions and prevent detection by the network. *Steganography Attacks* can also be launched by hiding information within the MAC fields of the 802.15.4 protocol [9].

3. Classification of IEEE 802.15.4 PHY Layer Attacks

While [3–7] focus on explaining the operation, intent, and impact of each of the *Radio Jamming* attacks individually, we provide our own two novel classifications in order to illustrate the dependencies and interrelationships between these attacks, as illustrated in Figure 1. It is important to note that the two presented classifications only consider *Radio Jamming* attacks. *Message Manipulation* and *Steganography Attacks* are not considered.

3.1. Classification A. In this classification, we define two broad categories of *Radio Jamming* attacks with respect to the number of target channels within the related frequency band over which jamming signals are transmitted. *Wide-Band Denial* targets all channels within the related frequency band, whereas all variants of *Pulse-Band Denial* target a single channel.

3.2. Classification B. This classification classifies attacks with respect to the fields of the transmitted frame which are corrupted due to adversary jamming. We infer from O'Flynn [4] that each of the *Radio Jamming* attacks explained in the previous section can corrupt a data frame using one of the following two techniques.

3.2.1. PHY Payload Corruption. In this method, the adversary aims to jam a few or all bytes of the PHY payload, thus corrupting the frame and indirectly leading to an incorrect Frame Check Sequence (FCS) at the receiver. Upon frame reception, the receiver discards the frame due to its incorrect FCS [4].

3.2.2. FCS Corruption. Contrary to *PHY Payload Corruption*, *FCS Corruption* involves the direct jamming of one or both bytes of the frame's two-byte FCS field while leaving the PHY payload as it is. This still leads to the received frame not being accepted by the receiver node [4].

It is worth noting that, for each of the two aforementioned frame corruption techniques, an adversary is able to detect the point in time at which the transmission of a frame's PHY payload or FCS starts by examining the frame's PHY layer header, which contains the length of the transmission frame. The frame content targeted for corruption by the adversary depends on the point in time at which the adversary starts emitting jamming energy, as well as how long the adversary continues to jam throughout the entire duration of frame transmission [4].

4. Comparative Analysis of IEEE 802.15.4 PHY Layer Attacks

In this section, we first define the criteria which will be used to evaluate each of the PHY layer attacks discussed in the previous section. Following this, we perform a detailed comparison between the attacks with respect to the defined evaluation criteria.

TABLE 1: Comparative analysis of IEEE 802.15.4 PHY layer attacks.

Attack	Energy efficiency	Effectiveness	Stealthiness	Primary security goals	
				Integrity	Availability
Wide-Band Denial	Low [3]	High	Low [3]		
Constant Jamming	Low [6]	High	Low [6]		
Deceptive Jamming	Low [6]	High	Moderate [6]		
Random Jamming	Moderate [6]	Moderate [6]	Moderate [6]	×	√ [7]
Interrupt Jamming	High [5]	High [5]	High [4]		
Activity Jamming and Scan Jamming	Moderate [5]	High	Moderate [5]		
Node-Specific Denial and Message-Specific Denial	High	High	High		
Message Manipulation Attacks	No enough information			√ [8]	×
Steganography Attacks	Dependent			×	×

4.1. Evaluation Criteria. We define four criteria for the purpose of PHY layer attack evaluation.

4.1.1. Energy Efficiency. An adversary launching an energy-efficient attack consumes minimum energy in order to disrupt network communications, thus preventing drainage of the adversary node's limited battery power [5].

4.1.2. Effectiveness. An effective attack is one that yields maximum disruption of network communications throughout the length of its deployment.

4.1.3. Stealthiness. A stealthy attack is an unintrusive attack which is launched with minimum probability of detection by the network [5].

4.1.4. Security Goal Violation. Security goals are used to assess a WPAN's level of security. Kumar et al. [10] define two classes of security goals: primary and secondary. Since no secondary goals are violated by any of the discussed attacks, we only focus on primary security goals.

4.2. Comparison between IEEE 802.15.4 PHY Layer Attacks. Table 1 compares between the IEEE 802.15.4 PHY layer attacks with respect to energy efficiency, effectiveness, stealthiness, and the primary security goals violated by each attack. Unreferenced entries in Table 1 indicate our own suggestions and inferences.

It is noted that Table 1 exhibits a direct proportionality between energy efficiency and stealthiness for almost all of the discussed PHY attacks. In other words, the lower the amount of energy consumed by an adversary during the attack (the higher the energy efficiency), the lower the probability of attack detection by the network within which the attack is launched (the higher the stealthiness).

4.2.1. Energy Efficiency, Effectiveness, and Stealthiness. The differences between the discussed IEEE 802.15.4 PHY layer attacks with respect to energy efficiency, effectiveness, and stealthiness are as follows.

(a) Radio Jamming. The following observations for all *Radio Jamming* attacks and their variants with respect to the above three evaluation criteria are noteworthy.

(i) *Wide-Band Denial. Wide-Band Denial* is not an energy-efficient attack because an adversary consumes a relatively large amount of power in order to constantly transmit jamming signals of high power over the entire frequency band. On the other hand, this attack is considered to be highly effective at disrupting network communications because it spans all channels of the related frequency band. Since a continuous jamming signal is present, this attack is very unstealthy and easy to detect [3].

(ii) *Constant Jamming.* Recall that the *Constant Jamming* variant of *Pulse-Band Denial* targets only a single channel within the frequency band. If all devices within the attacked network are expected to operate within a single common frequency channel, then the energy efficiency, effectiveness, and stealthiness characteristics of this attack converge to those of *Wide-Band Denial.*

(iii) *Deceptive Jamming.* This *Pulse-Band Denial* variant is as energy-efficient and as effective as *Constant Jamming* but is slightly more difficult to detect. This is because the deceptive use of packets instead of radio signals for the purpose of jamming confuses the receiver as to whether the received packets are being sent by a legitimate or illegitimate network node [6].

(iv) *Random Jamming.* By randomly alternating between sleeping and jamming modes, the *Random Jamming* variant of *Pulse-Band Denial* saves more energy, thus making it stealthier than its previous counterparts. The fact that jamming is no longer a continuous occurrence makes *Random Jamming* more effective and harder to detect [6].

(v) *Interrupt Jamming.* In this variant of *Reactive Jamming,* an adversary only starts jamming upon detection of a frame's preamble and SFD, making it a highly energy-efficient attack. Despite its low energy consumption, *Interrupt Jamming* is still highly effective, as an adversary only enters sleep state in the absence of radio activity [5]. *Interrupt Jamming* is very difficult to detect because jamming never extends beyond the length of frame transmission [4].

(vi) *Activity and Scan Jamming*. Due to their reliance on RSSI and CCA output, these two *Reactive Jamming* variants are known for displaying high rates of false positives in detecting network activity. False positives result either from the detection of channel noise or from the high latencies incurred by RSSI and CCA mechanisms in detecting channel activity. Channel noise detected by RSSI and CCA mechanisms can be either background noise or noise resulting from channel access contention over channels which are common between IEEE 802.15.4 Wireless Personal Area Networks (WPANs) and 802.11 Wireless Local Area Networks (WLANs). In order to compensate for this inaccurate detection of network activity, an adversary launching either attack must transmit more frequent jamming signals, thus consuming more energy and increasing the probability of detection [5]. Since Wood et al. [5] do not mention the possibility of false negative generation by *Activity* and *Scan Jamming*, it is safe to assume that network activity is always detected by both variants. From this, we infer that *Activity* and *Scan Jamming* are both highly effective at disrupting network communications.

(vii) *Node-Specific and Message-Specific Denial*. For these two *Reactive Jamming* variants, jamming is only initiated upon detection of a frame whose header indicates a specific destination address or frame type. As such, we consider both variants to be very energy-efficient. We also infer that both variants are highly effective at corrupting the targeted traffic and are very difficult to detect for the same reason as *Interrupt Jamming*.

(viii) *PHY Payload and FCS Corruption*. The reason for not including these two frame corruption techniques in Table 1 is because we consider them to be methods of launching the *Radio Jamming* attacks shown in Table 1, but they are not actual attacks themselves. Wood et al. [5] argue that energy savings with *FCS Corruption* are greater than with *PHY Payload Corruption*. This is because *FCS Corruption* involves jamming for at most two bytes of the intended frame's FCS field, whereas *PHY Payload Corruption* involves higher energy consumption due to the use of a longer jamming signal for a few or all bytes of the intended frame's payload. In our opinion, both methods of frame corruption are equally effective, as they both intend to directly or indirectly corrupt frame FCS, thus leading to the frame's rejection at the receiver.

(b) Message Manipulation Attacks. Wilhelm et al. [8] do not reveal any information regarding the energy efficiency, effectiveness, or ease of detection of *Symbol Flipping* and

Signal Overshadowing. However, in the results section of [8], it is stated that *Signal Overshadowing* has a 66.97% success rate in its deployment.

(c) Steganography Attacks. While Martins and Guyennet [9] make no explicit comments on this attack's energy efficiency, effectiveness, and stealthiness, we believe that there exists a tight relationship between these three criteria for this attack. The effectiveness of a *Steganography Attack* depends on the value and length of information exchanged among cooperating adversaries over the created hidden channel, as well as the effectiveness of the attacks launched by cooperating adversaries in response to the exchanged information. Recall that each bit of a secret message is embedded instead of the reserved bit of the PHY header field.

The longer the secret message that needs to be transmitted, the larger the number of frames exchanged over the network. Exchanging a larger number of frames leads to shorter lifetime of malicious adversaries due to higher energy consumption, in addition to higher probability of detection due to abnormal activity monitored by the network. Energy efficiency of PHY layer *Steganography Attacks* can be improved by transmitting some bits of a secret message in both PHY and MAC fields of transmitted frames, thus reducing the total number of exchanged frames [9].

4.2.2. Security Goal Violation. With reference to the security goals acquired from Kumar et al. [10], we find that the discussed PHY layer attacks target the violation of only two primary security goals, which are data integrity and network availability.

(a) Radio Jamming. Since all types of *Radio Jamming* attacks are launched with the intent of creating a DoS on the target network, all *Radio Jamming* attacks are considered to violate network availability [7]. A common misconception is that *Radio Jamming* attacks also target the integrity of data transmitted across the network because jamming corrupts, and thus modifies, either the payload or FCS fields of the frame if the modified frame is accepted by the receiver. This is not the case in *Radio Jamming* attacks, because the modified frame is rejected by the receiver due to an incorrect FCS.

(b) Message Manipulation Attacks. Contrary to *Radio Jamming*, *Message Manipulation Attacks* cause the acceptance of a modified message by the receiver, thus violating the integrity of the message [8]. *Message Manipulation Attacks* make no effort to target the availability of network resources.

(c) Steganography Attacks. Similar to *Message Manipulation Attacks*, *Steganography Attacks* do not target the violation of network availability. In our opinion, *Steganography Attacks* are not considered to violate the integrity of network transmissions. This is because adversaries merely use steganography to embed data within unused fields of transmitted frames, but no modifications are made to the existing frame data. This results in a correct FCS and frame acceptance at the receiver. As such, the original content of the frame and hence its integrity are retained.

5. Attacks on IEEE 802.15.4 MAC Layer

This section addresses attacks against the IEEE 802.15.4 MAC layer and is based on the research in [11].

5.1. Link Layer Jamming. Similar to *Radio Jamming* at the PHY layer, *Link Layer Jamming* is a MAC layer attack which is launched with the intent of creating a Denial of Service (DoS) against the network by disrupting the exchange of messages between transmitting and receiving network nodes. While *Radio Jamming* achieves its purpose by creating radio interference through the emission of radio signals, *Link Layer Jamming* involves the emission of packets rather than signals [12]. Two variants of *Link Layer Jamming* can be defined, both of which cause degradation and reduction of network performance and throughput [13]. These two variants differ with respect to the recurrence of packet transmission by the malicious adversary.

5.1.1. Random Jamming (Also Known as Blind Jamming [12]). In this variant, a malicious adversary emits packets of useless content at random time intervals and for no specific purpose. While this variant can be considered as a stand-alone attack, it is also the basis for the *One Random Attacker (ORA)* and *Two Random Attackers (TRA)* scenarios of the *Interference During CFP* variant of the *GTS Attack*, which are explained later.

5.1.2. Intelligent Jamming. An intelligent jammer emits packets of useless content at specific times for specific purposes [12]. In addition to acting as a stand-alone attack, this variant can also be used as the basis for launching more powerful and complicated types of attacks, such as *Acknowledgment (ACK) Attack*, *Man-In-The-Middle (MITM) Attack*, and the *One Intelligent Attacker (OIA)* and *Two Intelligent Attackers (TIA)* scenarios of the *Interference During CFP* variant of the *GTS Attack*, as explained later.

5.2. Node-Specific Flooding. Mišić et al. [13] describe an attack which involves the transmission of unnecessary packets whose destination addresses are set to the addresses of destination nodes targeted by malicious adversaries. The targeted nodes' power sources are eventually depleted due to excessive packet reception from the adversaries.

5.3. Back-Off Manipulation. The Carrier Sense Multiple Access with Collision Avoidance (CSMA-CA) [1] channel access mechanism is used to govern the rules for medium contention among network nodes in IEEE 802.15.4 networks. For beacon-less networks, unslotted CSMA-CA is used. For beacon-enabled networks, slotted CSMA-CA is used during the Contention Access Period (CAP) of each superframe duration. A malicious adversary can manipulate CSMA-CA rules in such a way that the adversary constantly uses a short back-off period instead of selecting a random back-off period from its contention window. In doing so, the adversary hijacks channel access by ensuring that it is always granted higher priority to access the channel than legitimate nodes, which use larger back-off periods [6]. This attack increases both waiting time of legitimate nodes during channel access and power consumption of nodes during the reception of adversary data [13].

Mišić et al. [13] mention two variants of *Back-Off Manipulation*, the difference between which is with respect to the methods they use to accomplish the common objective explained above.

5.3.1. Battery Life Extension (BLE) Pretense. BLE mode ensures the conservation of power for nodes operating on battery power. A malicious adversary can take advantage of this CSMA-CA feature by falsely pretending to run in BLE mode in order to acquire a smaller initial contention window size than the other legitimate nodes. This reduces the range of values from which the adversary can select its back-off period and ensures that its probability of accessing the medium is much higher than legitimate nodes.

5.3.2. Constant Back-Off Exponent (BE). A malicious adversary can choose not to increment its BE after a failed transmission attempt. Maintaining a constant BE prevents contention window size from being increased, thus increasing probability of channel access.

5.3.3. Random Number Generator (RNG) Tampering. Another way of increasing the odds of channel access is for a malicious adversary to modify its RNG in such a way that ensures that the back-off periods selected by the adversary are much smaller than those selected by legitimate nodes.

5.3.4. Back-Off Countdown Omission. Mišić et al. [13] describe an attack which involves the complete omission of the random back-off countdown by a malicious adversary. We consider this omission to be the same as the complete omission of the entire CSMA-CA protocol. The effect of this would be the ability of the adversary to transmit its packets more frequently than legitimate network nodes, thus causing collisions between the adversary's packets and legitimate network packets, resulting in the same DoS outcome as *Link Layer Jamming*.

5.4. Clear Channel Assessment (CCA) Manipulation. Mišić et al. [13] describe two attacks which can be launched against the CCA procedure of the CSMA-CA protocol. CCA is the process of initiating packet transmission if the channel is sensed idle for 2 successive back-off periods. Two variants of the CCA Manipulation attack can be defined as follows.

5.4.1. Clear Channel Assessment (CCA) Reduction. In this attack variant, if the adversary senses that the channel is idle for only 1 back-off period (not 2), it initiates packet transmission, giving channel access more quickly and frequently to adversaries than to legitimate network nodes.

5.4.2. Clear Channel Assessment (CCA) Omission. Rather than reducing the number of back-off periods during which CCA is performed, an adversary may choose to omit the

CCA procedure altogether in order to immediately start transmitting whenever the random back-off countdown is over. This could potentially cause collisions if the channel is not idle, leading to a DoS effect as in *Link Layer Jamming*.

5.5. Same-Nonce Attack.

Consider a node providing the access control service in secured operating mode. In this case, if two entries within this node's Access Control List (ACL) possess the same key and nonce, a malicious adversary obtaining the cipher texts pertaining to these two entries will be able to infer useful information about the transmitted data, as explained in [6].

5.6. Replay-Protection Attack.

Replay-protection is an IEEE 802.15.4 mechanism which causes a node to drop a frame if its sequence number is equal to or less than the sequence number of a preceding frame received by that same node. An adversary can send frames with large sequence numbers to targeted legitimate nodes, causing frames with smaller sequence numbers from other legitimate nodes to be dropped [6].

5.7. Acknowledgment (ACK) Attacks.

In IEEE 802.15.4 as well as in other types of networks, ACK frames are sent between network nodes in order to confirm successful frame transmission. For some types of frames, an *Acknowledgment Request* field is present, which is set to 1 if an acknowledgment is required upon frame receipt or 0 if no acknowledgment is required [1].

In this subsection, we explain two variants of the *ACK Attack*.

5.7.1. ACK Spoofing.

An adversary can perform *Intelligent Link Layer Jamming*, as described in Section 5.1.2, in order to prevent legitimate data from correctly being received by the receiver. The adversary then sends back a forged ACK on the receiver's behalf with the correct expected sequence number to the sender, thus preventing data retransmission by tricking the sender into thinking that the frame has successfully reached the receiver [3].

5.7.2. ACK Dropping.

In this variant, although the transmitted data is correctly received by the receiver, *ACK dropping* uses *Intelligent Link Layer Jamming* to jam the true ACK that is sent back by the receiver to the sender. Unlike *ACK Spoofing*, the malicious adversary sends no forged ACK in place of the jammed ACK. As such, the sender and receiver nodes' power and bandwidth are wasted during retransmissions up to a maximum number of retransmissions [6].

5.8. Man-In-The-Middle (MITM) Attack.

This attack is an extension of the *ACK Spoofing* variant of the *ACK Attack* described in Section 5.7.1. Following the transmission of the forged ACK by the adversary to the sender node, the adversary also transmits an altered version of the original data frame to the receiver and receives a true ACK frame from the receiver by pretending to be the original sender. *ACK* and *MITM Attacks* are also referred to as *Interception Attacks* [4].

5.9. Guaranteed Time Slot (GTS) Attacks.

In beacon-enabled networks, the PAN coordinator reserves Guaranteed Time Slots (GTS) within the Contention Free Period (CFP) of each superframe duration in order to guarantee channel access for network nodes running time-critical applications with real-time delivery, low latency, or specific bandwidth requirements. A maximum of 7 GTS can be assigned at any one time, with each GTS possibly occupying more than one superframe slot within the superframe's CFP. Allocation and deallocation of GTS are performed by the PAN coordinator on a first-come-first-serve basis [1]. *GTS Attacks* are launched against the network by misusing the GTS management scheme [12].

Since there is no method of verifying of sensor nodes' identifiers (IDs), Jung et al. [14] define two categories of variants of the IEEE 802.15.4 *GTS Attack* as follows.

5.9.1. Existing Identities in the PAN.

In this category, a *GTS Attack* is launched when malicious adversaries spoof the IDs of existing legitimate nodes in the PAN. Two variants for this category are defined in [12] as follows.

(a) DoS against Data Transmissions during CFP. This variant requires the adversary to passively eavesdrop on network traffic in order to collect information about the IDs of legitimate nodes and their allocated GTS. The adversary can then use this collected data to spoof the IDs of the legitimate nodes and to send GTS deallocation requests on their behalf to the PAN coordinator. This leads to the termination of channel access rights previously granted to the legitimate nodes during their previously assigned GTS.

(b) False Data Injection. While the *DoS During CFP* variant collects information about the IDs of nodes that have already been allocated GTS by the PAN coordinator, *False Data Injection* collects information about the IDs of legitimate PAN nodes that have not yet been allocated any GTS during the superframe's CFP. Using the collected information, the adversary pretends to be one of the unallocated nodes by spoofing its ID and sends a GTS allocation request on its behalf to the PAN coordinator. Finally, the adversary injects false traffic into the network during its falsely assigned GTS.

5.9.2. Nonexisting Identities in the PAN.

Rather than spoofing the IDs of legitimate nodes within the PAN, a malicious adversary can use its own or other nonexisting IDs to conduct either of the two attack variants contained within this category [4].

(a) DoS against GTS Requests. For this variant, a malicious adversary collects information about the GTS list, which contains both allocated and free GTS. Following this, the adversary keeps sending GTS allocation requests to the PAN coordinator until all 7 slots in the GTS list are filled up. Contrary to the *False Data Injection variant*, no

ID spoofing is involved, as the adversary sends allocation requests using its own or other nonexisting IDs to the PAN coordinator.

(b) Stealing Network Bandwidth. This attack variant is identical to the previous *DoS against GTS Requests* variant with the addition that the adversary also injects false data into the network during the assigned GTS. This variant is harder to detect than the previous *DoS against GTS Requests* variant because the PAN coordinator recognizes that the allocated slots are indeed being used for transmitting data and thus does not drop the allocated slots.

Sokullu et al. [12] identify four additional variants of the *GTS Attack*. We include these additional variants within a category of our choosing, which we call *Interference During CFP*.

5.9.3. Interference during CFP. In this type of *GTS Attack*, a malicious adversary collects information about the beginnings and ends of GTS which have been assigned to legitimate network nodes by the PAN coordinator. The adversary then creates interference by using *Link Layer Jamming* during these assigned slots with the intent of corrupting ongoing transmissions. The four variants defined in [12], which fall into this category, are as follows.

(a) One Intelligent Attacker (OIA). In the *OIA* scenario, a malicious adversary corrupts the communication with the maximum GTS length, either by corrupting only the GTS's first superframe slot or by corrupting all of the superframe slots contained within the GTS.

(b) One Random Attacker (ORA). In this scenario, a malicious adversary attacks the GTS of a randomly selected communication.

(c) Two Intelligent Attackers (TIA). This attack variant is considered as an extension to the *OIA* attack variant, with one malicious adversary attacking the communication with the largest GTS length and a second adversary attacking the communication with the second largest GTS length. This requires collaboration between the two adversaries.

(d) Two Random Attackers (TRA). As an extension to the *ORA* scenario, two malicious adversaries can attack the GTS of two randomly selected communications. Due to the random nature of this attack, it is possible for both adversaries to target the same communication.

5.9.4. DoS against CAP Maintenance. CAP maintenance involves the use of a number of preventative actions in order to ensure that the length of the CAP period of each superframe does not fall below a predefined threshold known as *aMinCAPLength* [1]. An adversary can launch an attack against CAP maintenance by constantly sending GTS requests, even when the superframe has no available capacity and/or the length of the CAP is about to fall below *aMinCAPLength*. This causes the length of the CAP to momentarily fall below *aMinCAPLength*, thus reducing

the amount of time which member nodes have in order to contend for channel access [1]. While this variant of the *GTS Attack* has not been previously mentioned in any references, its discovery is inspired by the brief statement made by Jung et al. [14] that preventative actions of CAP maintenance are ineffective if a malicious node constantly sends either GTS requests or data at the assigned GTS during the CFP.

5.10. PANId Conflict Attack. IEEE 802.15.4 defines a conflict resolution procedure, which is initiated when two PAN coordinators residing within the same Personal Operating Space (POS) have the same coordinator ID, also referred to as *PANId* [1]. A malicious adversary can abuse the conflict resolution procedure by transmitting fake *PANId* conflict notifications to the targeted PAN coordinator in order to initiate conflict resolution, thus momentarily delaying or even preventing communication between member nodes and the PAN coordinator [6].

5.11. Ping-Pong Effect. The *Ping-Pong Effect* is an attack which is launched with the aim of causing packet loss and service interruption, reducing node performance, and increasing energy consumption and network load. As per its name, this attack causes fast, repeated, and undue handovers of nodes between the coordinators of different PANs.

Balarengadurai and Saraswalhi [7] explain that a *Ping-Pong Effect* can be launched via the manipulation of one or both of the following two network parameters.

5.11.1. Membership Degree. A node switches to a new PAN coordinator if the membership degree to the new PAN coordinator is greater than its membership degree to its current PAN coordinator.

5.11.2. Election Possibility. A new node is elected as the PAN coordinator if its election possibility is higher than the election possibility of the current coordinator. Election possibility is determined with respect to factors such as mobility and remaining battery capacity.

5.12. Bootstrapping Attack. O'Flynn [4] explains an attack which forces a targeted network node to become unassociated with its PAN at a time of the adversary's choosing by launching any of the PHY or MAC layer attacks aimed at causing DoS. The next time that the legitimate node wants to rejoin the network, either the adversary passively eavesdrops on the association process in order to collect valuable bootstrapping information that it can use to perform its own association with the PAN, or the adversary can perform a *MITM Attack* in order to intervene with and thus prevent the association of the legitimate node with the PAN.

5.13. Steganography Attacks. Steganography Attacks are explained in Section 2 and can be launched by hiding information within the PHY and/or MAC frame fields of the 802.15.4 protocol.

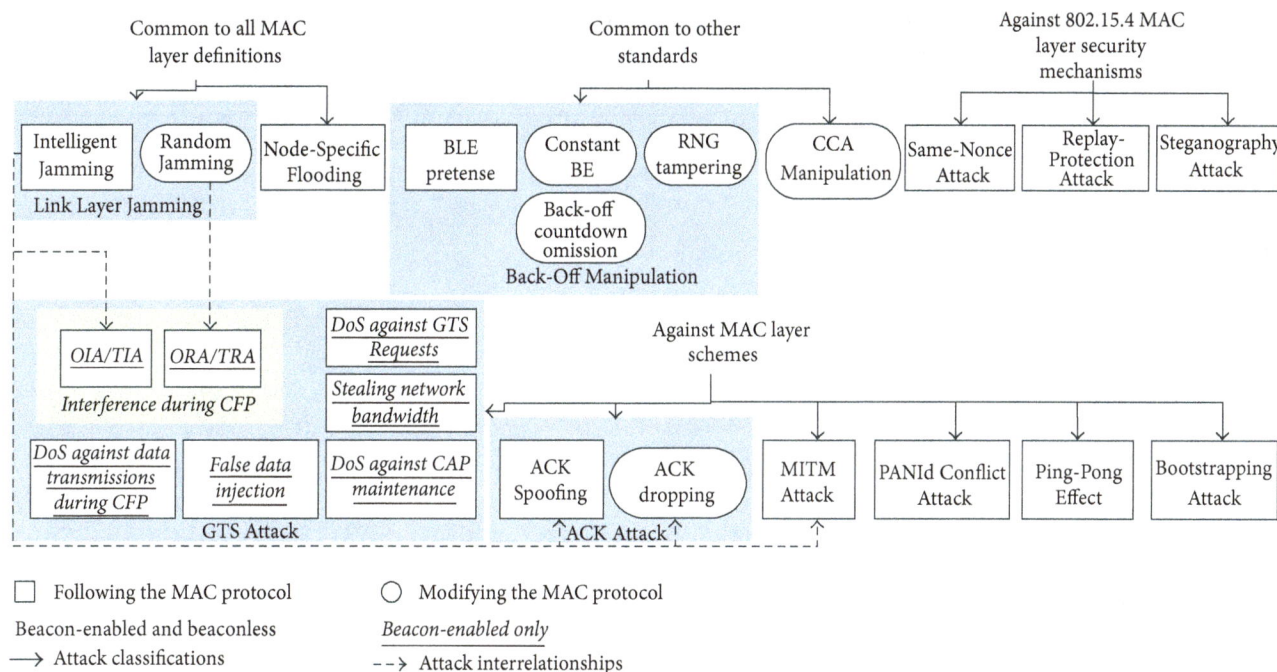

FIGURE 2: Classification of IEEE 802.15.4 MAC layer attacks.

6. Classification of IEEE 802.15.4 MAC Layer Attacks

The explained MAC layer attacks cannot be classified using one single deterministic classification. Therefore, we present two classifications which include some of the most important methods of classifying IEEE 802.15.4 MAC layer attacks obtained from external references. Novel extensions to these classifications are also presented. Figure 2 illustrates both classification methods, including our proposed extensions and the interrelationships between related attacks, which denote the attacks used to facilitate the launching of other attacks.

6.1. Classification A. Sokullu et al. [6] classify IEEE 802.15.4 MAC layer attacks into the following three main classes.

6.1.1. Common to All MAC Layer Definitions. Link Layer Jamming can be launched against all MAC layer definitions of all existing standards. We extend this class of attacks by adding *Node-Specific Flooding*.

6.1.2. Common to Other Standards. Back-Off Manipulation and CCA Manipulation attacks can be launched against both IEEE 802.15.4 Wireless Personal Area Networks (WPANs) and IEEE 802.11 Wireless Local Area Networks (WLANs) due to their similar CSMA-CA and Distributed Coordination Function (DCF) channel access protocols, respectively.

6.1.3. Against 802.15.4 MAC Layer Security Mechanisms. This class contains specific variants of some general attacks applied against IEEE 802.15.4 MAC layer security mechanisms. *Same-Nonce Attack* targets the access control service, and *Replay-Protection Attack* targets the replay-protection mechanism. *Steganography Attacks* are also included.

In addition to the above three classes of attacks, we extend this classification by including the following additional class of attacks.

6.1.4. Against 802.15.4 MAC Layer Schemes. This class refers to attacks applied against IEEE 802.15.4 MAC layer schemes. Contrary to Sokullu et al. [6], we argue that acknowledgments are considered to be an implemented MAC layer scheme and not a security mechanism. As such, we include *ACK and MITM Attacks* here. We also include the following additional attacks; *PANId Conflict Attack* targets the PANId conflict resolution procedure, *GTS Attacks* target the GTS management scheme, and *Ping-Pong Effect* and *Bootstrapping Attack* target PAN association.

6.2. Classification B. In this classification, we classify attacks with respect to conformance to MAC protocol rules and mode of network operation.

6.2.1. Conformance to MAC Protocol. Mišić et al. [13] classify attacks into those which either follow the MAC protocol to the letter or modify its rules.

6.2.2. Mode of Network Operation. All MAC layer attacks can be launched against both beacon-less and beacon-enabled networks except for all *GTS Attack* variants, as GTS are only present during the CFP of each superframe in beacon-enabled networks.

TABLE 2: Comparative analysis of IEEE 802.15.4 MAC layer attacks ($\sqrt{}$ for cause, $\sqrt{}\sqrt{}$ for effect).

Attack	DoS intent				Primary security goals			
	Exhaustion	Collision	Unfairness	Sleep	Confidentiality	Integrity	Authenticity	Availability
Link Layer Jamming	√√	√	√	×	×	×	√√	√
Node-Specific Flooding	×	√	√	×	×	×	√√	√√
BLE pretense, constant BE, RNG tampering, and CCA reduction	×	√√ [7]	×	×	×	×	√√	×
Back-off countdown omission and CCA omission	√	√√ [7]	×	×	×	×	√√	×
Same-Nonce Attack	×	×	×	√√	×	×	×	×
Replay-Protection Attack	×	√√	√	×	×	×	√√	√
ACK spoofing attack	×	√√	×	×	×	√√	√√	×
ACK dropping attack	×	√	√	×	×	×	√√	√√
MITM Attack	×	√√	×	×	√√	√√	√√	×
PANId Conflict Attack	×	√	√	×	×	×	√√	√√
DoS against data transmissions during CFP	×	√	×	×	×	√√	√√	×
DoS against GTS Requests	×	√	×	×	×	×	√√	√√
False Data Injection	×	√	×	×	√√	√√	√√	√√
Stealing network bandwidth	×	√	×	×	√√	×	√√	√√
DoS against CAP maintenance	×	√	×	×	×	×	√√	√√
Interference during CFP	√√	√	√	×	×	×	√√	√
Ping-Pong Effect	×	√	√	×	×	×	√√	√√
Bootstrapping Attack	×	√√	×	×	×	×	√√	√
Steganography Attack	×	√	×	×	×	×	×	×

7. Comparative Analysis of IEEE 802.15.4 MAC Layer Attacks

In this section, we perform a detailed comparison between the attacks with respect to the following evaluation criteria.

7.1. Evaluation Criteria

7.1.1. DoS Intent.
The primary intent of most MAC layer attacks is to cause a DoS against a specific part of or the entire network [12].

(a) Exhaustion Attacks. One form of DoS involves the depletion of the already-constrained power, bandwidth, memory, and/or storage resources of legitimate network nodes [7].

(b) Collision Attacks. An adversary can corrupt legitimate packets by initiating transmission during ongoing legitimate packet transmissions [7].

(c) Unfairness Attacks (Also Known as Misbehavior Attacks [4]). This attack ensures that an adversary is granted the same priority as or higher priority than legitimate nodes with respect to utilization of network resources, such as bandwidth and channel access. The latter causes starvation of legitimate nodes from network resources [7].

(d) Sleep Attacks. A Sleep Attack manipulates a targeted node's duty cycle (the percentage of time during which the node remains in active state). If the attack causes the targeted node's duty cycle to increase above average, it is also referred to as a Battery Exhaustion Attack [15].

7.1.2. Security Goal Violation.
As in Section 4, MAC layer attacks are compared with respect to their violations of the security goals obtained from Kumar et al. [10]. This paper only focuses on primary security goals, as no secondary goals are violated by any of the discussed attacks.

7.2. Comparison between IEEE 802.15.4 MAC Layer Attacks.
Table 2 illustrates the comparison between IEEE 802.15.4 MAC layer attacks with respect to both DoS intent and primary security goal violation. We analyze each of the MAC layer attacks and their variants from a cause and effect perspective. The cause ($\sqrt{}$) of an attack is the primary intent with which the attack is launched, whereas the effect ($\sqrt{}\sqrt{}$) of an attack refers to an unplanned repercussion of launching the attack in question.

(a) Data Confidentiality. Same-Nonce Attack is the only attack considered to violate data confidentiality, as it enables an adversary to decrypt ciphered network transmissions.

(b) Data Integrity. We consider attacks that corrupt only the payload field of the frame, while preserving the value of the original frame's Frame Check Sequence (FCS) field, to violate data integrity. As such, Steganography Attacks are not considered to violate the integrity of network transmissions.

(c) Data Authenticity. Attacks which involve the spoofing of legitimate node IDs violate authenticity.

(d) Network Availability. Any attack variant which has at least one DoS intent as its cause, as illustrated in Table 2, is considered to violate network availability.

8. Conclusion

This work constitutes a detailed survey on IEEE 802.15.4 PHY and MAC layer attacks.

In the first part of this survey, we extensively discussed 802.15.4 PHY layer attacks. The purpose and operation of each attack and its variants were explained. We presented two classifications for *Radio Jamming* attacks in particular. *Classification A* divided *Radio Jamming* attacks into *Wide-Band Denial* and *Pulse-Band Denial*, depending on the number of channels targeted by each within the related frequency band. *Classification B* classified *Radio Jamming* attacks into *PHY Payload Corruption* and *FCS Corruption* categories, depending on the method used for frame corruption. We explained how both frame corruption techniques in *Classification B* could be applied to any of the *Radio Jamming* attacks mentioned in *Classification A*. We concluded our discussion of 802.15.4 PHY layer attacks with a comparative analysis of the discussed PHY layer attacks, including their multiple techniques and variants. Attack differences were evaluated with respect to four evaluation criteria: energy efficiency, effectiveness, stealthiness, and violation of two particular primary security goals, which were data integrity and network availability.

IEEE 802.15.4 MAC layer attacks were addressed in the second part of this survey. We started off by clarifying the purpose and method of operation of each attack and its variants. A novel variant of the *GTS Attack*, which we called *DoS against CAP maintenance*, was proposed. Two classifications of the MAC layer attacks were presented. *Classification A* divided the attacks into four classes: common to all MAC layer definitions of all existing network standards, common to only a subset of other standards, launched against 802.15.4 MAC security mechanisms, and launched against implemented 802.15.4 MAC schemes. *Classification B* classified MAC attacks based on their conformance to MAC layer protocol rules and network mode of operation. Finally, a comparative analysis between all MAC layer attacks, as well as their multiple techniques and variants, was performed with respect to DoS intent and primary security goal violation.

Competing Interests

The authors declare that they have no competing interests.

References

[1] IEEE Standard for Information Technology, "Local and metropolitan area networks—specific requirements—part 15.4: wireless Medium Access Control (MAC) and Physical Layer (PHY) specifications for low rate Wireless Personal Area Networks (WPANs)," IEEE 802.15.4-2006, 2006.

[2] S. C. Ergen, *ZigBee/IEEE 802.15.4 Summary*, 2004, http://users.ece.utexas.edu/~valvano/EE345L/Labs/Fall2011/Zigbeeinfo.pdf.

[3] P. Jokar, H. Nicanfar, and V. C. M. Leung, "Specification-based intrusion detection for home area networks in smart grids," in *Proceedings of the IEEE 2nd International Conference on Smart Grid Communications (SmartGridComm '11)*, pp. 208–213, Brussels, Belgium, October 2011.

[4] C. P. O'Flynn, "Message denial and alteration on IEEE 802.15.4 low-power radio networks," in *Proceedings of the 4th IFIP International Conference on New Technologies, Mobility and Security (NTMS '11)*, pp. 1–5, IEEE, Paris, France, February 2011.

[5] A. D. Wood, J. A. Stankovic, and G. Zhou, "DEEJAM: defeating energy-efficient jamming in IEEE 802.15.4-based wireless networks," in *Proceedings of the 4th Annual IEEE Communications Society Conference on Sensor, Mesh and Ad Hoc Communications and Networks (SECON '07)*, pp. 60–69, IEEE, San Diego, Calif, USA, June 2007.

[6] R. Sokullu, I. Korkmaz, O. Dagdeviren, A. Mitseva, and N. R. Prasad, An Investigation of IEEE 802.15.4 MAC Layer Attacks, http://netos.ube.ege.edu.tr/source/publications/wpmc07.pdf.

[7] C. Balarengadurai and S. Saraswalhi, "Comparative analysis of detection of DDoS attacks in IEEE 802.15.4 low rate wireless personal area network," in *Proceedings of the International Conference on Modelling Optimization and Computing*, vol. 38, pp. 3855–3863, April 2012.

[8] M. Wilhelm, J. B. Schmitt, and V. Lenders, "*Practical Message Manipulation Attacks in IEEE 802.15.4 Wireless Networks*," 2015, http://www.lenders.ch/publications/conferences/Pilates12.pdf.

[9] D. Martins and H. Guyennet, "Attacks with steganography in PHY and MAC layers of 802.15.4 protocol," in *Proceedings of the 5th International Conference on Systems and Networks Communications (ICSNC '10)*, pp. 31–36, Nice, France, August 2010.

[10] J. V. Kumar, A. Jain, and P. N. Barwal, "Wireless sensor networks: security issues, challenges and solutions," *International Journal of Information and Computation Technology (IJICT)*, vol. 4, no. 8, pp. 859–868, 2014.

[11] Y. M. Amin and A. T. Abdel-Hamid, "Classification and analysis of IEEE 802.15.4 MAC layer attacks," in *Proceedings of the 11th International Conference on Innovations in Information Technology (IIT '15)*, pp. 74–79, Dubai, United Arab Emirates, November 2015.

[12] R. Sokullu, O. Dagdeviren, and I. Korkmaz, "On the IEEE 802.15.4 MAC layer attacks: GTS attack," in *Proceedings of the 2nd International Conference on Sensor Technologies and Applications (SENSORCOMM '08)*, pp. 673–678, Cap Esterel, France, August 2008.

[13] V. B. Mišić, J. Fung, and J. Mišić, "MAC layer security of 802.15.4-compliant networks," in *Proceedings of the 2nd IEEE International Conference on Mobile Ad-hoc and Sensor Systems (MASS '05)*, pp. 847–854, IEEE, Washington, DC, USA, November 2005.

[14] S. S. Jung, M. Valero, A. Bourgeois, and R. Beyah, "Attacking Beacon-enabled 802.15.4 networks," in *Security and Privacy in Communication Networks*, vol. 50 of *Lecture Notes of the Institute for Computer Sciences, Social Informatics and Telecommunications Engineering*, pp. 253–271, Springer, Berlin, Germany, 2010.

[15] F. Amini, J. Misic, and R. Eskicioglu, *Simulation and Evaluation of Security and Intrusion Detection in IEEE 802.15.4 Network*, University of Manitoba, Winnipeg, Canada, 2008.

10

A Movement-Efficient Deployment Scheme based on Information Coverage for Mobile Sensor Networks

Wei Cheng, Yong Li, Yi Jiang, and Xipeng Yin

School of Electronics and Information, Northwestern Polytechnical University, Xi'an, Shaanxi 710072, China

Correspondence should be addressed to Wei Cheng; pupil119@126.com

Academic Editor: Chi Chung Ko

Covering the surveillance region is a key task for wireless sensor networks. For mobile sensors, the deployment of sensors at appropriate locations is the key issue for sufficient coverage of surveillance area, and efficient resource management of sensor network. Previous studies most utilize physical coverage model for developing deployment schemes, in that the sensor resources may be wasted during the deployment for coverage. In this paper, a novel movement-efficient deployment scheme for mobile sensor networks is proposed, which adapts the information coverage model and the classical potential field method. The performances of the proposed scheme are evaluated comparing to two other schemes in aspects of coverage rate, coverage hole rate, ideal moving distance, and actual moving distance through extensive simulations. Simulation results show that the proposed scheme performs better than these related schemes in both coverage and movement efficiency.

1. Introduction

Mobile sensor networks are sensor networks in which sensors can move under their own control or under the control of the environment. A mobile sensor network is composed of some distributed mobile sensors to collect data. For the efficiency of wireless sensor networks, the appropriate scheme of deployment for mobile sensors is to maximize sensing coverage, which is the key task for a wireless sensor network [1].

Recently there is a lot of research about autonomous mobile robot deployment. Batalin and Sukhatme [2] proposed the local dispersion scheme to obtain a better coverage on whole surveillance region. Then a most famous approach, potential field theory scheme, was first proposed by Khatib [3] and then it was adapted for achieving elegant path planning algorithm of mobile robots [4] and multirobot manipulation [5]. Howard et al. [6] and Poduri and Sukhatme [7] employed the potential field-based schemes for area coverage in mobile sensor networks, which had been commonly used for mobile robotics. Tan et al. [8] proposed a floor-based scheme on deployment of mobile sensors. The main idea is to compartmentalize the surveillance region to floors with common height and make mobile sensors stay in floor line in the way in order to reduce the overlap area of the sensing disks.

Wang et al. [9] developed some schemes based on Voronoi diagram to maximize coverage of sensing. Once a coverage hole is discovered locally, the target location is estimated to guide the mobile sensors where to move at next round. The Voronoi diagram is used to discover the coverage holes and to develop a series of movement-assisted schemes for sensor deployment. Lee et al. [10] proposed an approach dealing with coverage hole by utilizing local Voronoi diagrams. While mobile sensors are deployed randomly, a sensor calculates the centroid for its local Voronoi polygon to get the next location where it will move to it in next time step. The sensors move round after round while the coverage is increased gradually. Thereafter, Han et al. [11] proposed another improved scheme which adapt centroid-directed virtual force. The same as other existing schemes, this scheme makes mobile sensors move around from high-density region to low-density region and avoid obstacles in surveillance region.

For coverage model, these previous works most utilize binary model that the sensing quality is constant within sensing range r_s while sensing quality is none outside sensing range. In this case, a given point is deemed to be covered if its physical distance to a sensor is shorter than sensing range. This concept of coverage is commonly regarded as *physical* coverage. However, Wang [12, 13] proposed a concept of *information coverage* based on estimation theory, in that sensors could cooperate to make estimates on sensing data at a given location of surveillance area. This new concept can reduce the requirements on sensor density for area coverage.

Although information coverage model has been used in sensor scheduling [14] and barrier coverage [15], it is rarely employed in the coverage issue for mobile sensor deployment. In this work, we will propose a movement-efficient scheme for mobile sensor deployment, which adapts sensing model based on information coverage and classical potential field method, so as to achieve the coverage requirement more practically and efficiently for mobile sensor networks. The main contribution of this paper is that we propose a novel scheme based on potential field for mobile sensor deployment, which adapts information coverage model. The proposed scheme can further improve coverage rate and avoid coverage hole effectively. Furthermore the proposed scheme could also increase movement efficiency in coverage process for mobile sensors.

The rest of this paper is organized as follows. Section 2 provides an overview on related works. In Section 3, the network assumptions, adversary model, and evaluation metrics are presented. Section 4 provides the detailed description of the strategies and our proposed deployment scheme. Then in Section 5 we evaluate the performance of the scheme by numeric simulations. Finally, Section 6 concludes this paper.

2. Related Work

2.1. Potential Field Method. According to potential field theory, a potential field is usually defined in a space so that its minimum could be achieved while all elements of space are nicely at target configuration [4–6, 16]. As objective is on the minimum ideally, all obstacles and borders in this space are all assumed to produce a high potential hill, so that, under this configuration of potential field, mobile sensor could move from the location with higher potential to location with lower potential. Sensors have to cover a target region, which may contain borders and obstacles, and to preserve communication capability between neighbor sensors. The mobile sensor gets attracted from the target locations while being repelled by borders and obstacles simultaneously. Finally the sum of all these potential fields determines the way of movements for the mobile sensor. For coverage issue, the overall potential field is achieved by adding the repulsive potentials that are from all obstacles and borders in surveillance area and the attractive potential from the target:

$$U(x, y) = U_{\text{att}}(x, y) + U_{\text{rep}}(x, y). \tag{1}$$

The related scheme proposed for maximizing the area coverage based on potential field could be presented as follows: each sensor would find the closest neighbor of its own by computing the distance between all the sensors. Then sensors would attempt to maintain a distance between their closest neighbor and themselves as communication range. The movements of sensors would turn into stable state when they all achieve this distance. Then every sensor would have one

communication link leastwise which is established between its closest neighbor and itself.

2.2. Voronoi Diagram Method. The Voronoi diagram is one of the basal tools for resolving coverage problem in wireless sensor networks. A Voronoi diagram for a set of seed points divides space into a number of regions. Each region corresponds to one of the seed points, and all the points in one region are closer to the corresponding seed point than to any other seed points. This is a key property for the sensing coverage since if sensor could not sense an expected event occurring in its region of polygon, none of the other sensors could sense it any more. Consequently, every sensor should operate the task of sensing within its local Voronoi polygon.

Utilizing property of Voronoi diagram, the area coverage could be transformed into coverage issue of every Voronoi polygon and reduce complexity of issues [17]. In Lee's scheme [10] and Han's scheme [11], every sensor utilizes the information locally to compute its local Voronoi polygon and then obtain the location with maximal effectiveness within local Voronoi polygon to enhance the local coverage rate and fix coverage holes at the most, so these schemes can be carried out currently and distributed by all mobile sensors. The centroid of the polygon is intersection point of all lines which divide the polygon into two portions of equal region. Thus this point could be the average of all points in the polygon. For coverage issue, centroid of local Voronoi polygon could be used as the point which is close to the location with maximal effectiveness so that these could efficiently increase the coverage of surveillance area. Assuming a polygon where n vertices (x_i, y_i) are presented clockwise and y coordinates of all vertices are nonnegative, then the closed-form expression of centroid (C_x, C_y) can be calculated as follows:

$$C_x = \frac{1}{6S_A} \sum_{i=0}^{n-1} (x_i + x_{i+1})(x_i y_{i+1} - x_{i+1} y_i),$$

$$C_y = \frac{1}{6S_A} \sum_{i=0}^{n-1} (y_i + y_{i+1})(x_i y_{i+1} - x_{i+1} y_i), \tag{2}$$

where area S_A of this polygon can be expressed as

$$S_A = \frac{1}{2} \sum_{i=0}^{n-1} (x_i y_{i+1} - x_{i+1} y_i). \tag{3}$$

Lee et al. [10] first proposed a novel scheme which deals with coverage hole by utilizing local Voronoi polygon. This scheme can be demonstrated by Figure 1; when all sensors are deployed randomly at first, each sensor computes centroid of its own local Voronoi polygon as possible future location and then moves towards this location. After several movements and iterations, the locations of all sensors are changed step by step; thus the whole coverage of sensing is refined gradually.

Han et al. [11] utilized three virtual forces from three potential fields independently on every sensor for coverage issue: the first is the repulsive potential U_{br} from surrounding borders and external obstacles, the second one is the repulsive potential U_{cov} from neighbor sensors, and the last one is

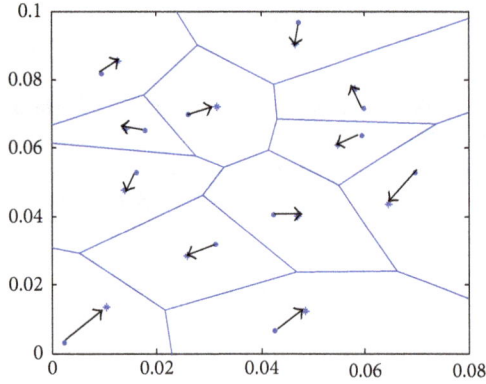

FIGURE 1: Snapshot of Centroid-Directed Potential Field.

the attractive cover potential U_v which could draw sensors towards the centroid of their local Voronoi polygons. Each of these potentials has its own direction and magnitude that depend on the relative locations of borders, obstacles, neighbor sensors, or centroid.

Thus mobile sensor is driven by U which is the sum of the three potential fields:

$$U = K_{br} U_{br} + K_{cov} U_{cov} + K_v U_v, \quad (4)$$

where K_{br}, K_{cov}, and K_v are corresponding positive scaling factors for each potential field.

3. Problem Formulation

3.1. Network Assumptions. We consider A as the surveillance area on a two-dimensional space. The borders of surveillance area can be regarded as wall-like obstacles. Given N mobile sensors which gather in one place at the beginning, and a target surveillance region, the goal of our work is to maximize the sensing coverage with less coverage hole. However, since sensor movements would cause significant energy consumption, we wish all sensors would move as less as possible to prolong the network lifetime.

To address this problem, the following assumptions will be made as follows.

(1) All of sensors have the same sensing range r_s and also communication range r_c. While sensors are within r_c of a sensor, they are counted as the neighbors of sensor.

(2) The mobile sensors could take omnidirectional movement and move at variable velocity in different rounds.

(3) Sensors could know their locations by some scheme and obtain locations of neighbor sensors by communication to each other.

(4) Sensors are assumed to be provided with equipment which could determine the relevant range of borders and obstacles.

(5) No mobile sensor will fail during the coverage process.

3.2. Evaluation Metrics

3.2.1. Coverage Rate. This metric is most fundamental for the performance of sensor deployment. This metric could be expressed as follows:

$$\text{rate}_{cov} = \frac{\bigcup_{i=1}^{N} C_i}{S_A}, \quad (5)$$

where C_i denotes the exact area covered by sensor i, the number of sensors is N, and S_A is the total area of the surveillance region.

3.2.2. Coverage Hole Rate. For simplicity of cognizance for coverage hole, we believe a point is in the coverage hole for given surveillance area if it satisfies the following conditions: (1) this point is not covered by any sensors in the surveillance area; (2) this point is not in the triangle whose vertexes are three nearest sensor's locations for this point. So the coverage hole rate can be expressed as

$$\text{rate}_{hole} = \frac{\bigcup_{i=1}^{M} H_i}{S_A}, \quad (6)$$

where H_i is the region of coverage hole i and M is the number of coverage holes.

3.2.3. Ideal Moving Distance. With the mobile sensors, we define ideal moving distance first as the direct distance between initial location and the final location. Then total ideal moving distance over all mobile sensors could be denoted as

$$L_{id,tot} = \sum_{i=1}^{N} \sqrt{\left(x_{Initial,i} - x_{Final,i}\right)^2 + \left(y_{Initial,i} - y_{Final,i}\right)^2}. \quad (7)$$

3.2.4. Actual Moving Distance. As the description above, we define actual moving distance as accumulative distance traveled in rounds from the initial location to the final location. When $l_i(t)$ is accumulative distance of sensor i at round t, actual moving distance over all mobile sensors after time duration T could be exactly expressed as

$$L_{tot} = \sum_{i=1}^{N} \sum_{t=1}^{T} l_i(t). \quad (8)$$

4. Strategies and Proposed Scheme

4.1. Information Coverage Concept. Considering K distributed sensors, every sensor takes a measurement on an unknown parameter θ of a phenomenon at some location and time [12, 13]. Let d_k, $k = 1, 2, \ldots, K$, denotes the distance between sensor k and the location with that phenomenon. The parameter θ is assumed to decay with range, while it is θ/d^α at distance d, where α is the decay exponent and $\alpha > 0$. Then the measurement z_k for the parameter θ, on sensor k, may be disturbed by an additive noise n_k. Thus

$$z_k = \frac{\theta}{d_k^\alpha} + n_k, \quad k = 1, 2, \ldots, K. \quad (9)$$

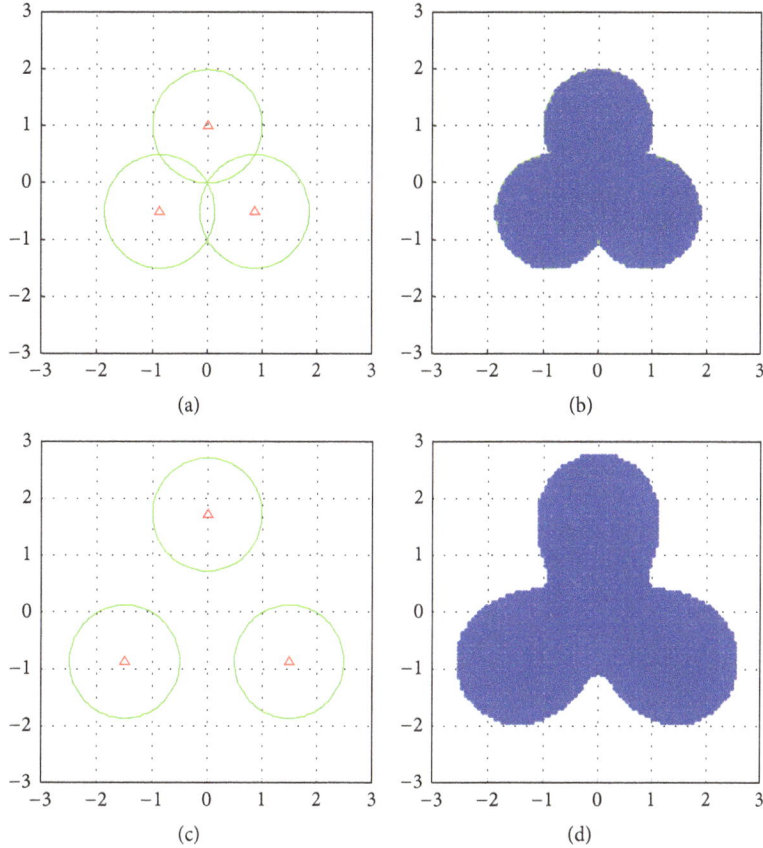

FIGURE 2: (a) Deployment for physical coverage. (b) Case of physical coverage. (c) Deployment for $(3, \varepsilon)$ information coverage. (d) Case of $(3, \varepsilon)$ information coverage.

The goal of parameter estimators is to estimate θ by these noisy measurements. Let $\hat{\theta}_K$ and $\tilde{\theta}_K = \hat{\theta}_K - \theta$, respectively, denote the estimate and estimation error of θ by K sensors.

We can observe that the estimation $\hat{\theta}_K$ and error $\tilde{\theta}_K$ are both random variables. The probability that the absolute value of $\tilde{\theta}_K$ is less than a constant A could be used to evaluate how well a given point is sensed; the larger this probability is, the more reliable the estimator is. Once it is larger than a scheduled threshold ε, that is, $\Pr\{|\tilde{\theta}_K| \leq A\} \geq \varepsilon$, this could be defined as *information coverage* with K cooperative sensors: a point is regarded as (K, ε)-covered if there are K sensors to estimate the parameter at a point so that $\Pr\{|\tilde{\theta}_K| \leq A\} \geq \varepsilon$, where $0 \leq \varepsilon \leq 1$. When K measurements are available, some standard estimators, such as the best linear unbiased estimator (BLUE), could be used to estimate parameter and obtain minimum mean squared error (MSE).

The key advantage of adapting this notion is to decrease sensor density and cover surveillance region completely. Assuming that all noises are independent and Gaussian, summation of these noises would still be Gaussian with zero mean and variance:

$$\tilde{\sigma}_K^2 = \sum_{k=1}^{K} a_K^2 \sigma_K^2, \tag{10}$$

where $a_k = B_K / d_k^\alpha \sigma_k^2$ and $B_K = \sum_{k=1}^{K}(1/d_k^{2\alpha}\sigma_k^2)$. Assume that all these noises have the same variance $\sigma_k^2 = \sigma^2$, $k = 1, 2, \ldots, K$. Hence there is

$$\Pr\left\{|\tilde{\theta}_K| \leq A\right\} = 1 - 2Q\left(\frac{A}{\sigma\sqrt{\left(\sum_{k=1}^{K} d_k^{-2\alpha}\right)^{-1}}}\right), \tag{11}$$

where $Q(x) = (1/\sqrt{2\pi}) \int_x^\infty \exp(-t^2/2)dt$.

Define sensing range r_s and the range where estimation error would be equal to threshold ε. Thus there is

$$Q\left(\frac{A}{r_s^\alpha \sigma}\right) = \frac{1}{2}(1 - \varepsilon). \tag{12}$$

For simplicity, we set r_s as the unit for range while $A = \sigma$ and then calculate threshold ε accordingly; hence $\varepsilon = 0.683$.

4.2. Information Coverage Based Deployment. In Wang et al.'s work [12], they introduced some regular sensor deployments for static sensors. Figure 2 demonstrates the physical coverage and information coverage when deployment pattern is tiling, that is, all equilateral triangles. In this case, with regular tiling and $\alpha = 1$, the information coverage on this area is much larger than that of a physical coverage without coverage hole.

There also was some other tiling which was square or hexagon in this work [12], whereas this equilateral triangles' deployment is more suitable for potential fields design because of its simplest form, such that distances between sensors are all equal in each tiling.

4.3. Potential Function Design. For developing a scheme based on this deployment of $(3, \varepsilon)$ information coverage for mobile sensors, two potential functions need to be designed. The first one would drive the mobile sensor to maximize the coverage region and the second one would prevent all mobile sensors from the borders and obstacles in surveillance area [6, 16]. These two potentials could be defined as cover potential and boundary potential.

The principal goal of cover potential is to drive the sensors to spread as long as the maximum region is covered while the communication connectivity could be still maintained. The distance between sensor and its closest neighbor is mainly used for computing this potential function. Thus this potential function could be formatted as

$$U_{\text{cov}} = \sum_{i=1}^{n} (d_i - d_{\text{CR}})^2, \qquad (13)$$

where d_i is the distance between sensor i and its closest neighbor.

We define d_{CR} as sensor distance of $(3, \varepsilon)$ information coverage, which can be computed based on deployment of equilateral triangles as follows. For $(3, \varepsilon)$ information coverage, there is

$$1 - 2Q\left(\frac{A}{\sigma\sqrt{\left(\sum_{k=1}^{K} D_k^{-2\alpha}\right)^{-1}}}\right) = \varepsilon. \qquad (14)$$

Since $K = 3$ and $D_k = D$ for $(3, \varepsilon)$ information coverage, where D is the distance between the vertex and geometric center, of this equilateral triangle, and $A = \sigma$, $\alpha = 1$, thus

$$\frac{1}{\sqrt{\left(\sum_{k=1}^{3} D^{-2}\right)^{-1}}} = Q^{-1}\left(\frac{1}{2}(1 - \varepsilon)\right), \qquad (15)$$

where $Q^{-1}(x)$ is the inverse Q function; then we can deduce that

$$D = \frac{\sqrt{3}}{Q^{-1}((1/2)(1 - \varepsilon))}. \qquad (16)$$

Considering both coverage and connectivity, we define D_v as the distance between the vertexes of this equilateral triangle; then $D_v = 2\cos(\pi/6) \cdot D$ and d_{CR} can be expressed as

$$d_{\text{CR}} = \begin{cases} D_v & \text{if } D_v < r_c \\ r_c & \text{if } D_v \geq r_c. \end{cases} \qquad (17)$$

Based on the potential function in (13), each sensor would attempt to hold on a distance as d_{CR} from its nearest neighbor.

Once the distance between sensors is larger than d_{CR}, the cover potential U_{cov} would produce an attractive force on sensors; while the distance between sensors is smaller than d_{CR} it would produce repulsive force. This scheme could drive spreading of sensors in the surveillance area so that the covered area could be maximized.

However, there are possibilities that sensors will move beyond the border of surveillance area, and this may reduce the efficiency of the coverage. Considering this issue, another potential function needs to be considered, which addresses the border of surveillance area and attempts to drive sensors within the surveillance area [6, 16]. If the border is regarded as an obstacle, then this could be regarded as a potential function of obstacle avoidance. Once the distance between sensor and the border or obstacle is smaller than the threshold value, this potential would try to keep the sensor away from border or obstacle at a limiting distance. It would produce a repulsive force that could drive sensors away from obstacles and borders. Once the distance between sensor i and border or obstacle is smaller than threshold d_{TH}, the border potential U_{br} would work immediately. So this potential function can be formatted as

$$U_{\text{br},i} = \begin{cases} \dfrac{\lambda}{d_{\text{br},i}} & \text{if } d_{\text{br},i} < d_{\text{TH}} \\ 0 & \text{if } d_{\text{br},i} \geq d_{\text{TH}}, \end{cases} \qquad (18)$$

$$U_{\text{br}} = \sum_{i=1}^{n} U_{\text{br},i},$$

where λ is a positive scaling factor. Then total potential function U can be computed by accumulating all of cover potentials and the border potentials, which can be expressed as

$$U = K_{\text{cov}} U_{\text{cov}} + K_{\text{br}} U_{\text{br}}, \qquad (19)$$

where K_{br} and K_{cov} are corresponding positive scaling factors for each potential field. The velocity of mobile sensor could be computed by working out the gradient of final potential function. The final potential function is the function of locations of mobile sensors, while the gradient of this potential function can be regarded as the time derivative of the location of each mobile sensor. Thus velocity can be expressed as

$$v(t) = -\nabla U(x, y) = \begin{bmatrix} -\dfrac{\partial U(x, y)}{\partial x} \\ -\dfrac{\partial U(x, y)}{\partial y} \end{bmatrix}. \qquad (20)$$

Each sensor calculates its new velocity and will move to next location according to this velocity, and then new locations can be obtained by central difference method.

5. Performance Evaluation

We would take comparison between the proposed scheme based on $(3, \varepsilon)$ information coverage and the existing schemes: potential field- (PF-) based scheme and Centroid-Directed Potential Field- (CDPF-) based scheme. We run

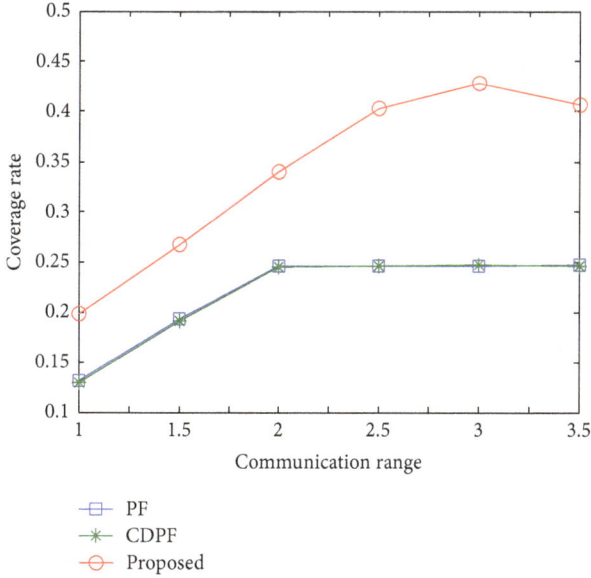

FIGURE 3: Coverage rate versus communication range (r_c/r_s).

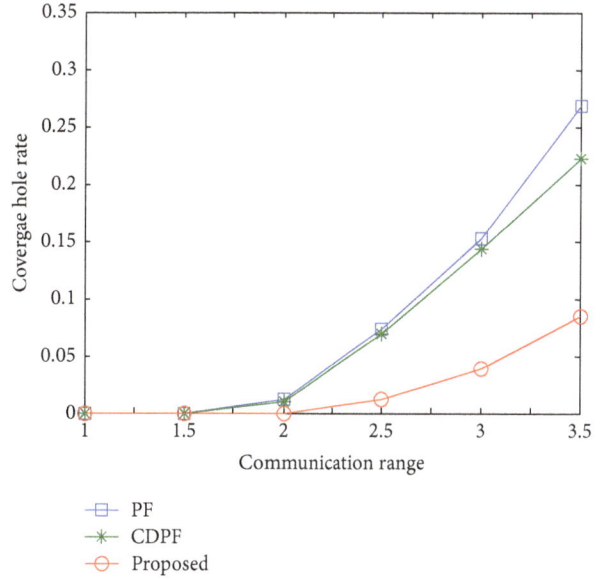

FIGURE 4: Coverage hole rate versus communication range (r_c/r_s).

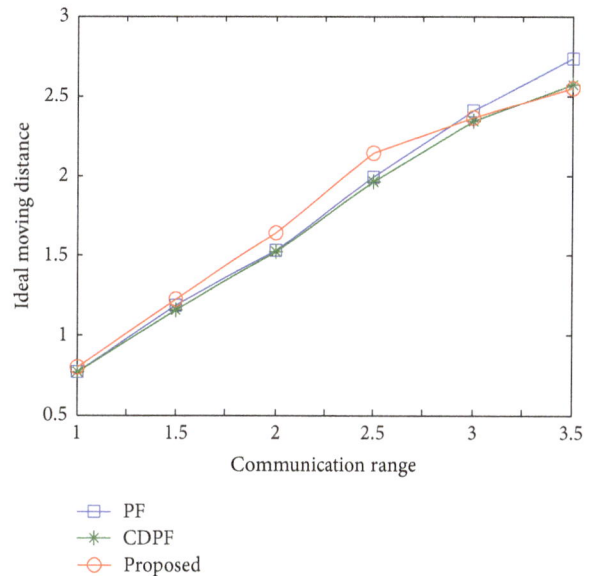

FIGURE 5: Ideal moving distance versus communication range (r_c/r_s).

each scheme in 3000 time stamps for one simulation and run it ten times for each parameter set to obtain average value. At the very beginning, we randomly deployed N sensors at the center of surveillance area, which has normalization boundary limits 1, so that the coordinate of mobile sensors (x_i, y_i) is always in $[0, 1]$. Therefore total area of the surveillance area $S_A = 1$. The default sensing range $r_s = 0.1$ while communication range of sensor r_c is at least one time more than it. The corresponding scaling factors K_{br} and K_{cov} are all equal to 1, while $K_v = 0.3$ for potential by centroid of the local Voronoi polygon, which is in Centroid-Directed Potential Field- (CDPF-) based scheme. Scaling factor for border potential $\lambda = 0.0009$, and the threshold distance $d_{TH} = 0.001$.

5.1. Impact of Communication Range. Deploy 8 mobile sensors in given surveillance area and make r_c vary from 1 to 3.5 times of r_s. As shown in Figures 3-4, we can observe that the proposed scheme can acquire almost two times larger coverage rate with minimum coverage hole rate, compared to other schemes. We can also observe that coverage rate increases with the communication range increases when the communication range is lower than three times of sensing range. While the communication range is more than three times of sensing range, the coverage rate is decreased and the coverage hole rate is quite large; that occurs because when communication range is very large, the probability of sensors that are far away becomes higher; thus the opportunity that sensors could cooperate with each other to achieve (K, ε) information coverage is reduced.

As shown in Figures 5-6, we can observe that the proposed scheme can acquire the best movement efficiency in terms of actual moving distance, compared to other schemes. Although the ideal moving distance of the proposed scheme is the highest, the actual moving distance is much lower than that of other schemes. That occurs because information coverage could avoid some movements such as coverage hole

fixing. When the model of information coverage is adopted, a physical coverage hole could be fixed by utilizing its neighbor sensors without taking any movement of other sensors.

5.2. Impact of Number of Sensors. We fix r_c as two times of r_s and the number of sensors is varied from 3 to 18. As shown in Figures 7-8, we could find that as the amount of sensors changes, the proposed scheme can always acquire the best coverage rate with minimum coverage hole rate, compared to other schemes. One thing that needs to be noticed is that when the number of sensors reaches a certain quantity, the coverage hole rate decreases while the number of sensors

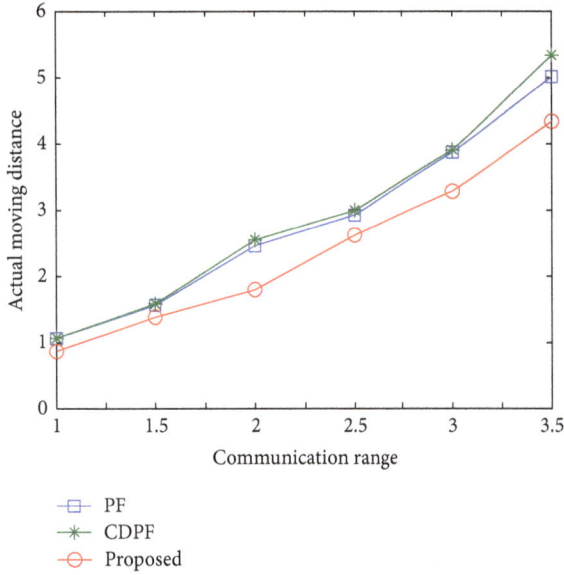

FIGURE 6: Actual moving distance versus communication range (r_c/r_s).

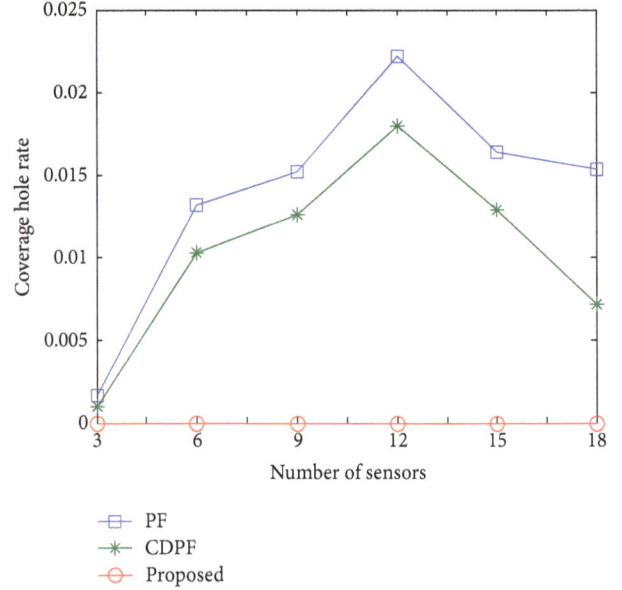

FIGURE 8: Coverage hole rate versus number of sensors.

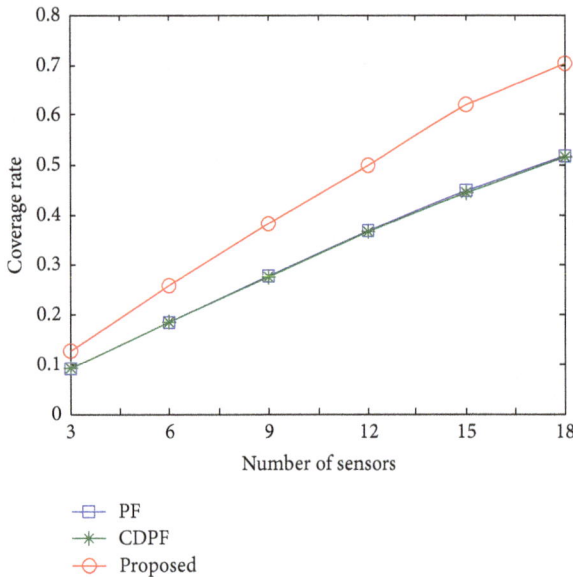

FIGURE 7: Coverage rate versus number of sensors.

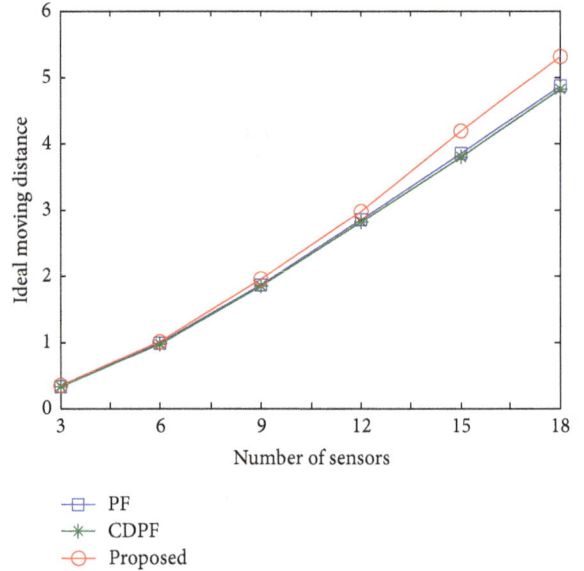

FIGURE 9: Ideal moving distance versus number of sensors.

increases for PF and CDPF schemes, as in Figure 8. That occurs because when the amount of sensors is quite large as mobile sensors spread, the probability of sensors that move back from the borders becomes higher, so that the opportunity of coverage hole refixing is increased. However, the coverage hole rate of the proposed scheme is always zero in this configuration, which proves that information coverage could fix coverage hole efficiently.

As shown in Figures 9-10, we can also observe that the proposed scheme can acquire the best movement efficiency compared to other schemes, which indicate that information coverage could improve both coverage rate and movement efficiency.

5.3. Cost of Proposed Scheme. Although the model of information coverage based deployment scheme could improve the performance of coverage for mobile sensors, there is still a cost that the communication cost becomes higher since sensors need to cooperate to achieve information coverage. Therefore, the network-wide trade-off between using sensor movements and using sensor commutation is worthy of further studies.

6. Conclusions

We propose movement-efficient deployment scheme based on potential field method for mobile sensors, which adapts

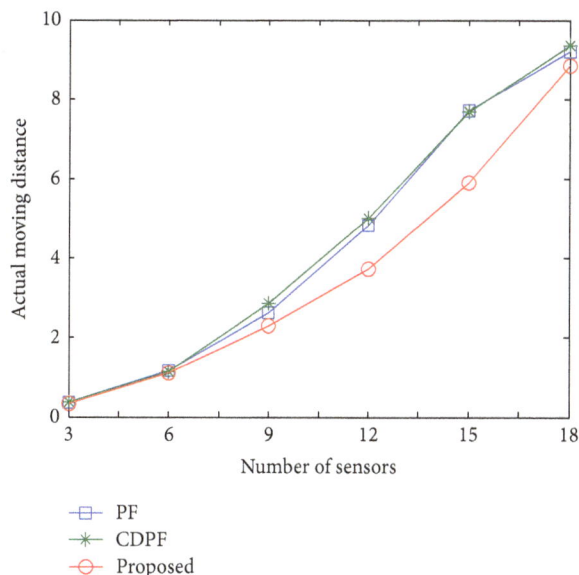

FIGURE 10: Actual moving distance versus number of sensors.

the information coverage model. Since the information coverage model could save sensor resources during deployment, the proposed scheme can improve coverage performance and movement efficiency. Numeric simulations show that the proposed scheme outperforms several previous approaches developed for deployment of mobile sensor networks.

Conflict of Interests

The authors declare that there is no conflict of interests regarding the publication of this paper.

Acknowledgments

The authors would like to thank Ghananeel A. Nighojkar in Wichita State University for his dissertation and simulation codes, which have motivated and helped this work, and would like to thank the reviewers for their helpful advice and comments. This work is supported by National Natural Science Foundation of China under Grants nos. 61301092 and 61401360; Fundamental Research Funds for the Central Universities under Grant no. 3102014JCQ01055; Natural Science Basis Research Plan in Shaanxi Province of China under Grant no. 2014JQ2-6033; and China Postdoctoral Science Foundation under Grant no. 2012M512026.

References

[1] T. Clouqueur, V. Phipatanasuphorn, P. Ramanathan, and K. K. Saluja, "Sensor deployment strategy for target detection," in *Proceedings of the 1st ACM International Workshop on Wireless Sensor Networks and Applications (WSNA '02)*, pp. 42–48, ACM, Atlanta, Ga, USA, September 2002.

[2] M. A. Batalin and G. S. Sukhatme, "Spreading out: a local approach to multi-robot coverage," in *Distributed Autonomous Robotic Systems 5*, chapter 10, pp. 373–382, Springer, Tokyo, Japan, 2002.

[3] O. Khatib, "Real-time obstacle avoidance for manipulators and mobile robots," *The International Journal of Robotics Research*, vol. 5, no. 1, pp. 90–98, 1986.

[4] R. Shahidi, M. Shayman, and P. S. Krishnaprasad, "Mobile robot navigation using potential functions," in *Proceedings of the IEEE International Conference on Robotics and Automation*, pp. 2047–2053, IEEE, Sacramento, Calif, USA, April 1991.

[5] P. Song and V. Kumar, "A potential field based approach to multi robot manipulation," in *Proceedings of the IEEE International Conference on Robotics and Automation*, pp. 1217–1222, IEEE, Washington, DC, USA, May 2002.

[6] A. Howard, M. J. Matarić, and G. S. Sukhatme, "Mobile sensor network deployment using potential fields: a distributed, scalable solution to the area coverage problem," in *Distributed Autonomous Robotic Systems 5*, chapter 8, pp. 299–308, Springer, Tokyo, Japan, 2002.

[7] S. Poduri and G. S. Sukhatme, "Constrained coverage for mobile sensor networks," in *Proceedings of the IEEE International Conference on Robotics and Automation*, pp. 165–171, New Orleans, La, USA, May 2004.

[8] G. Tan, S. A. Jarvis, and A.-M. Kermarrec, "Connectivity-guaranteed and obstacle-adaptive deployment schemes for mobile sensor networks," in *Proceedings of the 28th International Conference on Distributed Computing Systems (ICDCS '08)*, pp. 429–437, IEEE, Beijing, China, June 2008.

[9] G. Wang, G. Cao, and T. L. Porta, "Movement-assisted sensor deployment," in *Proceedings of the IEEE Conference on Computer Communications (INFOCOM '04)*, vol. 4, pp. 2469–2479, IEEE, March 2004.

[10] H.-J. Lee, Y.-H. Kim, Y.-H. Han, and C. Y. Park, "Centroid-based movement assisted sensor deployment schemes in wireless sensor networks," in *Proceedings of the IEEE 70th Vehicular Technology Conference Fall (VTC '09)*, pp. 1–5, Anchorage, Alaska, USA, September 2009.

[11] Y.-H. Han, Y.-H. Kim, W. Kim, and Y.-S. Jeong, "An energy-efficient self-deployment with the centroid-directed virtual force in mobile sensor networks," *Simulation*, vol. 88, no. 10, pp. 1152–1165, 2012.

[12] B. Wang, W. Wang, V. Srinivasan, and K. C. Chua, "Information coverage for wireless sensor networks," *IEEE Communications Letters*, vol. 9, no. 11, pp. 967–969, 2005.

[13] B. Wang, K. C. Chua, V. Srinivasan, and W. Wang, "Information coverage in randomly deployed wireless sensor networks," *IEEE Transactions on Wireless Communications*, vol. 6, no. 8, pp. 2994–3004, 2007.

[14] X. Deng, B. Wang, N. Wang, W. Liu, and Y. Mo, "Sensor scheduling for confident information coverage in wireless sensor networks," in *Proceedings of the IEEE Wireless Communications and Networking Conference (WCNC '13)*, pp. 1027–1031, Shanghai, China, April 2013.

[15] G. Yang and D. Qiao, "Barrier information coverage with wireless sensors," in *Proceedings of the 28th Conference on Computer Communications (IEEE INFOCOM '09)*, pp. 918–926, IEEE, Rio de Janeiro, Brazil, April 2009.

[16] G. A. Nighojkar, *Maximizing the area coverage of a mobile sensor network in a constrained environment [Ph.D. thesis]*, Wichita State University, Wichita, Kan, USA, 2006.

[17] A. Boukerche and F. Xin, "A Voronoi approach for coverage protocols in wireless sensor networks," in *Proceedings of the 50th Annual IEEE Global Telecommunications Conference (GLOBECOM '07)*, pp. 5190–5194, IEEE, Washington, DC, USA, November 2007.

11

A Novel WLAN Roaming Decision and Selection Scheme for Mobile Data Offloading

Nam Nguyen, Mohammad Arifuzzaman, and Takuro Sato

Graduate School of Global Information and Telecommunication Studies, Waseda University, Building No. 29-7, 1-3-10 Nishi-Waseda, Shinjuku-ku, Tokyo 169-0051, Japan

Correspondence should be addressed to Nam Nguyen; namnh@moegi.waseda.jp

Academic Editor: Francisco Martínez

The existing IEEE and 3GPP standards have laid the foundation for integrating cellular and WiFi network to deliver a seamless experience for the end-users when roaming across multiple access networks. However, in recent studies, the issue of making roaming decision and intelligently selecting the most preferable Point of Service to optimize network resource and improve end user's experience has not been considered properly. In this paper, we propose a novel cellular and WiFi roaming decision and AP selection scheme based on state of the art, 3GPP TS24.312 and IEEE 802.11u, k standards. Our proposed scheme assists the mobile nodes to decide the right timing to make roaming decision and select preferable point of service based on the operator's policies and real-time network condition. We also introduce our simulation model of a heterogeneous network with cellular and WiFi interworking as well as 3GPP ANDSF, TS24.312. It is a complete end-to-end system model from application to physical layer with considering user's mobility and realistic traffic model. The proposed scheme outperformed the conventional WiFi selection scheme in terms of dynamically steering mobile node's data traffic from macrocell to available Access Points. The proposed scheme increased the utilization and balanced the traffic load of access points and improved user's experienced throughput.

1. Introduction

In recent years, the proliferation of smart devices (smart phones and tablets) and mobile Internet applications is generating huge demand for mobile data traffic. In addition, we are shifting from voice-centric homogeneous network toward data-oriented heterogeneous network (HetNet). It is predicted that mobile data traffic will grow more than 1000 times in next decades, and billions of devices will be connected due to the popularity of Internet of Things by then [1]. Conversely, the licensed frequency band of cellular network is expensive and limited. Therefore, increasing the number of macrocell base stations is no longer an efficient method to boost network capacity. Therefore, offloading data traffic from the cellular network using alternative access network is becoming a major concern for network operators in order to avoid network congestion and degrading end-user's quality of experience (QoE). As the amount of data transfer increases, offloading is becoming increasingly important. The obvious choices for offloading are small cells and WiFi. The first option for data offloading is to deploy small-cells beneath the macrocell's coverage to increase the capacity as well as downlink and uplink throughput. However, small-cell technologies such as femtocell or picocell raise macrocell interference and cost-effective issues. On the other hand, WiFi is emerging as an effective alternative since it utilizes unlicensed frequencies, which causes less interference to macrocell. In addition, it is also very cost-effective compared to small-cell technology. For these reasons, more and more network operators are in favor of adopting WiFi on a large scale as the extended access network for their cellular network. WiFi will become increasingly important, playing the role of the third RAN, and will be the most reliable data offload technology in next-generation wireless network.

One of major challenges in WiFi offloading is to provide a smooth and seamless experience for end-user when roaming from cellular to WiFi APs and vice versa [2]. Most of researches in the literature are focused on this area [3–7].

However, there is another issue concerning how the mobile node decides the right timing to make roaming decision and selects the most preferable point of service anytime and anywhere. This issue becomes important to the operators deploying carrier WiFi AP for cellular offloading because their main concern is how to increase the utilization of available access points and proactively prevent network congestion for the macrocell. For example, currently, there are network operators such as AT&T US, KDDI Japan, deploying WiFi APs in crowded area in order to offload traffic and prevent congestion for cellular network. However, since the mobile users are usually not aware of the availability of WiFi AP, it is common that many of them might not switch their connection from cellular network to WiFi or the mobile terminals unconsciously connect to overloaded APs. As a result, the WiFi APs are typically underutilized and the end-users might experience bad QoS due to overload or congestion.

3GPP and IEEE have been actively working on the integration and interworking issues between 3GPP and non-3GPP wireless networks. For instance, 3GPP released a new network architecture that is a new all-IP core network architecture known as Evolved Packet Core (EPC) with a new entity called Access Network Discovery and Selection Function (ANDSF) TS 24.312 [8, 9] Mobile IP, Proxy IP as well as IEEE 802.11u [10] or Hotspot 2.0 (HS2.0) [11], and so forth. These new standards are the key enablers to make roaming between 3GPP and non-3GPP access network as smooth and seamless as it is in 3GPP cellular network. However, as far as we are concerned, the issue of how to decide the right timing to make roaming decision and select preferable point of service has not been addressed adequately by the existing standards. Since the key standards have not been finalized and optimized for dynamic data offloading purpose, the new amendments are expected in future releases of TS24.312 [9] and IEEE 802.11u [10]. For instance, in 3GPP ANDSF TS 24.312, the main function of ANDSF host entity is to provide policies or rules for discovering and selecting preferable access network or IP routing policies in case of multiple homing User Entity (UE). However, these policies are generally predefined and unchanged. If UE sole uses ANDSF's policies and does not consider any real-time network condition or measured information from UE, it can lead to undesirable network selection issues for UE [12]. For example, if the AP candidate is temporary overloaded or far away, it will degrade end-user's experienced QoS. On the other hand, IEEE 802.11u standard was mainly targeted on providing a seamless transition experience from WiFi to cellular network. Although it provides a standardized mean to let UE obtain additional AP and backhaul network condition information such as the channel load, AP's downlink/uplink capability for enhancing AP discovery and selection, it does not specify how the UE should use this information piece to find and select the preferable point of service.

In this research, we address the aforementioned issue by proposing a novel cellular-WiFi roaming decision and selection scheme specified for mobile data offloading purpose. Our proposed scheme takes advantage of state-of-the-art 3GPP (24.312 ANDSF) and IEEE (IEEE 802.11u, k) standards. By combining ANDSF's policies and network condition metrics measured by UE, the UE autonomously decides the right timing to make cellular-WiFi roaming decision and selects the most preferable point of access. In addition, we also present our simulation model of 3GPP ANDSF and vertical handoff, which is a system level end-to-end model from physical layer to application layer. The model features typical HetNet scenario with vertical handover between WiMAX base stations and WiFi APs, the ANDSF host, and the connection manager running on each mobile node. We also take into account user's mobility and realistic traffic model. During the simulation process, the ANDSF server and the UE's connection manager entity behave and interact with other nodes corresponding to real-time network events. For evaluation, we compare the performance of proposed scheme with that of the conventional WiFi selection (based on the RSS/SINR adopted in most of popular smart devices). The simulation result shows that our proposed scheme can proactively steer the UE's traffic from macrobase stations to available WiFi APs, which improves both network resource utilization and UE's experienced throughput.

2. Background and Related Work

2.1. 3GPP Access Network Discovery and Selection Function. 3GPP has considered the non-3GPP access interworking issue for cellular network since 3GPP release 8 [8]. 3GPP proposed a tight-coupling architecture that integrates non-3GPP access into 3GPP Evolved Packet Core (EPC). In addition, there is a new entity called Access Network Discovery and Selection Function (ANDSF) TS24.312 [9] appended to the EPC architecture. Its role is to provide the UEs with a set of rules for selecting preferable access network as well as IP routing policies for multihoming UE. Upon being provisioned with ANDSF's policies, the UE regularly checks the validity of the policies and selects the active rule with the highest priority. Validity criteria include date, time of day, or location. The UE can also use the valid routing policy (e.g., preferred RAT, specific preferable AP) to decide whether to route the ongoing traffic over 3GPP or WiFi. Researches on the ANDSF's functions and specification are ongoing and new enhancements and extensions for ANDSF's function are expected in future 3GPP releases.

2.2. IEEE 802.11u Standard. IEEE ratified an amendment for IEEE 802.11 standard to support WiFi and 3GPP network interworking known as IEEE802.11u [10]. This amendment was aimed at making roaming between WiFi and external network such as 3GPP network as smooth and seamless as roaming within 3GPP cellular network. Based on this amendment, WiFi Alliance also released specification for the next generation of WiFi AP also known as Hotspot 2.0 (HS2.0) [11] or WiFi Certified Passpoint. Beyond the security authentication enhancement, the main feature of HS2.0 in release 1 was to facilitate the WiFi AP discovery and selection procedures of supported UEs. Therefore, the HS2.0 provides UE various type of network condition information (AP, backhaul load condition, authentication type, connection capacity, etc.) prior to association through the Access Network Query Protocol (ANQP), and the Generic Advertisement Service

(GAS). These additional network condition metrics are beneficial to UE when selecting preferable AP as well as providing seamless handoff experience.

2.3. Related Works. In EPC architecture, the ANDSF is designated to deliver network selection policies from the operator for individual UE [8, 9]. Therefore, the operators can impose the policies to control when, where, and which access network that the UE can connect. However, one of the most important aspects of dynamic network selection is to consider the real-time network conditions as the criteria for selection. The conventional WiFi offloading strategy (adopted by most of smart device) is that WiFi always has higher priority than 3G/4G cellular. As a result, the UE switches to WiFi whenever AP is detected. Besides, WiFi AP selection is simply based on the Receiving Signal Strength (RSS) or SINR; thus the nearest AP is selected. However, in the dense area where there are large numbers of APs, we believe that RSS or SINR information alone is not enough to make intelligent WiFi selection. In such scenario, the selection scheme that is solely based on RSS can lead to bad quality of experience (QoE) for UE or network resource underutilization issue. Therefore, we need to take into account other factors to enhance roaming and point of service selection decision.

In [13] authors highlighted technical challenges in heterogeneous wireless networks underlying seamless vertical handover. The authors also presented a detailed survey on the vertical mobility management process with a focus on decision-making mechanisms. Handoff decision based on multiple inputs such as bandwidth, QoS, cost, and UE's velocity have been considered for years with complex system model and algorithms such as analytic hierarchy process (AHP), grey relational analysis (GRA) [14, 15]. However, the authors failed to describe how to implement their proposal in reality because some required inputs such as available bandwidth, cost of the service, UE's velocity, connection jitter, and packet loss are not available or difficult to obtain with the existing standards.

In [12] 4GAmericas provided a comprehensive view of the state of the art of the key enablers for integrating cellular network and WiFi. It explained the possibility of combining the advantages of ANDSF and HS2.0 could resolve the problems of macrocell and WiFi roaming. However, there was no detail description, numerical evaluation such as simulations or experiment's result provided. However, there was no detail description, numerical evaluation such as simulations or experiment's result provided. In [16], authors examined the mobility between different access technologies in heterogeneous wireless networks and focused on the case of interoperability issue. The quality of the service of mobility, the time required for the handover and the packet loss during handover, and so forth are also extensively analysed. There was 3GPP task group working on WLAN and 3GPP radio interworking at radio level. In [17], the authors tried to improve access network selection and traffic steering decision between 3GPP LTE network and WLAN. In their proposal, the cellular network provided additional Received Channel Power Indicator (RCPI) and/or (Received Signal to Noise Indicator) RSNI thresholds to the terminal so that it could

make roaming decision to WiFi when the LTE network condition was not favourable. In [18], the authors addressed solutions for WiFi offloading in LTE cellular networks when demands exceeded the capability of the LTE access. For evaluation, the authors compared the performance of each access technology using different network performance metrics. In [19], the authors proposed a novel network-assisted user-centric WiFi offloading model for maximizing per-user throughput in a heterogeneous network. In the proposed model, the network collects network information, such as the number of users in WiFi network and their traffic load. Subsequently, the network decided the specific portion of traffic to be transmitted via WiFi network so that the individual user's throughput could be maximized by offloading more traffic to WiFi. Through analysis, the authors investigated the effect of the WiFi offloading ratio on the per-user throughput. However, they did not consider user's mobility in their simulation model. In [20], the authors studied how much economic benefits can be anticipated thanks to delayed WiFi offloading method, by modeling the interaction between a single provider and users based on a two-stage sequential game. In this work, they first analytically proved that WiFi offloading is economically beneficial for both the provider and users. Their major focus was to understand how and how much users and the provider obtained the economic incentives by adopting delayed WiFi offloading and investigated the effect of different pricing and delay-tolerance. In [21], the authors investigated the performance improvement induced by adopting a hybrid cellular/WiFi communication architecture where the mobile users can be served by either the LTE eNodeB or a mobile WiFi AP. In this proposed scheme, the mobile WiFi APs are considered as relay entities that are wirelessly connected to the LTE eNodeB and share this broadband connection with other users over WiFi tethering or ad hoc network (using WiFi frequencies). Important performance metrics of the proposed hybrid scheme including the average bit error probability (ABEP), capacity, and outage probability were theoretically studied.

The IEEE 802.21 in [22], Media-Independent Handover (MIH), is a standardized framework proposed by IEEE, which facilitates vertical handover across heterogeneous networks. It defines a generic media-independent handoff (MIH) framework to support information exchange between network elements as well as a set of functional components to execute the roaming process. IEEE 802.21 specifies three media-independent services. Media-Independent Event Service (MIES), Media Independent Command Service (MICS), and Media-Independent Information Service (MIIS). The MIIS provides a data store of available networks and network parameters and defines standard query/response messages to access and retrieve such information for each available access network. The MIIS's role is similar to ANDSF. References [23, 24] provided an overview of vertical handoff approaches based on MIH framework. The work presented comprehensive solutions to ensure VHO between three types of different RATs: WiFi, WiMAX, and 3G, dealing with multiple parameters to make VHO decision. To the best of our knowledge, in recent years, researches related to MIH are gradually fading out from the academic studies and there is

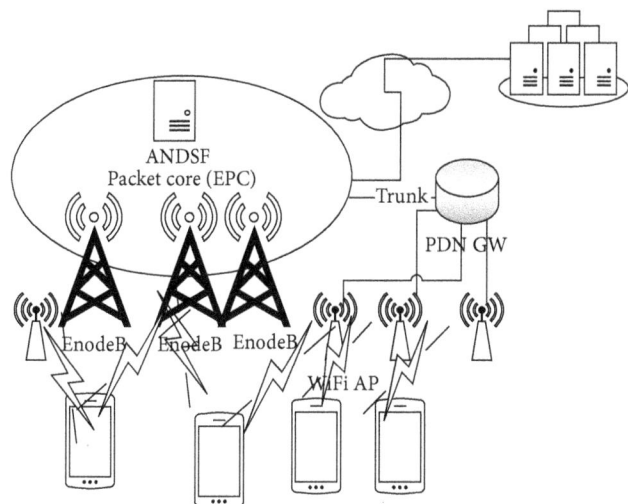

FIGURE 1: Heterogeneous network scenario with LTE cellular and WiFi interworking.

no initiative from industries to implement this standard in commercial products. The reason is that it required many changes and modifications to the MAC layer of existing IEEE 802.11, 802.16, 802.3 standards as well as 3GPP LTE, overall core network architecture, and protocols. In addition, the main function, MIIS service for network discovery, can be replaceable by ANDSF, which is more robust and popular. For these reasons, we consider using ANDSF instead of MIH to provide discovery and selection policy utilized in our scheme.

3. The Proposed Cellular and WiFi Roaming and Selection Scheme

3.1. The Proposed Scheme Formation Initiative. Our assumption is that we have a 3GPP network operator providing high-speed mobile data service using cellular (e.g., LTE) and WiFi networks. It also extends to cellular carriers and third party WiFi providers as long as they have interworking agreement and interconnected network. This is typical HetNet scenario in future wireless network because we anticipate that providing high-speed data service via multiple access networks will become popular in near future. Figure 1 shows the baseline scenario when users move from LTE network to WiFi network coverage. Depending on preference, in this situation, the UEs may decide to switch their access interface from LTE to WiFi. However, for a proactive HO roaming and selection decision scheme, whenever WiFi access network is available nearby, the operator should regulate every UE to offload some or all of its traffic through WiFi AP depending on the network condition. In order to effectively offload mobile traffic from macrocell to WiFi APs in heterogeneous network scenario in Figure 1, we need an entity from network side to regulate the cellular UEs when and where WiFi roaming is possible. In addition, it should also provide rightful selection rules for UE. For instance, only authorized APs (AP belonged to operator or authorized provider who has roaming agreement) can be selected. This selectively offloading traffic to legitimated WiFi networks gives mobile operators an opportunity to increase

their total network capacity to meet rising traffic demands and a way to extend network coverage and capacity to WiFi networks. For such scenario, the ANDSF [9] standard framework is a well suitable candidate since it is a 3GPP approach for controlling handover operation between 3GPP and non-3GPP access networks. Since release 8 of 3GPP, 3GPP has specified the ANDSF framework through which the network operator can provide a list of preferred access networks with intersystem mobility policies. However, the drawback of ANDSF's policies is its static characteristic since there is no mechanism to frequently update the policies. As described in [9], for prioritizing AP in selection, the operator assigns AP with an integer number from 1 to 255 (one is the highest priority) and the one with higher priority should be selected by the UE [9]. However, since the condition of the network may vary dynamically from time to time, if the UE only relies on the ANDSF selection rule for selecting new point of service, it can degrade the end-user experience in some cases (e.g., selected AP is too far away or congested one). Therefore, it is also necessary to take into account the real-time network condition at the UE side before making handoff decision. Among various network KPIs for network selection [14, 15] (such as bandwidth, delay, jitter, and latency), we consider the real-time load of WiFi AP is very important to deliver good user experience. Whether this metric is high or low could greatly affect user's experience because the channel load of AP correlates to the available bandwidth as well as delay or latency of the connection that the AP can offer. As far as we are concerned, the channel load information is first defined in IEEE 802.11k, e [25] and later included in the IEEE 802.11u or HS2.0 [11] specification. The UE can obtain this information piece via the beacon message or probe response from supported AP candidates. Since the HS2.0 is rolling out in near future, obtaining the real-time channel load of AP becomes feasible.

For these reasons, our proposed scheme for dynamic WiFi offloading is designed by following main points:

(i) Taking advantage of ANDSF's selection policies to regulate admission to access points, this requirement to prevent selecting unauthorized AP.

(ii) Using thresholds to control and select preferable AP candidate. Using thresholds to control roaming decision and select the preferable AP. The thresholds are managed by network operator and distributed to UEs via ANDSF MO. We use a defined object in ANDSF MO [9] so-called *MaximumBSSLoadValue* for traffic load threshold and we also propose an additional threshold for signal quality, *MinimumBSSRSSValue*.

(iii) Taking into account the load condition of AP obtained from AP candidates (we assume that the WiFi APs are compatible with either HS2.0 or IEEE 802.11k, u) to decide whether WiFi offloading is relevant and which AP is preferable.

3.2. Proposal for Enhancing ANDSF WLANSP Management Object. The ANDSF standard defined various policies such as intersystem mobility, Access Network Discovery Information through the Management Objects (MO), which is a tree based structure. In the latest release 12 [9], a new leaf of ANDSF MO

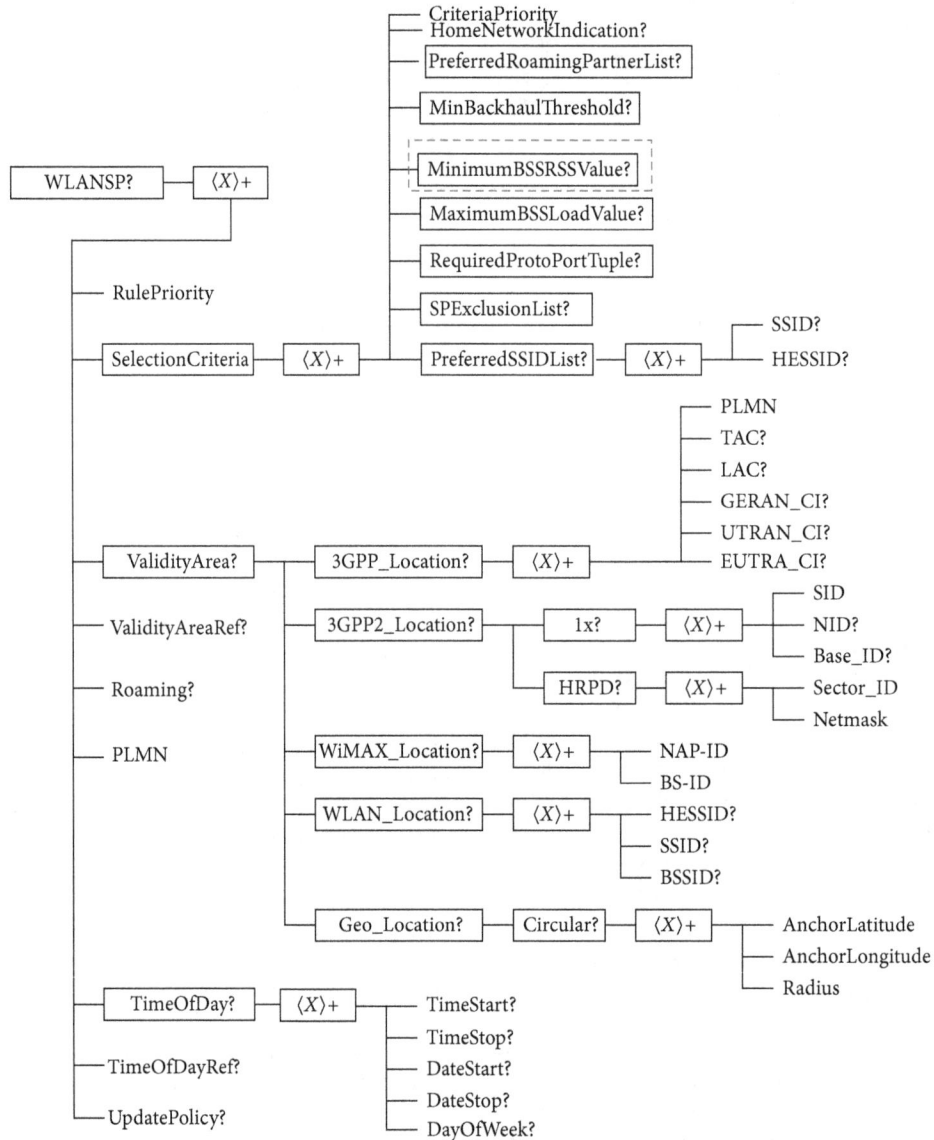

FIGURE 2: Example of enhanced ANDSF OM for WLANSP.

is added for WLAN selection so-called WLANSP (WLAN Selection Policy) (Figure 2). It is a set of operator-defined rules that determine how the UE selects a WLAN access network. The UE is provisioned with WLANSP rules from cellular operator.

Each WLANSP rule includes the following information:

(i) Validity conditions, that is, conditions indicating when the provided rule is valid (the validity conditions can include the time of day, geolocation, and network location (e.g., PLMN, Location Area).

(ii) One or more groups of WLAN selection criteria in priority order (each group contains one or more criteria that should be fulfilled by a WLAN access network in order to be eligible for selection).

The WLANSP MO can contain a set of one or many WLANSP rules. Each rule can have one or more WLAN selection

criteria defined in ANDSF/WLANSP/⟨X⟩/SelectionCriteria node including following attributes:

(i) CriteriaPriority.

(ii) HomeNetworkIndication.

(iii) PreferredRoamingPartnerList.

(iv) MinBackhaulThreshold.

(v) MaximumBSSLoadValue.

(vi) RequiredProtoPortTuple.

(vii) PreferredSSIDList.

(viii) SPExclusionList.

The UE can obtain these selection rules and use them for WiFi selection procedure. However, in our proposed scheme, for the sake of simplicity, we only consider the *MaximumBSSLoadValue* criterion for WiFi AP selection because it can serve as a load threshold in our selection scheme. If the load condition of AP candidate does not satisfy this threshold, it will be eliminated from the selection process. Therefore, this criterion is introduced in our scheme to preeliminated high load AP that cannot potentially provide desirable QoS. In addition, the threshold can also be used to control the number of UEs performing WiFi offloading. For example, if this load threshold value is properly set, only certain number of UEs can roam to WiFi so that the load threshold condition can be satisfied. The rest of cellular UEs is not allowed to offload because of the load threshold. In contrast, if the maximum load threshold is higher, more UEs can roam to WiFi. Therefore, network operator adjusts this threshold to control the percentage of cellular user roaming to WiFi for offloading.

Apart from the load metric, we also consider the receiving signal strength (RSS) as another important metric that can affect the UE's experienced QoS. If the AP is far away from the UE, it cannot sustain good QoS for the UE in spite of acceptable load condition. Therefore, in our proposed scheme, we add a RSS threshold to eliminate irrelevant AP candidate. Like the max load threshold, the operator utilizes this threshold to control the number of WiFi roaming UEs. For example, if the RSS threshold is high, only UEs that are near to the AP can roam to WiFi.

However, currently, such RSS threshold is not available in the ANDSF WLANSP MO. Therefore, we propose to append this selection policy node to the same OM leaf of *MaximumBSSLoadValue* in ANDSF WLANSP MO, $\langle X \rangle$/*WLANSP*/$\langle X \rangle$/*SelectionCriteria*/$\langle X \rangle$/*MinimumBSSRSSValue* so that this policy can be distributed to UEs in the same manner. Figure 2 shows the tree based structure of ANDSF WLANSP leaf and an example of how the additional selection node is added to existing ANDSF MO standard. If *MinimumBSSRSSValue* node is not present or the node is present and empty, the UE will not evaluate the node. If it is available, the UE can use this attribute in our scheme for AP selection.

3.3. Real-Time Channel Load for WiFi Access Point.

According to Hotspot 2.0 or IEEE 802.11u specification [11], the channel load metric of WiFi AP is defined as one of the new QoS network metrics appended to beacon message of WiFi AP. This channel load metric is inherited from legacy standard IEEE 802.11k [25] ratified in 2008.

The channel load indicates the channel occupancy status of AP. If the load is too high, it can significantly degrade the end-user experience since it increases the contention probability among cochannel UEs. According to [25], the channel utilization or channel load U_c is defined as the percentage of time, linearly scaled with 255 representing 100%, that the AP sensed the medium was busy, as indicated by either physical or virtual carrier sense (CS) mechanism. By this definition, this metric indicates the real-time load status of radio channel

$$U_c = \text{Integer}\left(\frac{t_{\text{busy}}}{T_M}\right) \times 255, \tag{1}$$

$$T_M = t_c \cdot t_b \cdot 1024. \tag{2}$$

The channel load is calculated by (1) and the *Channel_Busy_Fraction* t_{busy} is defined as the number of microseconds while the physical or virtual carrier sense mechanism in MAC layer keeps indicating busy channel status. The *Measurement_Duration* T_M represents the number of consecutive beacon intervals while the channel busy time is measured. T_M is calculated by (2) and t_c is the channel utilization measured interval representing the number of consecutive beacon intervals during which the channel busy time is measured. t_b is the beacon period value in IEEE802.11-time unit (TU). One TU is 1024 microseconds. The AP regularly calculates its channel load metric every T_M (μs) and the UEs obtain this channel load metric via either beacon message or probe response message.

3.4. Access Point QoS Indicator.

In our proposed scheme, we consider both the channel load and the RSS measured from the UE as the most significant metrics for selecting WiFi AP.

The *Receiving Signal Strength* (RSS) or SINR is the most widely used metric for conventional handover (HO) decision since it is easy to measure and directly relate to the radio channel quality. With strong RSS, the AP can use high channel modulation and coding scheme and the bit rate becomes higher accordingly. Therefore, it is evident that strong RSS is preferable.

The Channel Load. As aforementioned, this metric indicates the real-time load condition of the AP operating channel. In a wireless network, if the load is too high, it will result in poor quality of service (low throughput, high packet loss, or delay). This metric is even more important for WiFi since WiFi uses CSMA/CA for multiple access. The high channel load can result in high contention rate, which dramatically degrades active UE's throughput in the same channel. Therefore, lower load AP is preferable for good QoE

$$\overline{RSS}_i = \alpha \cdot RSS_i + (1 - \alpha) \cdot \overline{RSS}_{i-1}, \tag{3}$$

$$APQI_i = w_r \log_2\left(\frac{\overline{RSS}_i}{RSS\_MIN_i}\right) + w_l \log_2\left(\frac{1/Channel\_Load_i}{255}\right) \tag{4}$$

with $w_r + w_l = 1$.

Because the RSS and channel load have different unit and characteristic, we define a normalized metric known as APQI (Access Point Quality Indicator) in order to make it easier to evaluate and compare the metrics between AP candidates.

Firstly, since each of the above metrics has different unit and characteristics, we have to normalize them. In (4), $\overline{RSS}_i$ is the mean value of RSS, which is calculated as (3) while the *RSS_MIN* is minimum received signal strength of WiFi receiver specified by the Network Interface Controller (NIC) vendor. Each vendor of WiFi NIC has its own sensitive RSS value for the receiver. If the RSS is lower than this value, the data packet will be dropped. Therefore, the normalized term $\overline{RSS}_i/RSS\_MIN$ indicates how strong the received signal strength is compared to its minimum value. We use binary logarithmic function $\log_2(\overline{RSS_i}/RSS\_MIN_i)$ to reduce the impact of fluctuation of RSS input. By this definition, this value is independent of the vendor specified hardware specification and comparable among different vendor APs.

The *channel utilization* metric in (3) is defined as the percentage of time when the AP sensed that the radio channel kept indicating busy by either the physical or virtual carrier sense mechanism [25]. This percentage is linearly scaled of integer range [1–255] that corresponded to 0–100% channel load. By this definition, the term $Channel\_Load_i/255$ is the normalized value of the channel load (the $Channel\_Load_i$ is an integer number [1–255]). Here, we apply logarithmic functions, $\log_2(1/(Channel\_Load_i/255))$ to reduce the rapid fluctuation of $Channel\_Load_i$ input and make it comparable with normalized RSS. The $APQI_i$ is the $APQI$ of ith AP defined as in (4). The first term represents the benefit of signal quality of the radio link between the UE and the AP while the second term represents the benefit of real-time channel load condition of AP. In (4), the $APQI$ is inversely proportional to the channel load metric. It shows a trade-off between the signal quality and the traffic load metrics of AP candidate. The significance of each term depended on the weighted values w_r, w_l, for link quality and the channel load or channel utilization, respectively.

3.5. Proposed Roaming Decision and Selection Scheme for Mobile Data Offloading. Our roaming decision and selection scheme is a network assisted and UE driven scheme. The UE has a unique position, which allows it to receive both ANDSF's policies and the real-time network condition from AP candidates. Therefore, it is preferable to let the UE make the decision instead of any other entities in the network. The UE decides timing to make roaming decision and selects preferable point of service based on ANDSF's policies, UE's measured information, and channel load information from AP candidates. The proposed scheme is a decision-making procedure carried out by the connection manager (CM) of end-user's device. In this paper, the CM is a generic term referred to as a functional component that takes input from user preference, input from ANDSF, and APs, and performs connectivity management and the traffic steering, that is, binding user's application/flow to a radio. The CM decides whether cellular-WiFi roaming is relevant or not and which AP candidate is preferable for better QoS. In the proposed scheme, it is worth noting that the channel load metric of WiFi AP is required to decide the preferable candidate. However, it should not interfere with conventional homogenous cellular or WiFi handoff procedure.

Figure 3 illustrates the flowchart of proposed scheme.

Step 1. The CM monitors the QoS of current connection. If the QoS is degrading (the data rate drops to certain threshold), running application or user's preference requests for better QoS, it will trigger the process to find better point of service.

Step 2. The CM contacts the ANDSF server at the core network to fetch the list of legitimate WiFi APs and WLANSP selection rule corresponding to UE's location (cell ID of macrocell or SSID or GPS location if applicable). The ANDSF server returns a list of AP candidates that can be accessible from the UE's location. In addition, *MaximumBSSLoadValue* and *MinimumBSSRSSValue* thresholds are also obtained from ANDSF MO WLANSP.

Step 3. By using AP discovery information from ANDSF, the UE only scans the surrounding area to check whether the *candidate APs* are available because provisioned ANDSF's information may be out of date or unreachable from UE's location. By doing this step, UE eliminates unreachable AP and obtains the RSS and the channel load of each *candidate AP*. We assume that the *candidate APs* support either HS 2.0 or IEEE 802.11k, e. so that the real-time load and the RSS of each AP can be obtained from the beacon message or via ANQP protocol in case of HS.2.0.

Step 4. The CM obtains the list of APs that fulfil the load and RSS thresholds. The thresholds are specified by network carrier via ANDSF's WiFi selection policies. As aforementioned, the MaximumBSSLoadValue is used to select WiFi AP; if the load of AP is higher than this value, it will be eliminated. We introduce the load threshold in order to preeliminate unsuitable AP, which cannot guarantee a good QoS. The MinimumBSSRSSValue can serve as the RSS threshold to eliminate irrelevant AP candidates.

Step 5. If there is no qualified AP after Step 4, the UE will stay at the current network. The UE conducts the conventional homogenous handoff procedure if it is favourable. If these are qualified APs available, the CM calculates the APQI metric (4) for each AP candidate. If the UE is using WiFi, it will calculate the APQI of the associating AP. The AP with highest APQI is selected as the candidate for the next point of service. After deciding the preferable AP, the CM starts the handoff procedure to selected AP.

Step 6. If there is no better point of service, the UE will stay at the current network. The CM returns to Step 1.

The pseudocode of our proposed scheme is shown in Algorithm 1.

4. Simulation and Results

4.1. Simulation Model Preparation. Although there are several adoptions of ANDSF [9] standard for commercial solutions for network selection in heterogeneous network [26], as far as we are concerned, there is no simulation of ANDSF standard for academic studies. Therefore, in order to evaluate our proposed scheme, we developed a system model of HetNet with vertical handoff (VHO), ANDSF server/client, and connection manager running on each mobile node. The whole system model is developed by NCTUns 6.0, which is a powerful open source network simulator developed by National Chiao Tung University [27–30]. It can simulate varieties of popular wired and wireless network entities and protocols such as IEEE WiFi 802.11a/b/g, ad hoc networks, WiMAX IEEE 802.16 e/d/p, multiple-interface UE, GPRS,

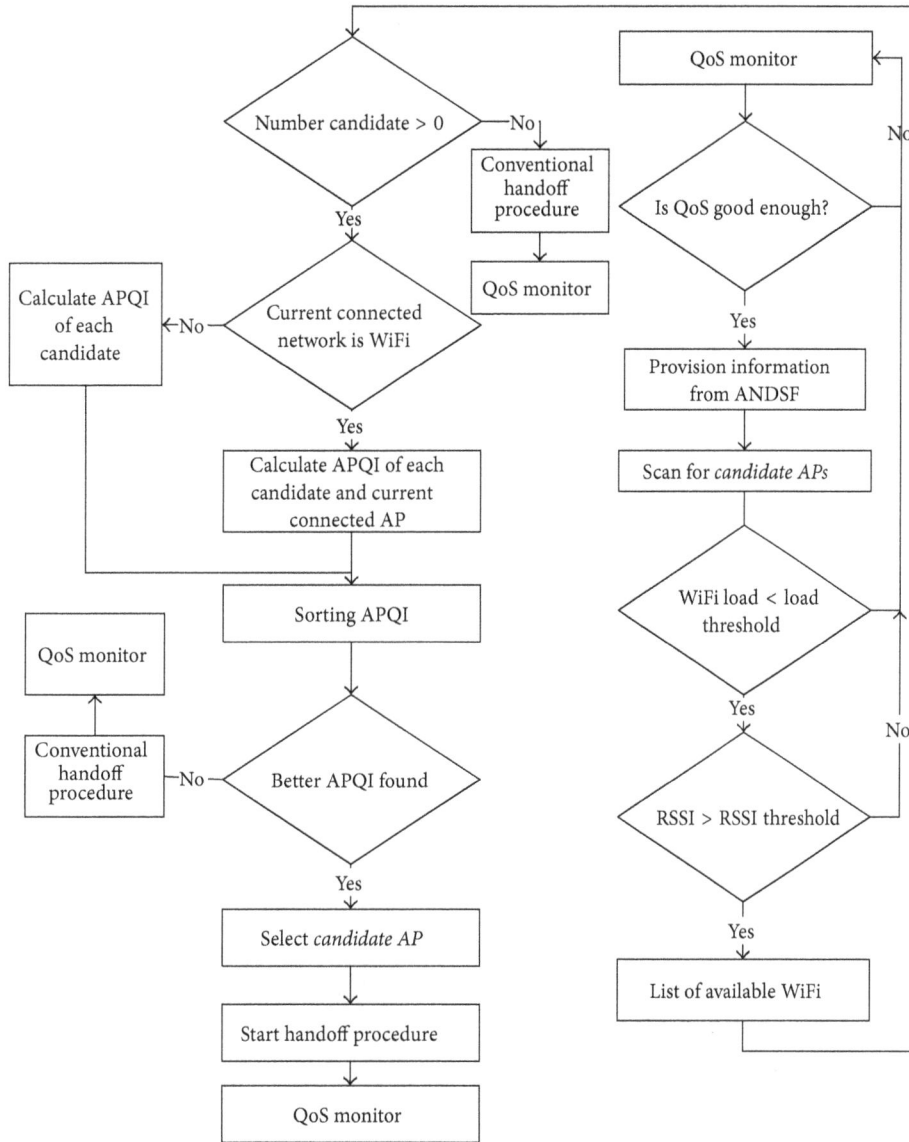

FIGURE 3: Proposed roaming decision and point of service selection scheme.

and satellite and vehicular networks. However, one of the key features of this simulator software is that it allows running user-defined C/C++ application on the simulated node. In our simulation model, we take advantage of this feature to simulate the ANDSF server and the connection manager (CM) for each UE node.

ANDSF server is a C program, which provisions information of available APs, load threshold, and RSS threshold. The CM is also C program running on each mobile node that executes our proposed scheme. It can obtain the RSS and the real-time channel load of AP candidate through MAC layer. The CM contacts the ANDSF host to fetch policies and thresholds via UDP socket. In addition, the CM can also obtain the RSS and the channel load of AP candidate from the MAC layer. As the simulation tool does not support IEEE802.11k, e or HS 2.0, we modify the MAC layer of the WiFi AP node to append the real-time channel load

information into the beacon message. In (1), the channel busy fraction is defined as the number of microseconds when the physical or virtual carrier sense indicates the channel is busy. In our simulation, we use the virtual carrier sense, Request to Send/Clear to Send (RTS/CTS) mechanism for detecting busy channel [25]. In [25] the default value of "*channelUtilizationBeaconIntervals*" in (1) is set to 50. However, this parameter is set to 10 in our simulation and the beacon broadcast interval is 100 ms.

The coefficients w_r and w_l are the weight values of RSS and channel load metric, respectively, used to calculate the APQI. In our simulation, we notice that if $w_r \gg w_l$, the system performs much like conventional WiFi selection based on RSS. Therefore, for the scope of this simulation, we choose $w_r = 0.4$ and $w_l = 0.6$.

Although the simulator software supports multiple-interface mobile UE and horizontal handover within WiMAX

Input: Set the parameters: *RSS, RSS_Min, Channel_load, MaximumBSSLoadValue, MinimumBSSRSSValue, w_r, w_l*
Output: AP candidate.
(1) **While** (1) **do** {
(2) **if** QoS degrading trigger **then**
(3) Obtain data from ANDSF server
  (a) Obtain AP candidates Information (SSID, operating channel)
  (b) Obtain Load and RSS thresholds *MaximumBSSLoadValue, MinimumBSSRSSValue*
(4) *qualified_AP_List = nil;*
for each AP_i in AP candidate List{
  (a) Obtain information of available AP_i information from physical Layer,
  (b) Obtain RSS_i, calculate $\overline{RSS}_i$ (3).
  (c) Obtain *channel_load_i* information.
  (d) **if** $\overline{RSS}_i$ > *MinimumBSSRSSValue* && *channel_load_i* < *MaximumBSSLoadValue* **then** *qualified_AP_List* add AP_i
}
If qualified_AP_List is empty **then** *return;*
(5) *Candidate_AP → APQI = calculate $APQI_1$ using* (4);
for each AP_i in AP candidate qualified_AP_List{
calculate $APQI_i$ using (4)
if AP_i → APQI > Candidate_AP → APQI **then** *Candidate_AP = AP_i;*
}
(6) **if** *Candidate_AP* != current AP **then**
return *Candidate_AP;*
}

ALGORITHM 1: Roaming decision and preferable AP selection.

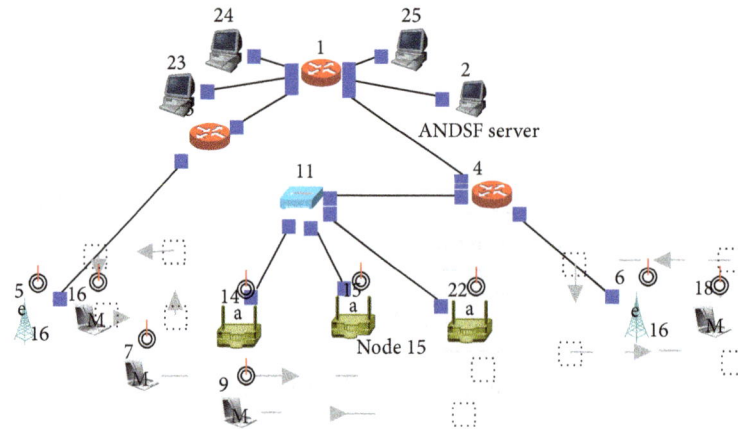

FIGURE 4: Simulation scenario setup for HetNet with interworking WiMAX BSs, WiFi APs, and Multiple Interfaces UEs.

or WiFi, it does not support VHO between WiMAX BS and WiFi AP. Therefore, we modify the source code of the simulation software to simulate the vertical handover between macrocell (WiMAX IEEE 802.16e) and WiFi IEEE 802.11a. It is worth noting that our proposed scheme is proposed for 3GPP cellular (LTE or UMTS) and WiFi. However, the simulation software NCTUns version 6.0 [27] does not support LTE. Therefore, we have to use WiMAX instead of LTE for our HetNet model. Since our proposed scheme does not consider any metric from the macrocell, using WiMAX instead of LTE for access network does not cause any difference in the simulation result. In addition, due to a problem related to mobile IP protocol of the simulator, the simulation of VHO is not a seamless handover. Therefore, the UE's connections are interrupted when performing the VHO.

4.2. Simulation Model Description. Figure 4 shows the screenshot of our simulation setup in the simulator and the simulation parameters are listed in Table 1. This setup is a typical WiFi deployment scenario to offload traffic from cellular network as we described in Figure 1. We arrange a typical WiMAX-WiFi interworking scenario with two IEEE 802.16e BS1, 2 (Nodes 5, 6), three 802.11a WiFi APs 1, 2, and 3 (Nodes 14, 15, and 22) and four multiple wireless interface UEs 1, 2, 3, and 4 (Nodes 7, 9, 16, and 18 consecutively) equipped with both WiMAX and WiFi 802.11a interfaces. All of WiMAX BSs and WiFi APs are interconnected via routers and switches, which is a simplified core network in Figure 1. We simulate the typical movement pattern of UEs when they move from the coverage of WiMAX to WiFi. During the simulation, UEs 1, 2, and 3 move from the

TABLE 1: Simulation parameter.

	WiMAX	WiFi
Technology	IEEE 802.16e	IEEE 802.11a
Coverage	1000 m	50 m
RX Thresh	−96 dBm	−82 dBm
Transmitting power	35 dBm	16.02 dBm
Bandwidth/QoS	5 Mb/s	9 Mb/s
Modulation scheme	OFDM 16QAM	OFDM 16QAM
Carrier frequency	2.3 GHz	5 GHz
channelUtilizationBeaconIntervals	NA	10
BS number	2	3
Propagation channel	Two-ray ground	Two-ray ground
Channel load threshold	NA	0.8
MS		
Node number		4
Mobility movement		Straight line
Multiple interface		WiMAX and WiFi
Speed		2–5 m/s
Traffic parameter		Greedy CBR
Miscellaneous		
Channel load broadcast interval		100 ms
w_r		0.4
w_l		0.6
α		0.5

coverage of BS1 toward WiFi coverage area while the UE4 (node 18) moves around BS2. All of the UEs have a CM application embedded with our proposed scheme to monitor and manage their connectivity. The ANDSF host (Node 2) resides at network side and it is reachable by all of UEs. The ANDSF communicates with the UE's CM to provide WiFi AP candidates as well as load and RSS threshold. In order to simulate WiFi roaming trigger event when QoS is not good enough, we randomly disrupt the WiMAX base station connection during the simulation.

At the beginning, the UEs are connected via WiMAX BSs; UEs 1, 2, and 3 (Nodes 7, 9, and 16) are associated with BS1 while UE4 (Node 18) is associated with BS2. At $T = 4$ s, the UEs start sending greedy CBR (Constant Bit Rate) toward the Correspondent Node (CN) (Nodes 2, 23, 24, and 25 in Figure 4). Figures 5 and 6 show the traffic of BS1, BS2. The throughput of BS1 and BS2 gradually increases and reaches the maximum bit rate at 1975 KB/s and 659 KB/s, respectively.

At $T = 10$ s, the CM of UE2 fetches information from ANDSF policies, which contains AP candidates 1, 2, and 3 (Nodes 14, 15, and 22). Subsequently, it scans the surrounding area to obtain the channel load and RSS of each AP candidate. The CM calculates the APQI of each AP according to (4). The UE2 picks AP1 because it has the highest APQI. The UE2 starts VHO procedure to AP1 at $T = 10$ s. As showed in Figures 5 and 7 the traffic of BS1 (red line) drops to 1316 KB/s at $T = 11$ s while the traffic of AP1 is gradually rising at $T = 13$ s when the VHO is completed.

At $T = 12$ s, the CM of UE1 (Node 9) detects the QoS of WiMAX BS degraded; it sends a request to ANDSF host to

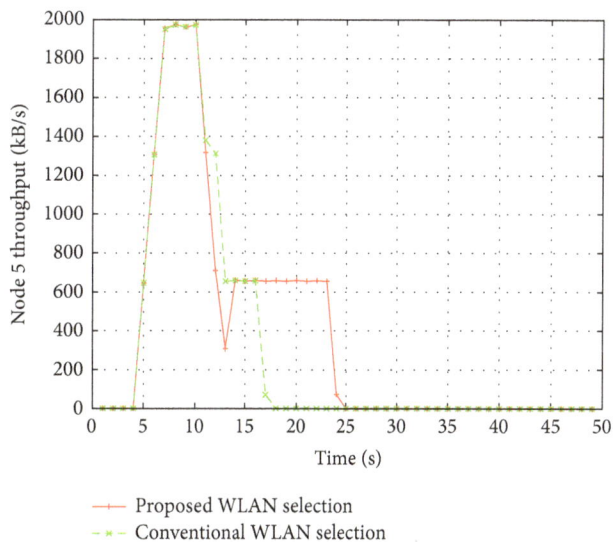

FIGURE 5: WiMAX BS1 (Node 5) throughput.

obtain the candidate list. The host returns the list of available APs 1, 2, and 3 (Nodes 14, 15, and 22). The UE1 carries out the same procedure as the UE2 to calculate the APQI of each AP candidate. The UE1 selects and starts VHO procedure to AP2 at $T = 12$ s. Figure 5 shows that the traffic of BS1 continual drops to 659.5 KB/s at $T = 14$ s. From Figure 8, at $T = 16$ s, the traffic of AP2 starts rising when the handover procedure of UE1 is completed.

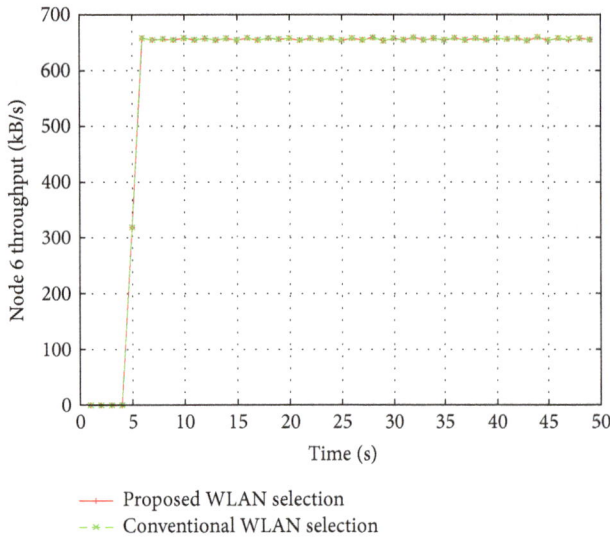

FIGURE 6: WiMAX BS2 (Node 6) throughput.

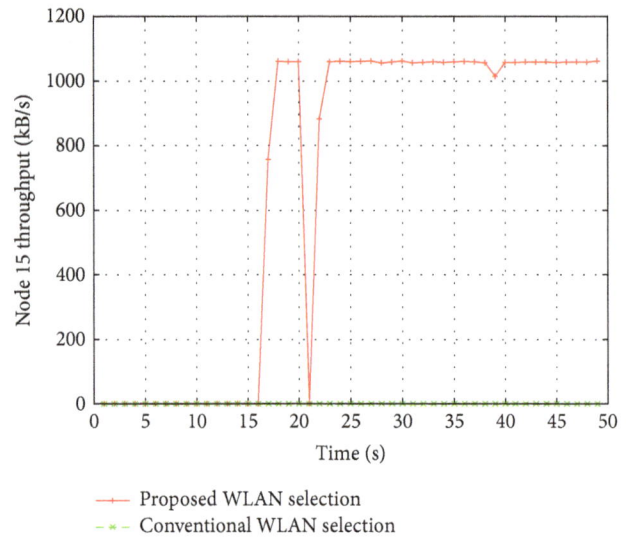

FIGURE 8: WiFi AP2 (Node 15) throughput.

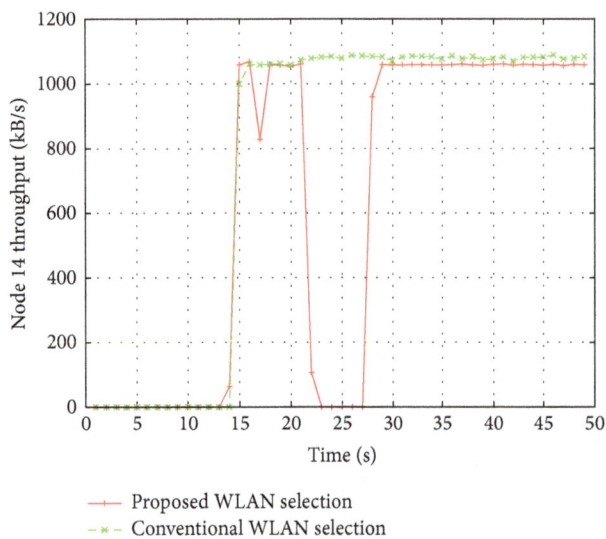

FIGURE 7: WiFi AP1 (Node 14) throughput.

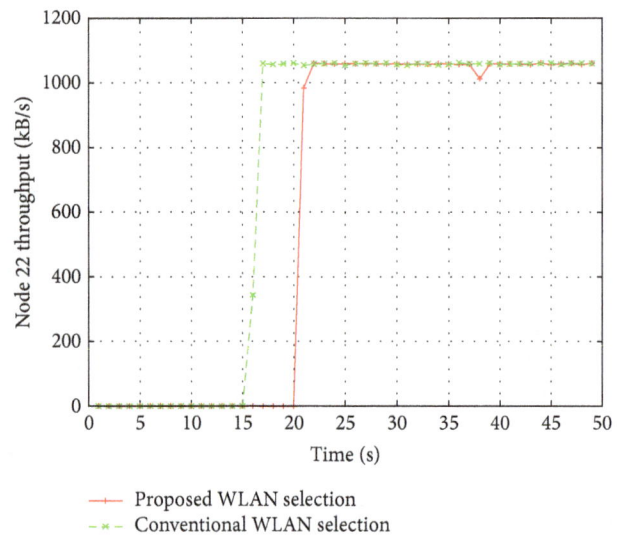

FIGURE 9: WiFi AP3 (Node 22) throughput.

At $T = 15$ s, the UE3 (Node 16) carries out the same procedure as UE1 and UE2. However, AP1 and AP2 are serving UE1 and UE2 and the channel load threshold is not satisfied. The UE3's CM decides to keep the connection with WiMAX BS1 since the specified APs are busy. As showed in Figure 4, from $T = 14$ s the throughput of BS1 remains at 650 KB/s which is the traffic of UE3.

At $T = 20$ s, the UE1 moves closer to the AP3. The RSS of AP3 satisfies the RSS threshold. The UE1's CM decides to switch the connection to AP3. In Figure 9, the traffic of AP3 is raising from $T = 20$ s while the traffic of AP2 in Figure 8 drops at the same time.

At $T = 21$ s, the CM of UE2 calculates the APQI of each AP candidate. The AP2 becomes a better candidate for UE2 since it remains idle and closer to UE2. The UE2 decides to select AP2 as the target AP for handoff. In Figure 7, the traffic

of AP1 drops and the traffic of AP2 increases at $T = 21$ s when the handoff process is going.

At $T = 24$ s, the UE3 calculates the APQI for each AP candidate. The AP1 becomes the best candidate for UE3 this time because it is idle and satisfies both channel load and RSS threshold. The UE3's CM decides to switch the connection from BS1 to AP1. In Figure 5, the traffic of BS1 drops to 0 KB/s at $T = 25$ s when all of UEs are transferred to Wi-F APs. In Figure 7, the traffic of AP1 increases from $T = 26$ s when handoff process is complete.

The UE4 moves back and forth around the BS2; however there is no available AP candidate at its location. Therefore, the UE4's CM decides to keep the connection with BS2 throughout the simulation.

Figure 10 shows the distribution of UEs at $T = 25$ s after transferring from WiMAX BS1; each WiFi AP serves one UE (Figure 11).

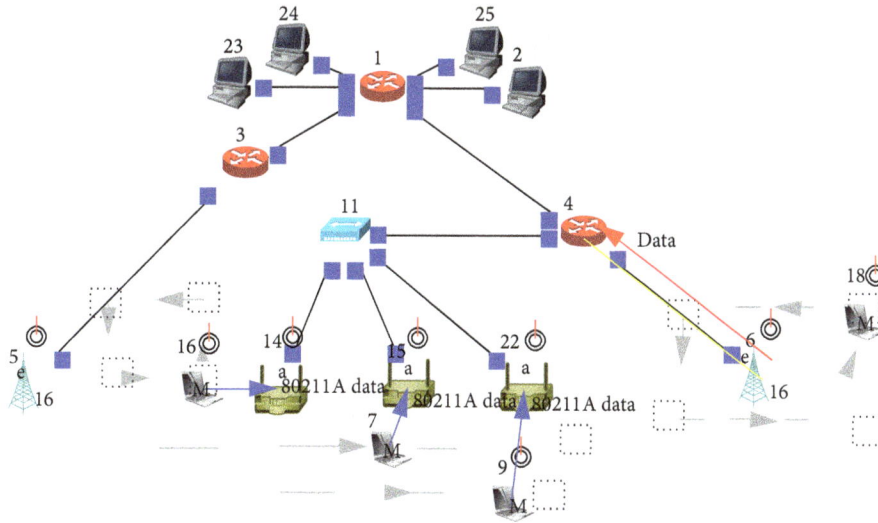

FIGURE 10: Screenshot from simulator for proposed scheme.

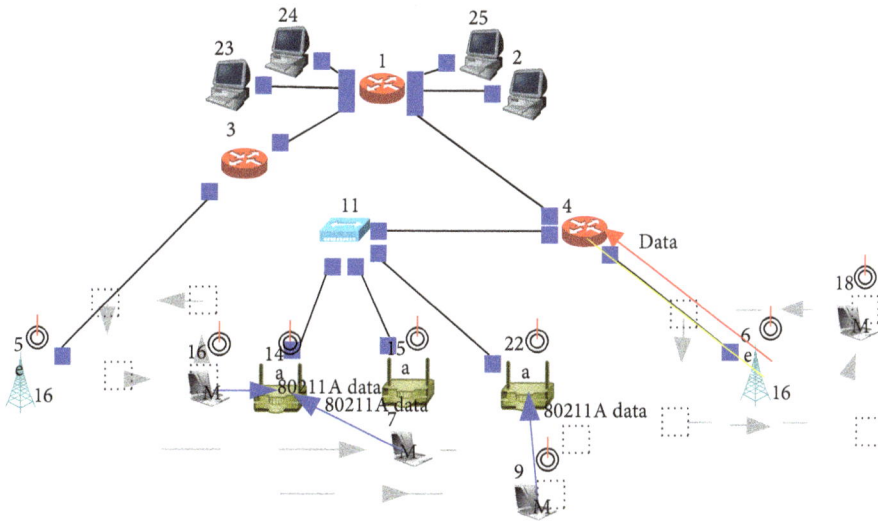

FIGURE 11: Screenshot from simulator with conventional scheme.

4.3. Simulation Result Discussion. In this section, we increased the density of the mobile nodes with random mobility. The simulation model is shown in Figure 12. The density of UEs gradually increases around WiFi APs when UEs move toward them. We compare the performance with that of the conventional WiFi selection scheme. As aforementioned, in the conventional selection based on RSS or SINR, if there are several of available APs, the nearest AP (regardless of AP's load status) will be selected and the connection will remain until the signal strength becomes unacceptable (below RX sensitive threshold).

At the beginning, the UEs are connected via WiMAX BSs. At $T = 4$ s, the UEs start sending greedy CBR (Constant Bit Rate) toward the Correspondent Node (CN) (Nodes 23, 24, and 25 in Figure 12). Figure 13 shows the traffic of UEs rising at $T = 4$ s.

At $T = 7$ s, the CM of UEs fetches discovery and selection policies from ANDSF, which contains AP candidates list,

signal strength, and AP's load thresholds corresponding to UEs location. Afterward, the UEs scan the surrounding area to obtain the channel load and RSS of each AP candidate. The AP candidates are evaluated based on their channel load and the received signal strength metrics. Only APs, which satisfy the load threshold and the signal strength threshold, are considered in the next steps. The CM of UEs calculates the APQI metric of each AP. The AP candidates are ranked by the APQI metric and the top AP is selected as AP candidate.

The numerical results show that the proposed scheme outperforms the conventional WiFi selection scheme in terms of overall system throughput or average UE's data throughput. With the knowledge of the network conditions and selection policies, UEs can offload their traffic more efficiently. With our proposed scheme, the UE can proactively decide the right timing for making WiFi roaming based on the policies from the network and UE's measured information. Furthermore, the UE can also dynamically select preferable

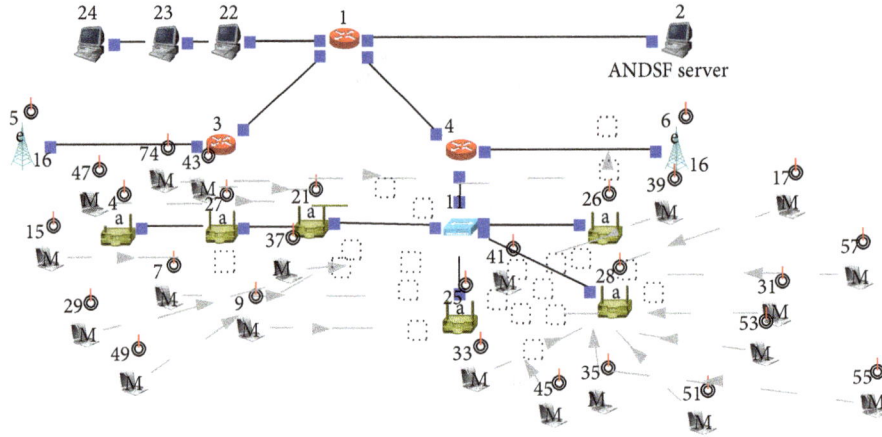

FIGURE 12: Simulation setup with 20 UEs and 6 APs and 2 WiMAX BSs.

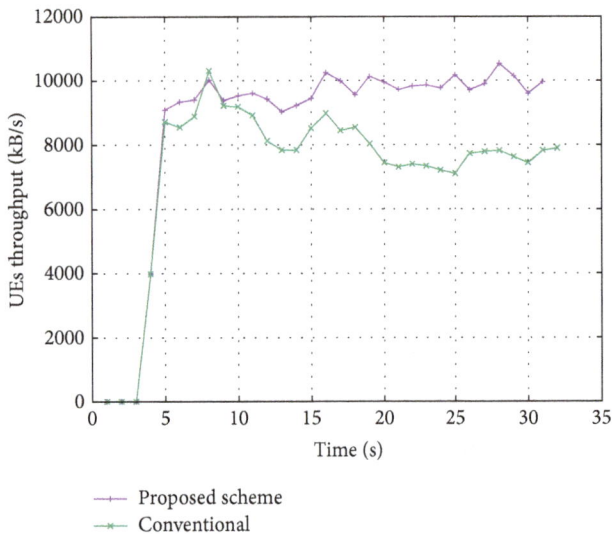

FIGURE 13: Overall system throughput comparison proposed scheme and conventional scheme.

AP (in terms of load condition and radio link condition) when moving. Therefore, the traffic of the macrocells is offloaded and distributed among available APs. It increases the utilization rate of available APs since all of APs are utilized while one AP is left unused in conventional case. In addition, we are also aware that proposed scheme has a drawback since it increases the number of handoffs. This drawback results from the fact that UEs are dynamically switched to preferable AP to optimize their throughput. Therefore, we have a trade-off between the number of handoffs and experienced throughput of mobile node. This problem will be considered in our future work.

5. Conclusion and Future Work

Our main contribution in this paper is that we proposed a novel WiFi roaming selection scheme for cellular data offloading. In our proposed scheme, we utilized the channel

load metric first defined in IEEE802.11k as well as IEEE 802.11u for the proposed scheme. As far as we are concerned, this channel load has not been considered in any related work. We defined a new metric, APQI, which was used to rank AP candidates based on their channel load status and signal strength. In addition, we utilized a series of signal level and AP load thresholds to eliminate unqualified candidates, which reduce the processing time and reduce unnecessary delay. We proposed to append signal strength threshold into the ANDSF OM WLANSP leaf so that UEs can easily fetch this information piece from ANDSF server for network selection. In addition, we also considered the practical implementation of our proposed scheme. Therefore, we reviewed the state of the art in 3GPP and IEEE and developed our scheme based on the ratified IEEE 802.11 k, u [11, 25] and 3GPP TS 23.312 [9] standard so that it can be applicable in future wireless network. In this paper, we also introduced our simulation model of ANDSF and vertical handover. Unlike previous works, our simulation model is an end-to-end model from application to physical layer. We took into account user's mobility and realistic traffic model. During the simulation process, the ANDSF server and the UE's connection manager entities behave and interact corresponding to real-time network events. By using this simulation model, we evaluated the proposed scheme in a typical heterogeneous network scenario with interworking macrocell (WiMAX) and WiFi.

We demonstrated how the proposed scheme performs in a typical HetNet scenario with WiFi APs and macrocells. The result showed that our proposed scheme dynamically steered the UE's traffic from macrocell and is distributed to available WiFi APs. As a result, both the overall system throughput and the utilization of available WiFi APs were improved. In addition, we observed better user's experienced throughput compared to conventional scheme. In our future work, we will further enhance the proposed scheme and evaluate the proposed scheme in more complex network scenario. In this paper, the network is assumed to support only data-like services. Other types of services such as voice or a combination of different services could also be studied.

A more realistic user behaviour model could be employed. The robustness of the results can be analysed by varying some system parameters such as hotspot radius and hotspot position. In addition, we will also consider the UE's context with different QoS requirements.

Conflict of Interests

The authors declare that there is no conflict of interests regarding the publication of this paper.

References

[1] W. Doug, "Announcing the Cisco Visual Networking Index Mobile Forecast, 2010–2015," April 2015, http://blogs.cisco.com/tag/mobile-data-forecast.

[2] D. Cavalcanti, D. P. Agrawal, C. Cordeiro, B. Xie, and A. Kumar, "Issues in integrating cellular networks, WLANs, and MANETs: a futuristic heterogeneous wireless network," *IEEE Wireless Communications*, vol. 12, no. 3, pp. 30–41, 2005.

[3] R. Chakravorty, P. Vidales, K. Subramanian, I. Pratt, and J. Crowcroft, "Performance issues with vertical handovers—experiences from GPRS cellular and WLAN hot-spots integration," in *Proceedings of the 2nd IEEE International Conference on Pervasive Computing and Communications (PerCom '04)*, pp. 155–164, March 2004.

[4] K. Murray, R. Mathur, and D. Pesch, "Intelligent access and mobility management in heterogeneous wireless networks using policy," in *Proceedings of the ACM 1st International Workshop on Information and Communication Technologies*, pp. 181–186, 2003.

[5] K.-S. Kong, W. Lee, Y.-H. Han, M.-K. Shin, and H. You, "Mobility management for all-IP mobile networks: mobile IPv6 vs. proxy mobile IPv6," *IEEE Wireless Communications*, vol. 15, no. 2, pp. 36–45, 2008.

[6] H. Fathi, S. S. Chakraborty, and R. Prasad, "Optimization of mobile IPv6-based handovers to support VoIP services in wireless heterogeneous networks," *IEEE Transactions on Vehicular Technology*, vol. 56, no. 1, pp. 260–270, 2007.

[7] C. Guo, Z. Guo, Q. Zhang, and W. Zhu, "A seamless and proactive end-to-end mobility solution for roaming across heterogeneous wireless networks," *IEEE Journal on Selected Areas in Communications*, vol. 22, no. 5, pp. 834–848, 2004.

[8] 3GPP. 3rd Generation Partnership Project, "Architecture enhancements for non-3GPP accesses (release 12)," Technical Specification 3GPP TS 23.402, 2014.

[9] Generation Partnership Project, "Access network discovery and selection function management object (release 12)," Technical Specification 3GPP TS 24.312; V12.4.0, 2014.

[10] IEEE, "IEEE standard for information technology—telecommunications and information exchange between systems—local and metropolitan area networks—specific requirements, part 11: wireless LAN Medium Access Control (MAC) and Physical Layer (PHY) specifications," IEEE Std 802.11, IEEE, 2012.

[11] Wi-Fi Alliance, "Hotspot 2.0 Technical Specification Package (Release 2); v1.0.0".

[12] 4G Americas, *Integration of Cellular and WiFi Networks*, 4G Americas, 2013.

[13] M. Zekri, B. Jouaber, and D. Zeghlache, "A review on mobility management and vertical handover solutions over heterogeneous wireless networks," *Computer Communications*, vol. 35, no. 17, pp. 2055–2068, 2012.

[14] M. Kassar, B. Kervella, and G. Pujolle, "An overview of vertical handover decision strategies in heterogeneous wireless networks," *Computer Communications*, vol. 31, no. 10, pp. 2607–2620, 2008.

[15] J. Fu, J. Wu, J. Zhang, L. Ping, and Z. Li, "A novel AHP and GRA based handover decision mechanism in heterogeneous wireless networks," in *Information Computing and Applications: First International Conference, ICICA 2010, Tangshan, China, October 15–18, 2010. Proceedings*, vol. 6377 of *Lecture Notes in Computer Science*, pp. 213–220, Springer, Berlin, Germany, 2010.

[16] H. Silva, L. Figueiredo, C. Rabadão, and A. Pereira, "Wireless networks interoperability-wifi wimax handover," in *Proceedings of the 4th International Conference on Systems and Networks Communications (ICSNC '09)*, pp. 100–104, IEEE, Porto, Portugal, September 2009.

[17] 3GPP Study on Wireless Local Area Network (WLAN) 3GPP Radio Interworking, "Technical specification group radio access network," 3GPP TR 37.834, 2013, (Release 12).

[18] D. H. Hagos and R. Kapitza, "Study on performance-centric offload strategies for LTE networks," in *Proceedings of the 6th Joint IFIP Wireless and Mobile Networking Conference (WMNC '13)*, pp. 1–10, IEEE, Dubai, United Arab Emirates, April 2013.

[19] B. H. Jung, N.-O. Song, and D. K. Sung, "A network-assisted user-centric WiFi-offloading model for maximizing per-user throughput in a heterogeneous network," *IEEE Transactions on Vehicular Technology*, vol. 63, no. 4, pp. 1940–1945, 2014.

[20] J. Lee, Y. Yi, S. Chong, and Y. Jin, "Economics of WiFi offloading: trading delay for cellular capacity," *IEEE Transactions on Wireless Communications*, vol. 13, no. 3, pp. 1540–1554, 2014.

[21] P. S. Bithas, A. S. Lioumpas, G. K. Karagiannidis, and B. S. Sharif, "Interference minimization in hybrid WiFi/cellular networks," in *Proceedings of the 19th IEEE International Workshop on Computer Aided Modeling and Design of Communication Links and Networks (CAMAD '14)*, pp. 198–202, IEEE, Athens, Greece, December 2014.

[22] IEEE, *IEEE Standard for Local and Metropolitan Area Networks. Part 21: Media Independent Handover Services*, IEEE, New York, NY, USA, 2009.

[23] M. Kassar, B. Kervella, and G. Pujolle, "An overview of vertical handover decision strategies in heterogeneous wireless networks," *Computer Communications*, vol. 31, no. 10, pp. 2607–2620, 2008.

[24] O. Khattab and O. Alani, "A survey on media independent handover (MIH) and IP multimedia subsystem (IMS) in heterogeneous wireless networks," *International Journal of Wireless Information Networks*, vol. 20, no. 3, pp. 215–228, 2013.

[25] IEEE, "IEEE standard for information technology—telecommunications and information exchange between systems—local and metropolitan area networks—specific requirements, part 11: wireless LAN Medium Access Control (MAC) and Physical Layer (PHY) specifications amendment 1: radio resource measurement of wireless LANs," IEEE Std 802.11k-2008, version 1.2.1, 2008.

[26] Green Packet Mobile data offload solution using ANDSF, April 2015, http://www.greenpacket.com/solutions/dynamic-WiFi-offload.html.

[27] S. Y. Wang, C. L. Chou, and C. C. Lin, "The design and implementation of the NCTUns network simulation engine," *Simulation Modelling Practice and Theory*, vol. 15, pp. 57–81, 2007.

[28] S.-Y. Wang, P.-F. Wang, Y.-W. Li, and L.-C. Lau, "Design and implementation of a more realistic radio propagation model for wireless vehicular networks over the NCTUns network simulator," in *Proceedings of the IEEE Wireless Communications and Networking Conference (WCNC '11)*, pp. 1937–1942, Cancun, Mexico, March 2011.

[29] S. Y. Wang and P. F. Wang, "NCTUns 6.0: a simulator for advanced wireless vehicular network research," in *Proceedings of the IEEE 71st Vehicular Technology Conference (VTC-Spring '10)*, pp. 1–2, Taipei, Taiwan, May 2010.

[30] S. Y. Wang and R. M. Huang, "NCTUns tool for innovative network emulations," in *Computer-Aided Design and Other Computing Research Developments*, pp. 155–183, Nova Science, 2009.

12

Joint Radio Resource Allocation and Base Station Location Selection in OFDMA based Private Wireless Access Networks for Smart Grid

Peng Du[1] and Yuan Zhang[2]

[1]*College of Automation, Nanjing University of Posts and Telecommunications, Nanjing 210023, China*
[2]*National Mobile Communications Research Laboratory, Southeast University, Nanjing, China*

Correspondence should be addressed to Yuan Zhang; y.zhang@seu.edu.cn

Academic Editor: George Tsoulos

This paper studies the base stations deployment problem in orthogonal frequency-division multiple access (OFDMA) based private wireless access networks for smart grid (SG). Firstly, we analyze the differences between private wireless access networks for SG and public cellular access networks. Then, we propose scheduling and power control based algorithms for the radio resource allocation subproblem and K-means, simulated annealing (SA), and particle swarm optimization (PSO) based algorithms for the base station (BS) location selection subproblem and iterate over these two sets of algorithms to solve the target problem. Simulation results show that the proposed method can effectively solve the target problem. Specifically, the combination of power control based resource allocation algorithm and PSO based location selection algorithm is recommended.

1. Introduction

It is critical that the underlying communication technology shall support efficient data exchange between various domains comprising smart grid (SG) [1]. This paper studies the orthogonal frequency-division multiple access (OFDMA) [2] based private wireless access networks for SG. Actually, wireless technologies can be used in the grid for monitoring, metering, and data gathering [3–17]. Specifically, SG devices such as switching station, distribution circuit, and distributed energy sites will produce various information data and send them periodically to the base station (BS) to realize the automation of power distribution and electricity information acquisition. In order to achieve this requirement, many technical problems need to be solved. Among them, this paper addresses the problem of how to optimize the deployment of BSs.

This problem has been extensively studied in the literature for the public cellular access networks. However, the scenario of private wireless access networks for SG is quite different from that of the public cellular access networks.

Property 1 (the locations of devices can be considered as fixed). Due to the peculiarity of SG, the locations of most power devices can be considered as fixed or quasi-fixed.

Property 2 (the uplink transmission is dominant [18, 19]). The smart grid network is an uplink dominated network, as the main data flow is from power devices to control center. The most often requirement of communications in SG is devices periodically reporting status monitoring data to the control center via BSs. Therefore, the direction of most data transmissions will be uplink.

Property 3 (the transmission rate requirement of devices can be considered as fixed [19]). The communications happening in the SG belong to the type of machine-to-machine (M2M) communications and the uplink data transmission rate requirement of each device can be considered as fixed or quasi-fixed.

Property 4 (the frequency separation requirement [19]). Wireless networks are more vulnerable than their wired counterparts due to the potential for direct access to the transport medium. Hence, security must be considered at every layer of the protocol stack in the private wireless access networks for SG. In addition to the authentication, authorization, and encryption considered at the application layer, the frequency separation mechanism at the physical layer shall also be considered, which is explained as follows. Actually, data produced by different types of devices in smart grid shall be transmitted to different destination systems. For example, as shown in Figure 1, the data produced by data terminal unit (DTU) shall be transmitted to the production service system, while the data produced by video terminal unit (VTU) shall be transmitted to the management service system. Due to the security purpose, the transmission paths used by different types of data shall be separated as much as possible. The separation can be achieved physically or logically. For example, four different approaches to construct the access network are illustrated in Figure 1, where the data paths in Figure 1(a) are the most separated; that is, the separation is achieved physically, while the data paths in Figure 1(d) are the least separated; that is, the separation can be achieved logically. Further, in addition to the separation for the wireline segment, data transmission over the wireless segment of data path shall also be separated for different types of devices. This requires that different types of devices shall use different frequency channels to transmit their data; that is, different types of devices sharing the same frequency channel will not be allowed. This is the frequency separation requirement considered in this work.

As a first step towards addressing the above issues, this paper investigates the problem of how to deploy BSs and allocate wireless resources so that the uplink transmission requirements are efficiently met. For this problem, we propose to decompose it into the resource allocation subproblem and the location selection subproblem and solve these two subproblems in an iterative fashion. The remainder of the paper is organized as follows. Section 2 formulates the joint resource allocation and location selection problem. Section 3 presents the overall framework to address this problem. Sections 4 and 5 propose resource allocation and location selection algorithms, respectively. Simulation results are reported in Section 6. Finally, we conclude in Section 7.

2. Problem Formulation

Consider a set of SG devices scattered in an area Ψ. Let $\mathcal{H}$ denote the set of devices. For each device $j \in \mathcal{H}$, let C_j denote the minimum uplink data rate requirement and P_j the uplink transmission power. The value of P_j shall satisfy $0 \leq P_j \leq P_{\max}$, where $P_{\max}$ is the upper bound. For convenience, let $C = \{C_j\}$ and $P = \{P_j\}$, respectively. Assume that all devices are classified into K different types. Let $\mathcal{H}_i$ denote the set of type-i SG devices, $1 \leq i \leq K$.

Assume that the private wireless access network consists of B BSs which are located in the area Ψ. Let $\mathbf{z}_b = (z_b^1, z_b^2)$ denote the deployment location of the bth BS, $1 \leq b \leq B$, where z_b^1 and z_b^2 are the horizontal and vertical ordinate of the deployment location, respectively. For convenience, let $\mathbf{z} = \{\mathbf{z}_b\}$. Not every location in Ψ can be the candidate location for

BS. Assume that Θ denotes the candidate BS location set in Ψ and the deployment location of BS can only be selected from the elements of Θ. That is, we restrict $\mathbf{z}_b \in \Theta$. In addition, let $\Omega = \{S_1, S_2, \ldots, S_B\}$ denote the relationship between SG devices and BSs, where S_b is the set of devices served by the bth BS. For simplicity and without loss of generality, we assume that the value of S_b is determined by the distance based rule. That is, a device will be served by the BS which is the closest to it.

Consider a OFDMA based private wireless access network. The radio resource is defined as follows. In the frequency domain, assume that the total bandwidth is divided into N channels. Let W and W_0 denote total and channel bandwidth in Hertz, respectively. In the time domain, assume that the time axes are organized into consecutive slots and L_0 consecutive slots constitute a frame. The basic resource unit for data transmission is a resource block (RB) which is defined as one channel in the frequency domain and one slot in the time domain, respectively. In each frame, assume that L slots can be used for uplink communications. Therefore, for each channel, there are L RBs which are allocatable. Finally, define binary variable $Y_{n,l,j}$ to denote the results of radio resource allocation which is valued 1 if the lth RB of the nth channel is allocated to device j and 0 otherwise. Each device shall be allocated a number of RBs to meet its minimum data rate requirement. For convenience, let $Y = \{Y_{n,l,j}\}$.

Given that the RB (n, l) has been allocated to device j, the received signal-interference-noise-ratio (SINR) experienced by device j on this RB at BS b can be written as

$$\gamma_{n,l,j} = \frac{P_j G_{jb}}{P_N + P_I},$$ (1)

where G_{bj} is the path loss from device j to BS b, P_N is the power of background noise, $P_I = \sum_{k \neq j, k \in \mathcal{D}_{n,l}} P_k G_{kb}$ is the power of interference, and $\mathcal{D}_{n,l}$ is the set of devices which share the same RB with device j. For simplicity and without loss of generality, we assume that the path loss mainly depends on the distance and can be calculated according to the formula $PL(x)$ for a distance separation of x meters and we assume that there is no interference between distant devices. Let $\widehat{C}_{n,l,j}$ denote the uplink data rate achieved by device j on RB (n, l) which is calculated by the Shannon formula as

$$\widehat{C}_{n,l,j} = \frac{W_0 \log\left(1 + \gamma_{n,l,j}\right)}{L_0}.$$ (2)

Then, the total data rate achieved by device j, denoted as $\widehat{C}_j$, can be calculated as

$$\widehat{C}_j = \sum_{(n,l):Y_{n,l,j}=1} \widehat{C}_{n,l,j}.$$ (3)

For convenience, let $\widehat{C} = \{\widehat{C}_j\}$.

Finally, the problem addressed in this paper can be formulated as, given the parameters Ψ, $\mathcal{H}$, C, K, $\{\mathcal{H}_i\}$, $P_{\max}$, B, Θ, W, W_0, N, L_0, and L, how to determine the values of deployment location $\mathbf{z}$, transmission power P, and

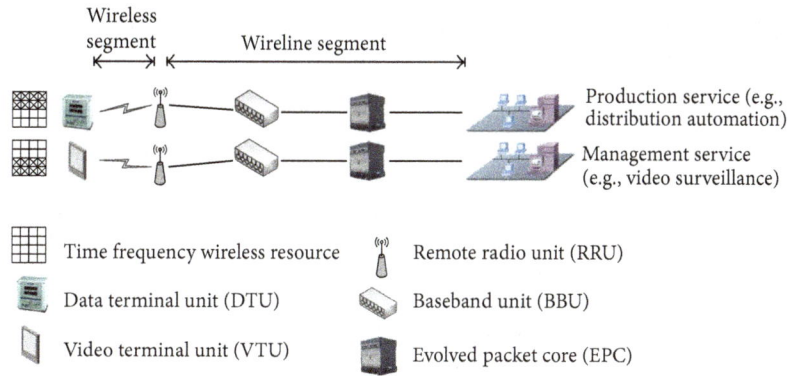

(a)

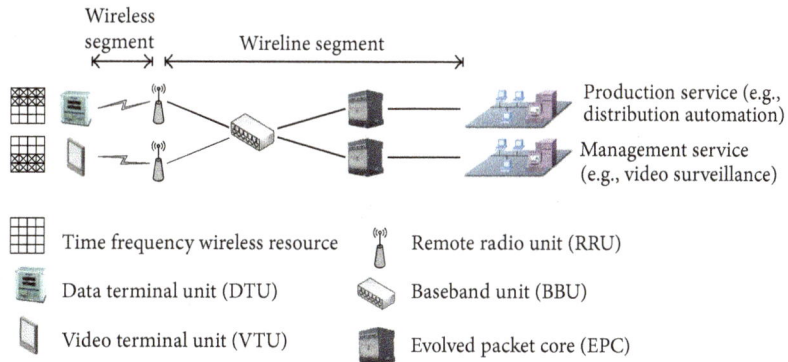

(b)

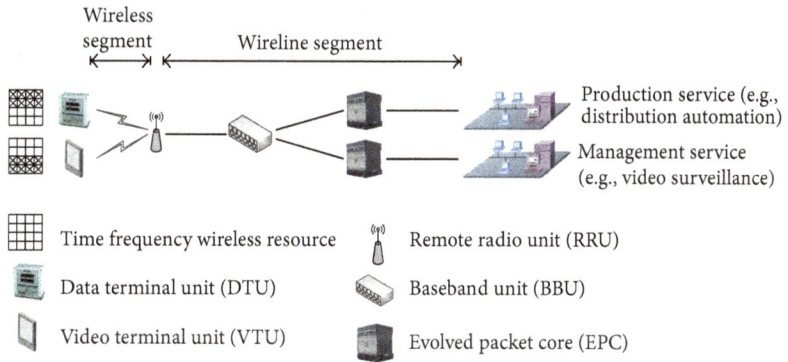

(c)

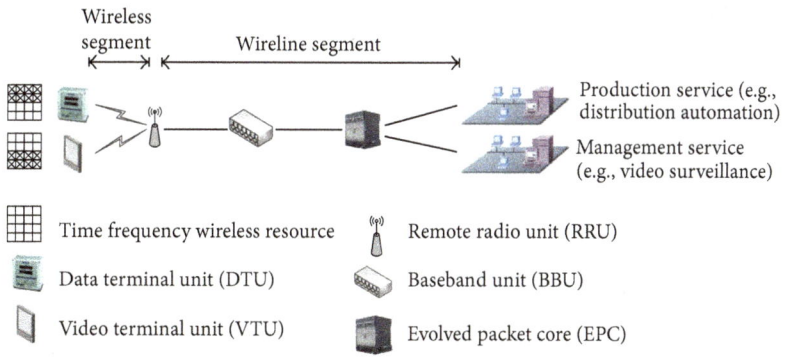

(d)

FIGURE 1: The separation requirement.

radio resource allocation Y, so that the achieved data rate $\widehat{C}$ approaches C as much as possible. The symbols used in this paper are summarized in List of Symbols.

3. The Framework

It is difficult to solve $\mathbf{z}$, P, and Y simultaneously. Therefore, we decompose the problem into two subproblems. The first is the location selection subproblem which determines $\mathbf{z}$; the second is the resource allocation subproblem which determines P and Y. Specifically, the resource allocation subproblem determines P and Y based on $\mathbf{z}$ produced by the location selection subproblem. Then, the payoff of the current $\mathbf{z}$ is calculated. Let V denote the payoff of a given $\mathbf{z}$.

The general expression of the payoff function can be written as

$$V = \sum_j \left(U_j\left(\widehat{C}_j\right) - I_j\left(P_j\right) \right), \qquad (4)$$

where $U_j(\cdot)$ is an increasing function representing the utility of device j and $I_j(\cdot)$ is also an increasing function representing the cost of device j. In this paper, we firstly let $U_j(\widehat{C}_j) = \widehat{C}_j/C_j$, where C_j is the minimum uplink data rate requirement of device j. Secondly, since the locations of devices in smart grid are fixed and the power can be supplied by alternating current adapter, we just let $I_j(P_j) = 0$. This is a difference between wireless communications for smart grid and for land mobile users. Therefore, we define the satisfaction ratio c_j of device j as the ratio between achieved data rate and required data rate; that is,

$$c_j = \frac{\widehat{C}_j}{C_j}, \qquad (5)$$

and we then define V as the sum of satisfaction ratio over all devices; that is,

$$V = \sum_j c_j, \qquad (6)$$

which is used to measure how good the given $\mathbf{z}$ is.

The problem can be solved by solving these two subproblems in an iterative fashion. The value of V for the current $\mathbf{z}$ will be fed back to the location selection subproblem for guided search of the better $\mathbf{z}$. The next two sections will solve these two subproblems in sequence.

4. Resource Allocation Methods

The task of resource allocation is to determine P and Y given $\mathbf{z}$. Two different methods based on different principles are presented. The first is scheduling based for which uplinks which are far away from each other are scheduled to share the same RB. The second is power control based for which the transmission power of each uplink is controlled so that uplinks which are not far away from each other can also share the same RB.

4.1. Scheduling Based Resource Allocation. This method consists of four steps, which are described in sequence as follows.

4.1.1. Uplink Transmission Power Setting. This subsection determines the transmission power P_j for each device j. As stated before, for this method, uplinks which are far away from each other (i.e., do not interfere with each other) will be scheduled to share the same RB. Therefore, for the scheduling based method, it can be expected that the interference power P_{I} in (1) is negligible. That is, we assume that there is no interference between distant devices. Thus, given the RB allocated to device j, the received SINR experienced by device j on this RB at BS b can be approximately written as

$$\gamma_j \approx \frac{P_j G_{jb}}{P_{\mathrm{N}}} \geq \Gamma, \qquad (7)$$

where device j is served by BS b (i.e., $j \in S_b$) and Γ is the minimum SINR requirement. Γ is a system parameter and common to all devices and RBs. Therefore, the uplink transmission power P_j can be set to

$$P_j = \min\left(P_{\mathrm{N}} \cdot \frac{\Gamma}{G_{jb}}, P_{\max} \right). \qquad (8)$$

That is, since in this method distant devices between which there is no interference are scheduled simultaneously, there is no power control and power is strictly a function of the target minimum SINR requirement.

4.1.2. Interference Graph Construction. The interference graph is used to indicate whether any two devices can reuse the same RB due to the interference between them. As indicated by Property 4, different types of devices shall transmit data over different channels. So we need to construct interference graph for each type, respectively. Let $\mathcal{G}_i(V_i, E_i)$ denote the interference graph for the ith type, $1 \leq i \leq K$, where V_i is the vertex set in which each vertex represents a device of the ith type and E_i is the edge set in which each edge $e_{j,k}$ represents devices j and k which cannot reuse the same RB. There are two rules to decide if edge $e_{j,k}$ exists. Assume that devices j and k are served by BS b_j and b_k, respectively. The first rule is if $b_j = b_k$, then edge $e_{j,k}$ exists. The second rule is if $b_j \neq b_k$ but the interference caused to each other is too large, then edge $e_{j,k}$ exists. Specifically, if the distance between device j and BS b_k is less than the interference radius R_j of device j or if the distance between device k and BS b_j is less than the interference radius R_k of device k, then edge $e_{j,k}$ exists.

The calculation of interference radius is as follows. For device j, the interference radius R_j is defined as the distance at which the received SINR is η, where η is the SINR requirement to ensure that the device does not cause nonnegligible interference to other uplinks that are out of the range of interference radius. According to (7), we have the equation for R_j as

$$\frac{P_j \cdot \mathrm{PL}\left(R_j\right)}{P_{\mathrm{N}}} = \eta, \qquad (9)$$

from which the value of R_j can be solved. After calculating the interference radius for each device, the interference graph $\mathcal{G}_i$

```
Require: z.
Ensure: 𝒢ᵢ, 1 ≤ i ≤ K.
  (1)  for i = 1 to K do
  (2)     for any two devices j and k in 𝒽ᵢ do
  (3)        if bⱼ = bₖ then
  (4)           connect vertexes j and k in 𝒢ᵢ;
  (5)        end if
  (6)        if dis(j, bₖ) < Rⱼ or dis(k, bⱼ) < Rₖ then
  (7)           connect vertexes j and k in 𝒢ᵢ;
  (8)        end if
  (9)     end for
  (10) end for
```

ALGORITHM 1: Interference graph construction.

can be constructed. The procedure is outlined in Algorithm 1, where $\mathrm{dis}(j, b)$ in line (6) represents the distance between device j and BS b.

4.1.3. Utility Function Calculation. For each i, $1 \le i \le K$, the utility function $F_{i,n}$ is defined as the sum of satisfaction ratio over all devices of the ith type given that a total of n channels have been allocated to them. To calculate $F_{i,n}$, define $\Delta F_{i,n}$ as the sum of satisfaction ratio over all devices of the ith type given that the nth channel has been allocated to them. Then the value of $F_{i,n}$ can be obtained according to

$$F_{i,n} = F_{i,n-1} + \Delta F_{i,n}, \tag{10}$$

where $F_{i,0} = 0$. Further, to calculate $\Delta F_{i,n}$, define $\Delta F_{i,n,l}$ as the sum of satisfaction ratio over all devices of the ith type given that the lth RB of the nth channel has been allocated to them. Then the value of $\Delta F_{i,n}$ can be obtained according to

$$\Delta F_{i,n} = \sum_{l=1}^{L} \Delta F_{i,n,l}. \tag{11}$$

Let $\mathcal{H}_{i,n,l}$ denote the set of devices of the ith type that share the lth RB of the nth channel. Then the value of $\Delta F_{i,n,l}$ can be calculated as

$$\Delta F_{i,n,l} = \sum_{j \in \mathcal{H}_{i,n,l}} \frac{\widehat{C}_{n,l,j}}{C_j}, \tag{12}$$

where $\widehat{C}_{n,l,j}$ can be obtained by (2). Finally, we say a device is feasible in slot l if the total power allocated to this device in this slot does not exceed $P_{\max}$. The procedure to calculate utility function is outlined in Algorithm 2, where the set $\mathcal{H}_{i,n,l}$ is determined in a heuristic manner in lines (4)–(10).

4.1.4. RB Allocation. This subsection presents the RB allocation algorithm. As indicated by Property 4, due to the security consideration, different types of devices shall use different frequency channels and different types sharing the same frequency channel are not allowed. This is the constraint which channel allocation shall satisfy. The procedure of the

scheduling based RB allocation is outlined in Algorithm 3, where n_i denotes the number of channels which have been allocated to the ith type. Specifically, after the type which is allocated to the nth channel has been selected in line (3), the RBs of the nth channel shall be allocated according to $\{\mathcal{H}_{i,n,l}\}$ which has been obtained in Algorithm 2, as shown in line (6).

4.2. Power Control Based Resource Allocation. This method consists of four steps, which are described in sequence as follows.

4.2.1. Grouping. Let S_b^i denote the set of type-i devices which are served by BS b, $1 \le i \le K$. The value of S_b^i can be derived from the value of S_b, which can be derived from the value of **z**. Let $\mathcal{H}_i = \{\mathcal{H}_{i1}, \ldots, \mathcal{H}_{ig}, \ldots, \mathcal{H}_{iG_i}\}$ denote the grouping for the ith type, where $\mathcal{H}_{ig}$ is the set of devices of the ith type which can share the same RB and G_i is the number of groups. The procedure of grouping is outlined in Algorithm 4.

4.2.2. Uplink Transmission Power Control. Since all devices in $\mathcal{H}_{ig}$ share the same RB, the received SINR in (1) can be rewritten as

$$\gamma_{n,l,j} = \frac{P_j G_{jb}}{P_{\mathrm{N}} + \sum_{k \ne j, k \in \mathcal{H}_{ig}} P_k G_{kb}} \ge \Gamma, \tag{13}$$

where $j \in \mathcal{H}_{ig}$ and Γ is the minimum SINR requirement. Similarly, Γ is a system parameter and common to all devices and RBs.

We propose an iterative update algorithm for finding the minimum transmission power satisfying the above equation. Specifically, for the tth iteration, the optimal power $P_j^{[t]}$ to be used by device j can be obtained by solving the following equation:

$$\frac{P_j^{[t]} G_{jb}}{P_{\mathrm{N}} + \sum_{k \ne j, k \in \mathcal{H}_{ig}} P_k^{[t-1]} G_{kb}} = \Gamma, \tag{14}$$

where $P_k^{[t-1]}$ is the power settings obtained at iteration $t - 1$. According to (14), the value of $P_j^{[t]}$ can be easily obtained using the bisection method [20]. Additionally, if the value of $P_j^{[t]}$ is greater than $P_{\max}$, it will be set as $P_{\max}$. The update of the values of transmission power proceeds in iterations until the power convergence.

4.2.3. Utility Function Definitions. The utility function $F_{i,n,g}$ is defined as the sum of satisfaction ratio over all devices in $\mathcal{H}_{ig}$ given that a total of n channels have been allocated to them. To calculate $F_{i,n,g}$, define $\Delta F_{i,n,g,l}$ as the sum of satisfaction ratio over all devices in $\mathcal{H}_{ig}$ given that the first l RBs of the nth channel have been allocated to them. Then the value of $F_{i,n,g}$ can be obtained according to

$$F_{i,n,g} = F_{i,n-1,g} + \Delta F_{i,n,g,L}, \tag{15}$$

Require: $\mathcal{G}_i$, $1 \leq i \leq K$.
Ensure: $\{F_{i,n}\}$ and $\{\mathcal{H}_{i,n,l}\}$.
  (1) **for** $i = 1$ to K **do**
  (2)   **for** $n = 1$ to N **do**
  (3)   **for** $l = 1$ to L **do**
  (4)   initialize $\mathcal{H}_{i,n,l} = \emptyset$;
  (5)   delete from $\mathcal{G}_i$ devices which are not feasible in slot l anymore;
  (6)   **while** $\mathcal{G}_i \neq \emptyset$ **do**
  (7)   determine device j^* with the lowest satisfaction ratio in $\mathcal{G}_i$;
  (8)   put j^* into $\mathcal{H}_{i,n,l}$;
  (9)   delete j^* and devices connected to it from $\mathcal{G}_i$;
  (10)   **end while**
  (11)   calculate $\Delta F_{i,n,l}$;
  (12)   recover $\mathcal{G}_i$;
  (13)   update satisfaction ratio of all devices in $\mathcal{G}_i$;
  (14)   **end for**
  (15)   calculate $\Delta F_{i,n}$;
  (16)   calculate $F_{i,n}$;
  (17)   **end for**
  (18) **end for**

ALGORITHM 2: Utility function calculation.

Require: $\{F_{i,n}\}$ and $\{\mathcal{H}_{i,n,l}\}$.
Ensure: $\{Y_{n,l,j}\}$.
  (1) initialize $n_i = 0$, $1 \leq i \leq K$;
  (2) **for** $n = 1$ to N **do**
  (3)   allocate the nth channel to the i^*th type with the minimum F_{i,n_i} and break the tie arbitrarily;
  (4)   let $n_{i^*} = n_{i^*} + 1$;
  (5)   **for** $l = 1$ to L **do**
  (6)   allocate the lth RB to devices in $\mathcal{H}_{i^*,n_{i^*},l}$;
  (7)   let $Y_{n,l,j} = 1$ for each $j \in \mathcal{H}_{i^*,n_{i^*},l}$;
  (8)   **end for**
  (9) **end for**

ALGORITHM 3: Scheduling based RB allocation.

Require: z.
Ensure: $\{\mathcal{H}_{ig}\}$.
  (1) **for** $i = 1$ to K **do**
  (2)   let $G_i = \max\{|S_1^i|, |S_2^i|, \ldots, |S_B^i|\}$;
  (3)   **for** $g = 1$ to G_i **do**
  (4)   **for** $b = 1$ to B **do**
  (5)   **if** $S_b^i \neq \emptyset$ **then**
  (6)   select any device j from S_b^i and put into $\mathcal{H}_{ig}$;
  (7)   delete device j from S_b^i;
  (8)   **end if**
  (9)   **end for**
  (10)   **end for**
  (11) **end for**

ALGORITHM 4: Grouping.

Require: $\{\mathscr{H}_{ig}\}$.
Ensure: $\{Y_{n,l,j}\}$.
(1)   initialize $n_i = 0$, $1 \leq i \leq K$;
(2)   **for** $n = 1$ to N **do**
(3)   allocate the nth channel to the i^*th type with the minimum F_{i,n_i} and break the tie arbitrarily;
(4)   let $n_{i^*} = n_{i^*} + 1$;
(5)   **for** $l = 1$ to L **do**
(6)   select $\mathscr{H}_{i^* g^*}$ which is feasible in slot l and has the minimum $\Delta F_{i^*,n_{i^*},g,l-1}$;
(7)   let $Y_{n,l,j} = 1$ for each $j \in \mathscr{H}_{i^* g^*}$;
(8)   calculate $\Delta F_{i^*,n_{i^*},g,l}$ for each g;
(9)   **end for**
(10)   calculate F_{i,n_i} for each i;
(11) **end for**

ALGORITHM 5: Power control based RB allocation.

where $F_{i,0,g} = 0$ and the value of $\Delta F_{i,n,g,L}$ can be obtained according to

$$\Delta F_{i,n,g,l} = \begin{cases} \Delta F_{i,n,g,l-1}, & \sum_{j \in \mathscr{H}_{ig}} Y_{n,l,j} = 0 \\ \Delta F_{i,n,g,l-1} + \sum_{j \in \mathscr{H}_{ig}} \dfrac{\widehat{C}_{n,l,j}}{C_j}, & \text{otherwise,} \end{cases} \quad (16)$$

where $F_{i,n,g,0} = 0$ and $1 \leq l \leq L$.

4.2.4. RB Allocation. This subsection presents the RB allocation algorithm. Similarly, different types of devices are not allowed to share the same frequency channel, which is the constraint which channel allocation shall satisfy.

For convenience, we define function $F_{i,n}$ as

$$F_{i,n} = \sum_{g=1}^{G_i} F_{i,n,g}. \quad (17)$$

In addition, we say a group $\mathscr{H}_{ig}$ is feasible in slot l if the total power allocated to each device $j \in \mathscr{H}_{ig}$ in this slot does not exceed $P_{\max}$. The procedure of the power control based RB allocation is outlined in Algorithm 5, where n_i also denotes the number of channels which have been allocated to the ith type. Specifically, after the type which is allocated to the nth channel has been selected in line (3), the RBs of the nth channel shall be allocated according to $\{\mathscr{H}_{ig}\}$ which has been obtained in Algorithm 4, as shown in line (6).

5. Location Selection Methods

The task of location selection is to search for the location $\mathbf{z}$. Three different location selection methods are presented. The first is K-means based [21]. This method is raw and is used as the benchmark in this work. The next two are simulated annealing (SA) based [22] and particle swarm optimization (PSO) based [23], respectively.

5.1. K-Means Based Location Selection. Initially, $\mathbf{z}_b = (z_b^1, z_b^2)$ is randomly selected from the candidate location set Θ as the deployment locations of BSs, where z_b^1 and z_b^2 are the horizontal and vertical ordinate of the deployment location, respectively. Then, we can obtain the corresponding $\Omega = \{S_1, S_2, \ldots, S_B\}$ which describes the relationship between SG devices and BSs. Next, the BS locations are updated as follows. Assume that the locations of device j are $\mathbf{x}_j = (x_j^1, x_j^2)$, where x_j^1 and x_j^2 are the horizontal and vertical ordinate of the location of device j, respectively. The new BS locations can be calculated as

$$z_b^h = \frac{1}{|S_b|} \sum_{j \in S_b} x_j^h, \quad (18)$$

where $1 \leq b \leq B$, $h \in \{1, 2\}$, and $|S_b|$ is the number of devices served by the bth BS. For each b, if the calculated $\mathbf{z}_b$ does not belong to Θ, it shall be set as the element in Θ which is the closest to the calculated value.

5.2. SA Based Location Selection. The location selection is to iterate over all candidate locations to find the best location that maximizes the satisfaction ratio. Since the enumeration is practically impossible, an algorithm with controllable complexity which can output a solution within the given time limit is desirable. We consider a stochastic local search algorithm which progressively traverses from one location to its neighbor in a probabilistic manner for finding the global optimal solution. Specifically, an algorithm based on simulated annealing is proposed, as outlined in Algorithm 6.

Beginning with an initial location, the variable $\mathbf{z}_{\text{best}}$ records the location with the highest payoff obtained so far as the algorithm proceeds. In lines (4) and (9), the resource allocation methods in Section 4 are used to determine the values of P and Y. At each iteration, a new location $\mathbf{z}_{\text{next}}$ among the neighborhood of current location $\mathbf{z}$ is chosen in line (8). The new location $\mathbf{z}_{\text{next}}$ is determined as follows. First, for the current $\mathbf{z}$, we can obtain $\Omega = \{S_1, S_2, \ldots, S_B\}$ and then calculate the satisfactory ratio of each S_b, $1 \leq b \leq B$. For each iteration only one BS location is changed. We choose BS b^* with the lowest satisfactory ratio to change the location. Specifically, we select a candidate BS location from Θ which is no more than d meters away from the original BS location

```
(1)   initialize c = 0;
(2)   initialize t = t_init;
(3)   initialize z;
(4)   determine the values of P and Y given z;
(5)   determine the value of V given z, P, and Y;
(6)   initialize z_best = z and V_best = V;
(7)   while c < c_max do
(8)       update z_next;
(9)       update P_next and Y_next given z_next;
(10)      update V_next given z_next, P_next, and Y_next;
(11)      if V_next > V then
(12)          update z = z_next and V = V_next;
(13)          if V_next > V_best then
(14)              update z_best = z_next and V_best = V_next;
(15)          end if
(16)      else
(17)          update z = z_next and V = V_next with probability e^{(V_next−V)/t};
(18)      end if
(19)      let c = c + 1;
(20)      let t = αt;
(21)  end while
(22)  return z_best.
```

ALGORITHM 6: SA based iterative procedure.

as the new BS location, where d is a parameter. If $\mathbf{z}_{next}$ yields a better payoff than $\mathbf{z}$, the search proceeds with $\mathbf{z}_{next}$ for the next iteration. Otherwise, $\mathbf{z}_{next}$ is still chosen with probability $e^{(V_{next}-V)/t}$ based on the concept of simulated annealing in line (17). In line (20), the temperature t decreases after each iteration according to an annealing schedule $t = \alpha t$, where $0 < \alpha < 1$ is also a parameter. Different values of c_{max}, α, and d can be set to control the speed of cooling.

5.3. PSO Based Location Selection.

In this subsection, a particle swarm optimization based algorithm is presented to search for the location. Assume that the swarm consists of M particles and the search space is B dimensional. Let $\mathbf{Z}_m = (\mathbf{z}_{m1}, \ldots, \mathbf{z}_{mb}, \ldots, \mathbf{z}_{mB})$ represent the position of the mth particle, where $\mathbf{z}_{mb}$ is a two-dimensional vector representing the deployment location of the bth BS. Let $\mathbf{v}_m = (\mathbf{v}_{m1}, \ldots, \mathbf{v}_{mb}, \ldots, \mathbf{v}_{mB})$ represent the velocity of the mth particle, where $\mathbf{v}_{mb} = (v^1_{mb}, v^2_{mb})$ is a two-dimensional vector for which v^1_{mb} and v^2_{mb} represent the horizontal and vertical velocity, respectively. Let $\mathbf{P}_m = (\mathbf{p}_{m1}, \ldots, \mathbf{p}_{mb}, \ldots, \mathbf{p}_{mB})$ represent the position of the best solution found by the mth particle and let $\mathbf{P}^* = (\mathbf{p}^*_1, \ldots, \mathbf{p}^*_b, \ldots, \mathbf{p}^*_B)$ represent the position of the best solution found by all particles during the search. The position of each particle is updated by using $\mathbf{Z}^{[t+1]}_m = \mathbf{Z}^{[t]}_m + \mathbf{v}^{[t+1]}_m$, where $\mathbf{Z}^{[t]}_m$ is the position of the mth particle at iteration t and $\mathbf{v}^{[t+1]}_m$ is the new velocity of the mth particle at iteration $t + 1$. The velocities of the particles are updated according to $\mathbf{v}^{[t+1]}_m = w\mathbf{v}^{[t]}_m + c_1\xi(\mathbf{P}^{[t]}_m - \mathbf{Z}^{[t]}_m) + c_2\eta(\mathbf{P}^{*[t]} - \mathbf{Z}^{[t]}_m)$, where $\mathbf{P}^{[t]}_m$ is the position of the best solution found by the mth particle at iteration t, $\mathbf{P}^{*[t]}$ is the position of the best solution found by all particles during the search so

far, and ξ and η are random values generated by the uniform distribution in the interval $[0, 1]$.

Additionally, for the PSO based algorithm, there are two types of collisions. For the first type, the particles could be attracted to regions outside the feasible search space Θ; for the second type, the velocity of particles could be too large. The anticollision mechanisms for preserving the feasibility of solution are as follows. For the first type of collision, if $\mathbf{z}_{mb} \notin \Theta$ occurs, we set $\mathbf{z}_{mb}$ randomly selected location in Θ. For the second type of collision, if it occurs, we set

$$v^h_{mb} = \begin{cases} v_{max}, & \text{if } v^h_{mb} > v_{max} \\ -v_{max}, & \text{if } v^h_{mb} < -v_{max}, \end{cases} \tag{19}$$

where $h \in \{1, 2\}$ and v_{max} is the velocity limit.

The procedure for PSO based algorithm is outlined in Algorithm 7, where c_{max} is the iteration limit.

6. Performance Evaluation

6.1. Parameter Setting.

Assume there are a total of $K = 3$ types of SG devices. In the case of no particular description, the required uplink data rate of each type is $C_1 = 100$ kbps, $C_2 = 400$ kbps, and $C_3 = 800$ kbps, respectively, and the number of devices of each type is 50, 50, and 50, respectively. We randomly distribute these devices in a circle region Ψ with a radius of 1200 meters. Further, we assume that Θ contains a total of 350 candidate BS locations which are also randomly generated in Ψ. Based on the simulation settings in [24, 25], wireless communication related parameters are set as follows. The maximum transmission power P_{max} is 20 dBm. The path loss formula is $PL(x) = 6 + 42.68 \log(x)$ dB for a distance

```
(1)  initialize c = 1;
(2)  for particle m = 1 to M do
(3)      initialize the velocity v_m in [-v_max, v_max];
(4)      initialize the position Z_m in Θ;
(5)      determine the value of V_m given Z_m;
(6)      initialize P_m = Z_m and V_best,m = V_m;
(7)  end for
(8)  calculate V_best = max {V_best,1, V_best,2, ..., V_best,M} and determine P*;
(9)  while c < c_max do
(10)     for particle m = 1 to M do
(11)         update the velocity v_m;
(12)         update the position Z_m;
(13)         determine the value of V_m given Z_m;
(14)         if V_m > V_best,m then
(15)             update V_best,m = V_m;
(16)             update P_m;
(17)         end if
(18)     end for
(19)     calculate V'_best = max {V_best,1, V_best,2, ..., V_best,M};
(20)     if V'_best > V_best then
(21)         update V_best = V'_best;
(22)         update P*;
(23)     end if
(24)     let c = c + 1;
(25) end while
(26) return P*.
```

ALGORITHM 7: PSO based iterative procedure.

separation of x meters. The total bandwidth W is 5 MHz and the bandwidth of each channel W_0 is 180 kHz. Assume that the power of background noise $P_N = N_0 W$, where the noise power spectrum density $N_0 = -174$ dBm/Hz. The minimum SINR requirement Γ is 3 dB, which is used in (7) and (13) to determine transmit power. The SINR requirement η is -2 dB, which is used in (9) to determine interference radius. Finally, the number of slots in each frame L_0 is 20. In the case of no particular description, assume that the number of usable slots L is also 20. For SA, there are three parameters t_{init}, α, and d. For t_{init} and d, based on the recommendations in [20, 26, 27], we set $t_{init} = 1000$ and $d = 30$. For α, we have run many simulation experiments to find an appropriate value of it. Simulation results show that the larger the value of α is, the better the supporting ratio is. Therefore, since the value of α shall be between 0 and 1, we set $\alpha = 0.99$. For PSO, there are five parameters M, V_{max}, w, c_1, and c_2. For M, w, c_1, and c_2, based on the recommendations in [27, 28], we set $M = 10$, $w = 0.7$, $c_1 = 2$, and $c_2 = 2$. For V_{max}, we have run many simulation experiments to find an appropriate value of it. Simulation results show that the value of V_{max} shall not be too small or too large. Specifically, if the value of V_{max} is too small, the convergence rate of PSO will be very slow; if the value of V_{max} is too large, PSO will oscillate and not converge. Therefore, after many simulation experiments, we have selected $V_{max} = 150$ to achieve acceptable convergence rate. Finally, for both algorithms, the iteration limit c_{max} is set to be 1000.

Combining different resource allocation and location selection algorithms, we have a total of six different schemes.

We evaluate the performance of above schemes for different parameter configurations. For each parameter configuration, we run simulation experiments for 1000 times and average the results.

6.2. Simulation Results. This subsection presents the performance evaluation results of the proposed schemes under different scenarios and the effects of various system parameters are evaluated and compared.

6.2.1. Convergence. We show in Figure 2 a typical trace of the progression of benefits for guided stochastic search in all schemes, where "PC" and "Sched" represent power control and scheduling based resource allocation algorithm, respectively. We can find that the payoff of the best location selection is increased gradually and will be converged to a constant value finally. Therefore, the curves in Figure 2 show that the proposed schemes are converged to a steady state. Additionally, we can observe that the solution quality and the required number of iterations to converge are significantly different from each other. Firstly, the final values of payoff for different schemes are different. Specifically, the "PC + PSO" scheme can achieve the highest payoff (i.e., 128.5030) among all schemes. Recall that the payoff is defined as the sum of satisfaction ratio over all devices where the satisfaction ratio of a device is defined as the ratio between achieved data rate and the required data rate. For this set of simulation experiments, since there are totally 150 devices (as stated in the beginning of Section 6.1), the value of payoff will not be higher than 150. Therefore, a payoff of 128.5030 means that

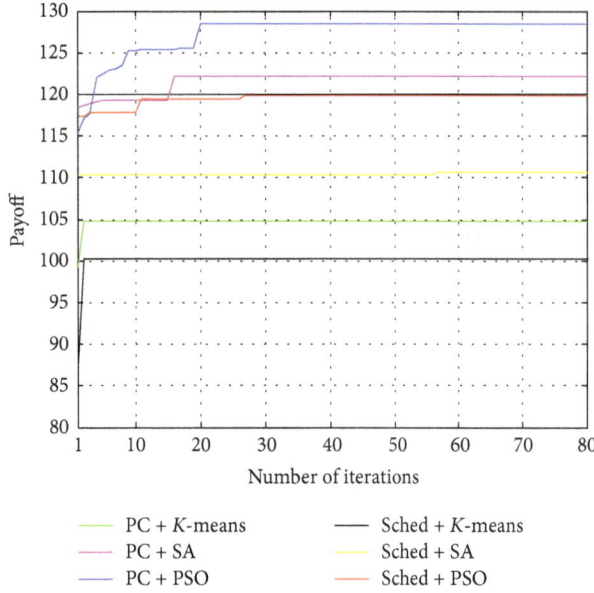

FIGURE 2: Convergence of the proposed schemes.

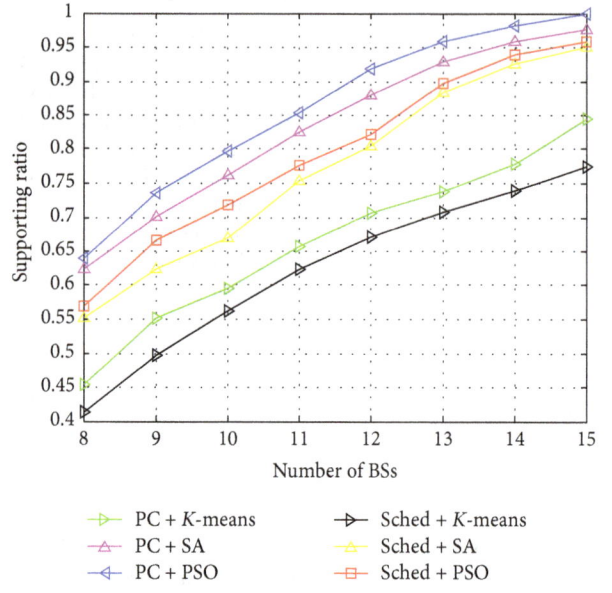

FIGURE 4: Impact of the number of BSs.

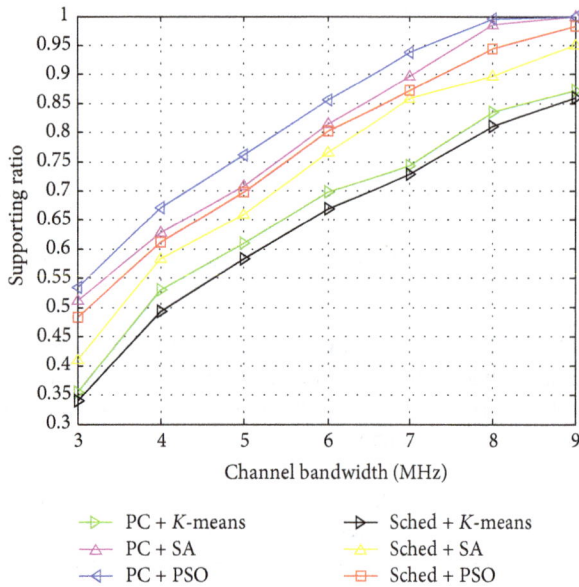

FIGURE 3: Impact of the number of channels.

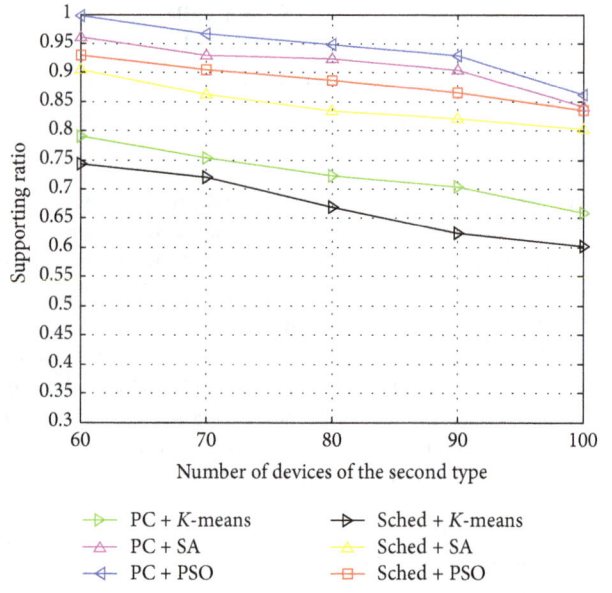

FIGURE 5: Impact of the number of devices.

most data rate requirements have been satisfied. Secondly, for K-means related schemes (i.e., the "PC + K-means" and "Sched + K-means" schemes), although their payoff is not high (i.e., 104.8572 and 100.1876), the required numbers of iterations to converge (i.e., 2 and 2) are much smaller than other schemes; that is, they converge much faster than other schemes. Therefore, we can conclude that different schemes can achieve different tradeoffs between solution quality and convergence rate.

For any device j, if its uplink data rate requirement is met (i.e., $\widehat{C}_j \geq C_j$), we say this device is satisfied. Further, we define the supporting ratio as the ratio between the number

of devices which have been satisfied and the total number of devices. In the following simulation experiments, we will evaluate the impact of the number of channels (i.e., the total bandwidth), the number of BSs, and the number of devices on the performance (i.e., the supporting ratio) of all these six schemes. Additionally, we would like to claim that all the values plotted in Figures 3, 4, and 5 are obtained after the algorithms have converged to a steady state.

6.2.2. Impact of the Number of Channels. The number of channels is equal to $\lfloor W/W_0 \rfloor$, where W is the total bandwidth. Figure 3 shows the supporting ratio of all proposed schemes when the total bandwidth W or, equivalently, the number of channels is varied. For this set of simulation experiments, there are totally 150 devices, for which the sum of data rate requirements is $50 \times C_1 + 50 \times C_2 + 50 \times C_3 = 65$ Mbps. We set the number of BSs B to be 10. It can be observed that when the total bandwidth (i.e., the number of channels) increases, the supporting ratio increases. Specifically, when the total bandwidth is 9 MHz (i.e., the number of channels is 50), the supporting ratio of the "PC + PSO" and "PC + SA" schemes is as high as 1 (i.e., the data rate requirements of all 150 devices have been satisfied), but the supporting ratio of the "PC + K-means" and "Sched + K-means" schemes is just 0.8713 and 0.8605, respectively (i.e., there are still $150 \times (1 - 0.8713) = 20$ and $150 \times (1 - 0.8605) = 21$ devices whose data rate requirements are not satisfied, resp.).

Finally, it can be observed that the "PC + PSO" scheme is the best among all other schemes, which will be validated again by the following simulation results. This is due to two aspects of reasons. For the first reason, SA and PSO are metaheuristics, which efficiently explore the search space to find near-optimal solutions. By searching over a large set of feasible solutions, they can find good solutions with less computational effort compared to simple heuristics (e.g., the K-means method). Therefore, SA and PSO are superior to K-means in finding good solutions. For the second reason, if two devices are close to each other, they could interfere with each other if they use the same RB. For the PC method, the transmission power of each device is controlled so that devices which are close to each other can also share the same RB; for the Sched method, only devices which are far away from each other can share the same RB. Since the PC method allows devices which are close to each other to transmit data simultaneously, it can admit more devices than the Sched method. On the other hand, for the PC method, since there exists interference among neighbor devices, each device will have to increase its transmission power to combat such interference to meet the minimum SINR requirement. This makes devices using the PC method consume more power resource than the Sched method. Therefore, the PC method can admit more devices than the Sched method via consuming more power.

6.2.3. Impact of the Number of BSs. Figure 4 shows the supporting ratio of all proposed schemes when the number of BSs B is varied. We can observe that when the number of BSs increases, the supporting ratio increases, since the average distance between devices and access points is shortened. Specifically, when the number of BSs is 15, the supporting ratio of the "PC + PSO" and "PC + SA" schemes is 1 and 0.9767, respectively (i.e., there are zero and $150 \times (1 - 0.9767) = 4$ devices whose data rate requirements are not satisfied, resp.), but the supporting ratio of the "PC + K-means" and "Sched + K-means" schemes is just 0.8447 and 0.7743, respectively (i.e., there are still $150 \times (1 - 0.8447) = 24$ and $150 \times (1 - 0.7743) = 34$ devices whose data rate requirements are not satisfied, resp.). Therefore, we can conclude that the "PC + PSO" scheme is the best one and for the simulated scenario at least 15 BSs shall be deployed so that

the supporting ratio of one can be achieved. For the following simulations, we will set the value of B to be 15.

6.2.4. Impact of the Number of Devices. Figure 5 shows the supporting ratio of all proposed schemes when the number of devices is varied. For convenience, let N_i denote the number of devices of the ith type. Let $N_{i,\text{init}}$ denote the initial value of N_i. As stated in the beginning of Section 6.1, we set $N_{1,\text{init}} = 50$, $N_{2,\text{init}} = 50$, and $N_{3,\text{init}} = 50$. We will collect the performance metrics (i.e., the supporting ratio) which is a function of (N_1, N_2, N_3). However, it is hard to visualize high-dimensional data when the dimension is greater than two. Therefore, we run the simulation for three times. For the jth ($j = 1, 2, 3$) run, we change the values of $N_j = N_{j,\text{init}} + n$ while keeping the values of other N_i ($i \neq j$) fixed to be $N_{i,\text{init}}$, where $n = 10, 20, 30, 40, 50$. Due to the limited space, we only plot the simulation results of the second run in Figure 5 where the horizontal axis represents the number of devices of the second type. We can observe that when the number of devices increases, the supporting ratio decreases, since the radio resource consumed by each type of devices increases. Specifically, when the number of devices of the second type is increased to be 100, there are totally $50 + 100 + 50 = 200$ devices, for which the sum of data rate requirements is $50 \times C_1 + 100 \times C_2 + 50 \times C_3 = 85$ Mbps. For this scenario, the supporting ratio of the "PC + PSO" and "PC + SA" schemes is still 0.8626 and 0.8420, respectively (i.e., there are $150 \times 0.8626 = 129$ and $150 \times 0.8420 = 126$ devices whose data rate requirements can be satisfied, resp.), but the supporting ratio of the "PC + K-means" and "Sched + K-means" schemes is only 0.6600 and 0.6015, respectively (i.e., there are only $150 \times 0.6600 = 99$ and $150 \times 0.6015 = 90$ devices whose data rate requirements have been satisfied, resp.). Comparing these curves, we can also conclude that the "PC + PSO" scheme is more preferable than other schemes.

7. Conclusions

In this paper, we study the joint BS location selection, transmission power control, and wireless channel allocation problem in OFDMA based private wireless access networks for smart grid. We transform the joint problem into channel allocation and site selection subproblems and solve these two subproblems iteratively. According to the simulation results, the combination of power control based resource allocation algorithm and PSO based location selection algorithm is recommended to solve the joint problem.

List of Symbols

Ψ: The area in which a set of SG devices is scattered
$\mathcal{H}$: The set of devices
C_j: The minimum uplink data rate requirement of device j
P_j: The uplink transmission power of device j
$P_{\max}$: The maximum uplink transmission power
K: The number of types of devices
$\mathcal{H}_i$: The set of SG devices of the ith type

B: The number of BSs

$\mathbf{z}_b$: The deployment location of the bth BS

Θ: The set of candidate BS locations

S_b: The set of devices served by the bth BS

W: The total bandwidth in Hertz

N: The number of channels into which the total bandwidth is divided

W_0: The channel bandwidth in Hertz

L_0: The number of slots in a frame

L: The number of slots which can be used for uplink communications in each frame

$Y_{n,l,j}$: The binary variable indicating whether the lth RB of the nth channel is allocated to device j

$\gamma_{n,l,j}$: The received SINR experienced by device j on the RB (n,l) at BS b

G_{bj}: The path loss from device j to BS b

P_{N}: The power of background noise

P_{I}: The power of interference

$\mathscr{D}_{n,l}$: The set of devices which share the same RB with device j

$\mathrm{PL}(x)$: The path loss for a distance separation of x meters

$\widehat{C}_{n,l,j}$: The uplink data rate achieved by device j on RB (n,l)

$\widehat{C}_j$: The total data rate achieved by device j

C: The set of all C_j

P: The set of all P_j

$\mathbf{z}$: The set of all $\mathbf{z}_b$

Ω: The set of all S_b

Y: The set of all $Y_{n,l,j}$

$\widehat{C}$: The set of all $\widehat{C}_j$

c_j: The satisfaction ratio of device j

V: The sum of satisfaction ratio over all devices

Γ: The minimum SINR requirement

$\mathscr{G}_i$: The interference graph for the ith type

V_i: The vertex set in $\mathscr{G}_i$

E_i: The edge set in $\mathscr{G}_i$

$e_{j,k}$: The edge which represents devices j and k cannot reuse the same RB

R_j: The interference radius of device j

η: The SINR requirement to calculate R_j

$\mathrm{dis}(j,b)$: The distance between device j and BS b

$F_{i,n}$: The sum of satisfaction ratio over all devices of the ith type given that a total of n channels have been allocated to them

$\Delta F_{i,n}$: The sum of satisfaction ratio over all devices of the ith type given that the nth channel has been allocated to them

$\Delta F_{i,n,l}$: The sum of satisfaction ratio over all devices of the ith type given that the lth RB of the nth channel has been allocated to them

$\mathscr{H}_{i,n,l}$: The set of devices of the ith type that share the lth RB of the nth channel

n_i: The number of channels which have been allocated to the ith type

S_b^i: The set of type-i devices which are served by BS b

$\mathscr{H}_{ig}$: The set of devices of the ith type which can share the same RB

G_i: The number of groups

$\mathscr{H}_i$: The set of all $\mathscr{H}_{ig}$

$P_j^{[t]}$: The power setting obtained at iteration t

$F_{i,n,g}$: The sum of satisfaction ratio over all devices in $\mathscr{H}_{ig}$ given that a total of n channels have been allocated to them

$\Delta F_{i,n,g,l}$: The sum of satisfaction ratio over all devices in $\mathscr{H}_{ig}$ given that the first l RBs of the nth channel have been allocated to them

$F_{i,n}$: The sum of $F_{i,n,g}$ over all groups

$\mathbf{x}_j$: The locations of device j.

Competing Interests

The authors declare that there is no conflict of interests regarding the publication of this paper.

Acknowledgments

This work was supported by the National Natural Science Foundation of China (no. 61571111).

References

[1] R. Ma, H.-H. Chen, Y.-R. Huang, and W. Meng, "Smart grid communication: its challenges and opportunities," *IEEE Transactions on Smart Grid*, vol. 4, no. 1, pp. 36–46, 2013.

[2] E. Dahlman, S. Parkvall, and J. Skold, *4G: LTE/LTE-Advanced for Mobile Broadband*, Academic Press, New York, NY, USA, 2013.

[3] X. S. Shen, "Empowering the smart grid with wireless technologies," *IEEE Network*, vol. 26, no. 3, pp. 2–3, 2012.

[4] H. Gharavi and B. Hu, "Multigate communication network for smart grid," *Proceedings of the IEEE*, vol. 99, no. 6, pp. 1028–1045, 2011.

[5] C. Gentile, D. Griffith, and M. Souryal, "Wireless network deployment in the smart grid: design and evaluation issues," *IEEE Network*, vol. 26, no. 6, pp. 48–53, 2012.

[6] Q.-D. Ho, Y. Gao, and T. Le-Ngoc, "Challenges and research opportunities in wireless communication networks for smart grid," *IEEE Wireless Communications*, vol. 20, no. 3, pp. 89–95, 2013.

[7] Z. Zhu, S. Lambotharan, W. H. Chin, and Z. Fan, "Overview of demand management in smart grid and enabling wireless communication technologies," *IEEE Wireless Communications*, vol. 19, no. 3, pp. 48–56, 2012.

[8] A. Abdrabou and A. M. Gaouda, "Uninterrupted wireless data transfer for smart grids in the presence of high power transients," *IEEE Systems Journal*, vol. 9, no. 2, pp. 567–577, 2015.

[9] P.-Y. Kong, "Wireless neighborhood area networks with QoS support for demand response in smart grid," *IEEE Transactions on Smart Grid*, vol. 7, no. 4, pp. 1913–1923, 2015.

[10] W.-Z. Song, D. De, S. Tan, S. K. Das, and L. Tong, "A wireless smart grid testbed in lab," *IEEE Wireless Communications*, vol. 19, no. 3, pp. 58–64, 2012.

[11] B. Fateh, M. Govindarasu, and V. Ajjarapu, "Wireless network design for transmission line monitoring in smart grid," *IEEE Transactions on Smart Grid*, vol. 4, no. 2, pp. 1076–1086, 2013.

[12] H. Gharavi and B. Hu, "Scalable synchrophasors communication network design and implementation for real-time distributed generation grid," *IEEE Transactions on Smart Grid*, vol. 6, no. 5, pp. 2539–2550, 2015.

[13] M. M. Aly and M. A. El-Sayed, "Enhanced fault location algorithm for smart grid containing wind farm using wireless communication facilities," *IET Generation, Transmission & Distribution*, vol. 10, no. 9, pp. 2231–2239, 2016.

[14] X. Wang and P. Yi, "Security framework for wireless communications in smart distribution grid," *IEEE Transactions on Smart Grid*, vol. 2, no. 4, pp. 809–818, 2011.

[15] T. Liu, Y. Liu, Y. Mao et al., "A dynamic secret-based encryption scheme for smart grid wireless communication," *IEEE Transactions on Smart Grid*, vol. 5, no. 3, pp. 1175–1182, 2014.

[16] B. Hu and H. Gharavi, "Smart grid mesh network security using dynamic key distribution with merkle tree 4-way handshaking," *IEEE Transactions on Smart Grid*, vol. 5, no. 2, pp. 550–558, 2014.

[17] F. Salvadori, C. S. Gehrke, A. C. de Oliveira, M. de Campos, and P. S. Sausen, "Smart grid infrastructure using a hybrid network architecture," *IEEE Transactions on Smart Grid*, vol. 4, no. 3, pp. 1630–1639, 2013.

[18] S. Chen, "A novel TD-LTE frame structure for heavy uplink traffic in smart grid," in *Proceedings of the 2014 IEEE Innovative Smart Grid Technologies-Asia (ISGT Asia '14)*, pp. 158–163, Kuala Lumpur, Malaysia, May 2014.

[19] Jiangsu Electric Power Company of China, "Electric power broadband wireless multi-service bearer network," White Paper, 2015.

[20] H.-Y. Hsieh, S.-E. Wei, and C.-P. Chien, "Optimizing small cell deployment in arbitrary wireless networks with minimum service rate constraints," *IEEE Transactions on Mobile Computing*, vol. 13, no. 8, pp. 1801–1815, 2014.

[21] J. Macqueen, "On convergence of K-means and partitions with minimum average variance," *Annals of Mathematical Statistics*, vol. 36, no. 3, pp. 1084–1090, 1965.

[22] H. Keinanen, "Simulated annealing for multi-agent coalition formation," in *Agent and Multi-Agent Systems: Technologies and Applications*, Lecture Notes in Computer Science, pp. 30–39, Springer, Berlin, Germany, 2009.

[23] J. Kennedy and R. Eberhart, "Particle swarm optimization," in *Proceedings of the IEEE International Conference on Neural Networks*, pp. 1942–1948, Perth, Australia, November 1995.

[24] 3GPP, "Further advancements for E-UTRA physical layer aspects," 3GPP TR 36.814, v9.0.0, 2010.

[25] 3GPP, "LTE coverage enhancements," 3GPP TR 36.824, v11.0.0, 2012.

[26] Y. Lu, Y. Lin, Q. Peng, and Y. Wang, "A review of improvement and research on parameters of simulated annealing algorithm," *College Mathematics*, vol. 31, no. 6, pp. 96–103, 2015.

[27] D. Wang, *Intelligent Optimization Methods*, Higher Education Press, Beijing, China, 2007.

[28] A. I. S. Nascimento and C. J. A. Bastos-Filho, "A particle swarm optimization based approach for the maximum coverage problem in cellular base stations positioning," in *Proceedings of the 10th International Conference on Hybrid Intelligent Systems (HIS '10)*, pp. 91–96, IEEE, Atlanta, Ga, USA, August 2010.

13

Anticollusion Attack Noninteractive Security Hierarchical Key Agreement Scheme in WHMS

Kefei Mao, Jianwei Liu, and Jie Chen

School of Electronic and Information Engineering, Beihang University, Beijing 100191, China

Correspondence should be addressed to Kefei Mao; owen.buaa@gmail.com

Academic Editor: Jit S. Mandeep

Wireless Health Monitoring Systems (WHMS) have potential to change the way of health care and bring numbers of benefits to patients, physicians, hospitals, and society. However, there are crucial barriers not only to transmit the biometric information but also to protect the privacy and security of the patients' information. The key agreement between two entities is an essential cryptography operation to clear the barriers. In particular, the noninteractive hierarchical key agreement scheme becomes an attractive direction in WHMS because each sensor node or gateway has limited resources and power. Recently, a noninteractive hierarchical key agreement scheme has been proposed by Kim for WHMS. However, we show that Kim's cryptographic scheme is vulnerable to the collusion attack if the physicians can be corrupted. Obviously, it is a more practical security condition. Therefore, we proposed an improved key agreement scheme against the attack. Security proof, security analysis, and experimental results demonstrate that our proposed scheme gains enhanced security and more efficiency than Kim's previous scheme while inheriting its qualities of one-round communication and security properties.

1. Introduction

Wireless Health Monitoring System (WHMS) is a dedicated network environment that supports the biometric information acquisition devices to gather people's health data anytime and anywhere [1]. Moreover, WHMS is a typical example of using wireless technologies to reduce medical expense and improve social benefits, such as detecting the lonely stroke patients timely [2, 3]. Security and privacy are the major concerns in medical activities, and WHMS is not an exception [4–6]. To provide privacy and security assurances in WHMS, it is important to provide security services by using cryptographic algorithms. Thus, obtaining cryptographic keys is an essential operation to achieve the security goals in WHMS. There are several key agreement schemes that have been proposed for WHMS applications [5, 7–10].

The noninteractive scheme is becoming a very active direction in the sensors networks [11–14] because sensor nodes have limited energy and processing and storage abilities. A noninteractive hierarchical key agreement scheme, called the Freshness-Preserving Noninteractive Hierarchical Key Agreement Protocol (FNKAP), was proposed by Kim [8] in 2014. The major advantages of the proposed scheme in Kim [8] go as follows. Firstly, there is only one-round communication to agree on a session key between two entities. Secondly, it is declared that the FNKAP achieves the patient anonymity and the session key confidentiality, and it can resist active and passive security attacks. However, we found that there is a flaw in the FNKAP when the physicians are not to be trusted. The scheme is not strong enough against the collusion attack where there are two adversaries who are a physician and a patient, separately. More precisely, in order to obtain a specific patient's electronic medical data, the adversary can pretend to be sick and become the same physician's patient with the victim in the real world. Then, the adversary bribes any other physician to get the private values of a physician. Finally, the adversary could calculate the session key and decrypts the victim's electronic health data freely. Note that a physician can casually expose the private values because the disclosed values are untraceable in Kim's scheme. As a result, this method of attack is reasonable and straightforward to implement.

The contributions of this paper are twofold. First, we illustrate that there is a weakness in FNKAP and introduce specific attack methods. Second, we propose an enhanced security hierarchical key agreement scheme with noninteracting for WHMS based on pairings. Security proof and analysis illustrate our scheme enhances security strength of FNKAP, and it can resist the collusion attack. Moreover, theoretical analysis results show that our scheme is more efficient than Kim's work.

The rest of this paper is organized as follows. We formalize a basic system structure for WHMS in Section 2, and we also give the security model and define the adversary's ability in the same section. We simply highlight Kim's scheme [8] in Section 3. The weakness of Kim's scheme is discussed in Section 4. We detail our enhanced security hierarchical key agreement scheme against the security attacks in Section 5. We present the analysis of our improvements regarding correctness and security in Section 6. We compare our scheme with Kim's scheme in terms of functionality and performance in Section 7. Finally, this paper is concluded in Section 8.

2. Preliminaries

We first illustrate the basic system structure of WHMS in this section. Moreover, we introduce security threats, security model, bilinear group, and mathematic assumption, separately. Basic notations are provided in Notations section.

2.1. Basic Structure. As depicted in Figure 1, a typical hierarchical key agreement for WHMS involves five types of parties. They are, namely, the u-Health Server (SV), the physicians (PH), the patients (PA), the gateways (GW), and the sensor nodes (SN). There is a hierarchical permission structure from the u-Health Server SV to the physicians $PH_i, i \in (1, 2, \ldots)$, to the gateway $GW_{i,j}, i, j \in (1, 2, \ldots)$ of patients $PA_{i,j}$ and to the sensor node $SN_{i,j,d}, i, j, d \in (1, 2, \ldots)$ of the gateway $GW_{i,j}$ [7, 8].

As the root authority, SV is responsible for managing the entities' authorities in WHMS. SV produces the private keys of entities such as GW, SN, and PH. When PA wants to use WHMS, his/her GW and SN should agree on the session keys with the physician (PH), separately. Similarly, when PH wants to send a diagnostic report to PA, he/she should also agree on a session key with PA.

2.2. Security Threats. Kim assumes that the physicians are trusted in paper [8]. However, we point out that the scheme should take the risk of the physician's corruption because it is more practical. In practice, not all physicians are trusted all the time. For example, as reported, the staff of a famous hospital sold the patient's personal medical data in USA [7], and 500 patients' medical information may have been compromised at a medical center in LA because an employee's laptop was stolen [15].

Thus, we assume a security model in which the adversary has the following abilities. First, the adversary can totally control the channel. Therefore, the adversary can eavesdrop, intercept, modify, replay, or inject any data via the channel. Second, the adversary can compromise the secure information from the physicians except for the victim's current physician. Third, the adversary can also compromise several sensor nodes and gateways except for the victim's current GW and SN.

We aim to achieve the following security goals under the above security threats.

Key Agreement. Two entities establish a session key which is only known by specific entities.

Anonymity and Untraceability. The identities of GW and SN should be kept confidential from the adversary and cannot be traced by the adversary.

Resistance Passive and Active Attacks. The scheme is secure against the passive and active attacks.

2.3. Security Model. Inspired by the security model for a noninteractive hierarchical key agreement scheme [11] and the original Bellare-Rogaway key exchange model [16], the security model of our scheme is stated as follows.

Participants. We model the scheme participants as a finite set U of fixed size with each A being a Probabilistic Polynomial Time (PPT) turing machine. Each scheme participant $A \in U$ may execute a polynomial number of protocol instances in parallel. We will refer to sth instance of principal A communicating with peer B as $\prod_{A,B}^s$.

Adversary Model. The adversary $\mathscr{A}$ is modeled as a PPT Turing machine and can be given all public parameters of the system, and he/she can access the oracle by issuing some specified queries:

(i) Send($\prod_{A,B}^s, D$). The adversary $\mathscr{A}$ sends the message D to the session s executed by A communicating with B. Since our proposal is a noninteractive scheme, the query does not need to be responded to.

(ii) Establish(A). The adversary $\mathscr{A}$ names a node A and obtains all the secret values held by the node. Neither of the patient's gateway and sensor nodes named in the test query or any of their ancestors can be established.

(iii) Reveal($\prod_{A,B}^s$). If the query is achieved, the system returns the session key to the adversary $\mathscr{A}$. The session between the target patient's facilities (a gateway and sensor nodes) and the physician cannot be revealed.

(iv) Test($\prod_{A,B}^s$). Only one query of this form is allowed for the adversary $\mathscr{A}$. The adversary $\mathscr{A}$ names ID_A and ID_B and executes this query at any time. Then, a number sk is returned as follows. A bit b is chosen at random in $\{0, 1\}$. If $b = 1$ then the adversary gets the secret key shared between the two nodes, and if $b = 0$ it gets

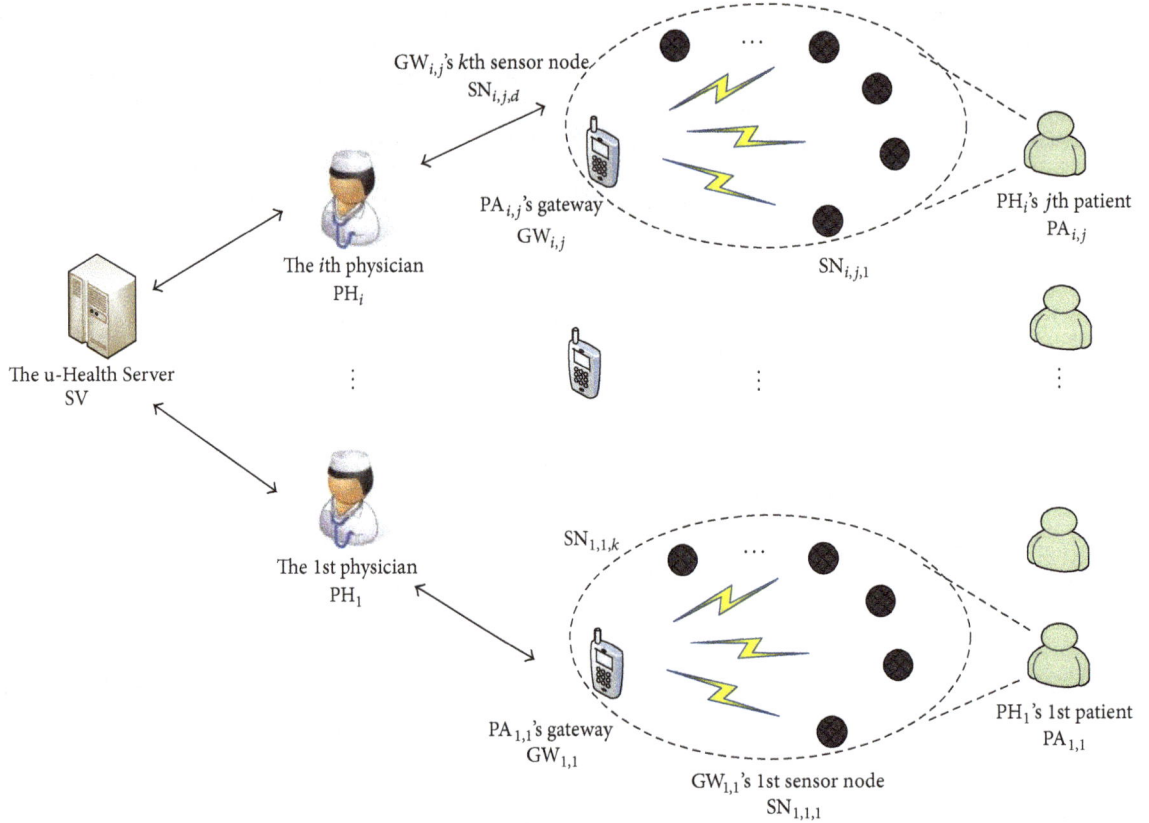

FIGURE 1: Basic hierarchical key agreement structure in WHMS.

a key chosen at random from the set of all possible shared keys.

Definition 1 (HKA-security). As a function of the security parameter k, we define the advantage $\text{Adv}_{\mathcal{A},\Sigma}^{\text{HKA}}$ of the PPT adversary $\mathcal{A}$ in an attacking scheme Σ as

$$\text{Adv}_{\mathcal{A},\Sigma}^{\text{HKA}} = \left| 2\text{Succ}_{\mathcal{A},\Sigma}^{\text{HKA}} - 1 \right|. \tag{1}$$

Here, $\text{Succ}_{\mathcal{A},\Sigma}^{\text{HKA}}$ is the probability that the adversary queries $\text{Test}(\prod_{A,B}^{s})$ and outputs a bit b^* such that b is used by the test query. We call a hierarchical key agreement scheme Σ to be HKA secure if for any PPT adversary the $\mathcal{A}$ function is negligible.

2.4. Bilinear Groups

Definition 2 (bilinear map). G_1 is an additive cyclic group of prime order q and G_2 is a multiplicative cyclic group of prime order p. The bilinear pairing is a map $\hat{e}: G_1 \times G_1 \to G_2$ with the following properties [17].

Bilinearity. For all $P, Q \in G_1$ and $a, b \in Z_q^*$, we have $\hat{e}(aP, bQ) = \hat{e}(P, Q)^{ab}$.

Nondegeneracy. The map does not send all pairs in $G_1 \times G_1$ to the identity in G_2.

Computability. There exists an efficient algorithm to compute $\hat{e}(P, Q)$ for all $P, Q \in G_1$.

2.5. Mathematic Assumption. The mathematic assumptions used in the paper are listed as follows.

Definition 3 (Elliptic Curve Discrete Logarithm Problem, ECDL problem). Suppose E is an elliptic curve over a finite field F_q. Given $Q, P \in E$ to find the $n \in Z_q^*$, $nP = Q$ is believed to be hard [18].

Definition 4 (Bilinear Diffie-Hellman Problem, BDH problem). BDH problem is defined as follows. There is a bilinear map $\hat{e}: G_1 \times G_1 \to G_2$. Given $(P, aP, bP, cP \in G_1)$ for $a, b, c \in Z_q^*$, to compute the $\hat{e}(P, P)^{abc} \in G_2$ is believed to be hard [17].

Definition 5 (Decisional Bilinear Diffie-Hellman Problem, DBDH problem). DBDH problem is defined as follows. There is a bilinear map $\hat{e}: G_1 \times G_1 \to G_2$. Given $(P, aP, bP, cP \in G_1)$ for $a, b, c, r \in Z_q^*$, to differentiate the $\hat{e}(P, P)^{abc} \in G_2$ and $\hat{e}(P, P)^r \in G_2$ is believed to be hard [17].

2.6. Notations. To provide a quick reference, the basic notations used in the paper are listed in Notations section.

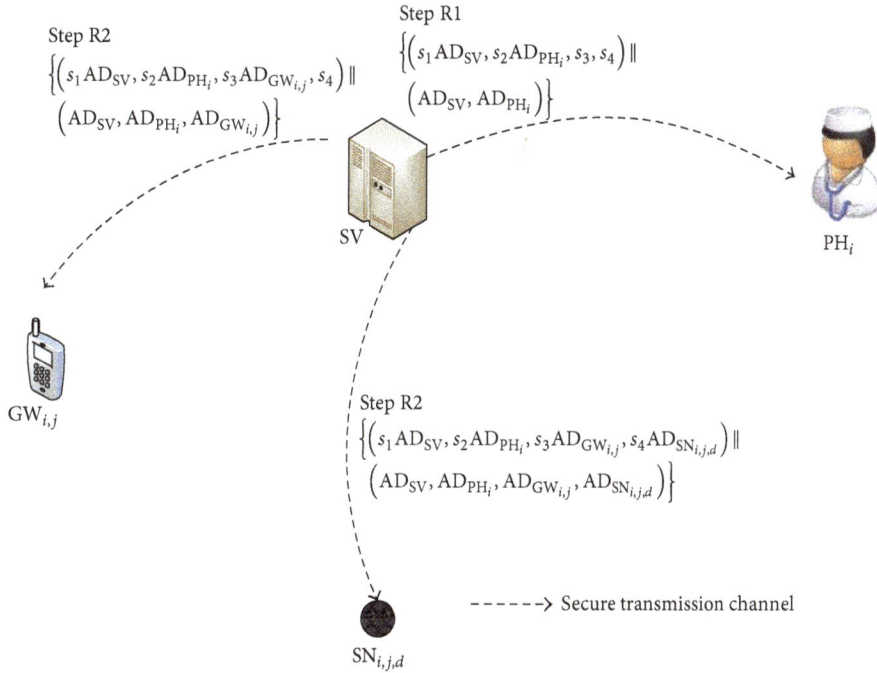

FIGURE 2: Physician and patient registration phase in Kim's paper [8].

3. Review of the Kim's Scheme

In this section, we briefly review Kim's key agreement scheme [8], which consists of three phases: System Initialization Phase, Physician and Patient Registration Phase, and Noninteractive Key Agreement and Secure Communication Phase.

3.1. System Initialization Phase. SV generates two groups G_1 and G_2 of prime order p with a bilinear map $\hat{e} : G_1 \times G_1 \rightarrow G_2$. Also, it chooses a cryptographic hash function $H : \{0, 1\}^* \rightarrow G_1$. After that, SV picks four random numbers $s_1, s_2, s_3, s_4 \leftarrow Z_q^*$ as the master private keys. Then, SV computes an amplified identity $\text{AD}_{\text{SV}} = H(\text{ID}_{\text{SV}})$ and a public key $s_1\text{AD}_{\text{SV}}$. Finally, SV keeps the master private keys and the amplified identity, securely.

3.2. Physician and Patient Registration Phase. Before providing service, the patient PA and his/her physician PH must register in SV. Here, the statement $A \Rightarrow B : M$ denotes that B receives a message M from A via a secure channel. The Physician and Patient Registration Phase is basically shown in Figure 2.

Step R1 ($\text{SV} \Rightarrow \text{PH}_i : \text{Reg}_{\text{PH}_i}$). When a physician PH_i wants to be a legal e-medical physician, he/she sends his/her identity ID_{PH_i} to SV via a secure channel. Then, SV validates the identity ID_{PH_i}. If the solution is positive, SV sends $\text{Reg}_{\text{PH}_i} = \{(s_1\text{AD}_{\text{SV}}, s_2\text{AD}_{\text{PH}_i}, s_3, s_4) \| (\text{AD}_{\text{SV}}, \text{AD}_{\text{PH}_i})\}$. Here, $\text{AD}_{\text{PH}_i} = H(\text{ID}_{\text{PH}_i})$. Finally, PH_i stores the received information, securely.

Step R2 ($\text{SV} \Rightarrow \text{GW}_{i,j} : \text{Reg}_{\text{GW}_{i,j}}$ and $\text{SV} \Rightarrow \text{SN}_{i,j,d} : \text{Reg}_{\text{SN}_{i,j,d}}$). When a patient $\text{PA}_{i,j}$ of PH_i wants to use the service

in the WHMS, he/she should register his/her gateway $\text{GW}_{i,j}$ and k sensor nodes $\text{SN}_{i,j,d}, 1 \le d \le k$ in SV. SV validates the identity ID_{PH_i} submitted by $\text{PA}_{i,j}$. If the solution is positive, SV receives the gateway's identity $\text{ID}_{\text{GW}_{i,j}}$ and the sensor nodes' identity $\text{ID}_{\text{SN}_{i,j,d}}, 1 \le d \le k$. Then, SV sends $\text{Reg}_{\text{GW}_{i,j}}$ and $\text{Reg}_{\text{SN}_{i,j,d}}$ to them via a secure channel. Here, $\text{Reg}_{\text{GW}_{i,j}}$ and $\text{Reg}_{\text{SN}_{i,j,d}}$ as follows:

$$\text{Reg}_{\text{GW}_{i,j}} = \left\{ \left(s_1\text{AD}_{\text{SV}}, s_2\text{AD}_{\text{PH}_i}, s_3\text{AD}_{\text{GW}_{i,j}}, s_4 \right) \| \right.$$

$$\left. \left(\text{AD}_{\text{SV}}, \text{AD}_{\text{PH}_i}, \text{AD}_{\text{GW}_{i,j}} \right) \right\},$$

$$\text{Reg}_{\text{SN}_{i,j,d}} \tag{2}$$

$$= \left\{ \left(s_1\text{AD}_{\text{SV}}, s_2\text{AD}_{\text{PH}_i}, s_3\text{AD}_{\text{GW}_{i,j}}, s_4\text{AD}_{\text{SN}_{i,j,d}} \right) \| \right.$$

$$\left. \left(\text{AD}_{\text{SV}}, \text{AD}_{\text{PH}_i}, \text{AD}_{\text{GW}_{i,j}}, \text{AD}_{\text{SN}_{i,j,d}} \right) \right\}.$$

Finally, $\text{GW}_{i,j}$ and $\text{SN}_{i,j,d}$ store their received information, securely.

3.3. Noninteractive Key Agreement and Secure Communication. In this phase, the sensor node $\text{SN}_{i,j,d}$ and the gateway $\text{GW}_{i,j}$ of the patient PA_i and the physician PH_i agree on a fresh session key for establishing a secure communication channel. Here, the statement $A \rightarrow B : M$ denotes that B receives a message M from A via a unsecure channel. The Noninteractive Key Agreement and Secure Communication is basically shown in Figure 3.

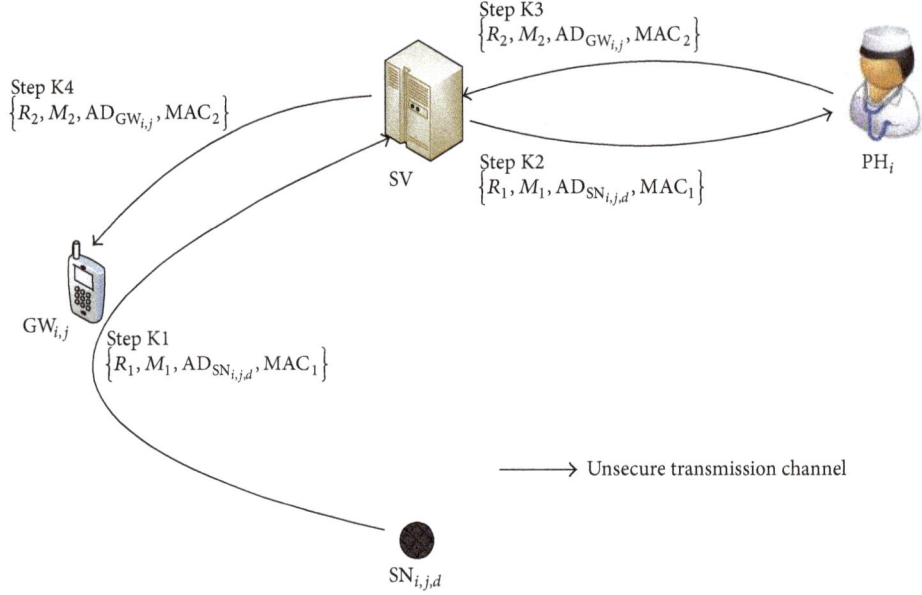

FIGURE 3: Noninteractive key agreement and secure communication in Kim's paper [8].

Step K1 ($\text{SN}_{i,j,d} \rightarrow \text{SV} : \{R_1, M_1, \text{AD}_{\text{SN}_{i,j,d}}, \text{MAC}_1\}$). $\text{SN}_{i,j,d}$ chooses a random number r_1 and computes $R_1 = r_1 \text{AD}_{\text{SN}_{i,j,d}}$. The fresh session key sk_1 is computed as follows:

$$
\text{sk}_1 = \hat{e}\left(s_1 \text{AD}_{\text{SV}}, \text{AD}_{\text{SV}}\right) \cdot \hat{e}\left(s_2 \text{AD}_{\text{PH}_i}, \text{AD}_{\text{PH}_i}\right)
$$
$$
\cdot \hat{e}\left(s_3 \text{AD}_{\text{GW}_{i,j}}, \text{AD}_{\text{PH}_i}\right) \tag{3}
$$
$$
\cdot \hat{e}\left(s_4 \text{AD}_{\text{SN}_{i,j,d}}, \text{AD}_{\text{PH}_i}\right)^{r_1}.
$$

Then, $\text{SN}_{i,j,d}$ computes $M_1 = E_{\text{sk}_1}(\text{Data}_i)$ and $\text{MAC}_1 = H(\text{sk}_1 \parallel R_1 \parallel M_1)$, where Data_i is the data collected by $\text{SN}_{i,j,d}$.

Step K2. When PH_i is authenticated by SV, he/she can check the data of the patient PA_i. PH_i computes the fresh session key sk_1' as follows:

$$
\text{sk}_1' = \hat{e}\left(s_1 \text{AD}_{\text{SV}}, \text{AD}_{\text{SV}}\right) \cdot \hat{e}\left(s_2 \text{AD}_{\text{PH}_i}, \text{AD}_{\text{PH}_i}\right)
$$
$$
\cdot \hat{e}\left(\text{AD}_{\text{PH}_i}, \text{AD}_{\text{GW}_{i,j}}\right)^{s_3} \cdot \hat{e}\left(\text{AD}_{\text{PH}_i}, R_1\right)^{s_4}. \tag{4}
$$

Then, PH_i computes $\text{MAC}_1' = H(\text{sk}_1' \parallel R_1 \parallel M_1)$. Only if MAC_1' is equal to MAC_1 does PH_i assure the correctness of sk_1'. Then, PH_i decrypts M_1 to get Data_i by using the key sk_1'.

Step K3 ($\text{PH}_i \rightarrow \text{GW}_{i,j} : \{R_2, M_2, \text{AD}_{\text{GW}_{i,j}}, \text{MAC}_2\}$). When PH_i wants to send the electronic health report to PA_i, he/she

chooses a random number r_2 and computes $R_2 = r_2 \text{AD}_{\text{PH}_i}$. PH_i computes the fresh session key sk_2 as follows:

$$
\text{sk}_2 = \hat{e}\left(s_1 \text{AD}_{\text{SV}}, \text{AD}_{\text{SV}}\right) \cdot \hat{e}\left(s_2 \text{AD}_{\text{PH}_i}, \text{AD}_{\text{PH}_i}\right)
$$
$$
\cdot \hat{e}\left(\text{AD}_{\text{PH}_i}, \text{AD}_{\text{GW}_{i,j}}\right)^{s_3} \tag{5}
$$
$$
\cdot \hat{e}\left(\text{AD}_{\text{PH}_i}, \text{AD}_{\text{GW}_{i,j}}\right)^{s_4 r_2}.
$$

In addition, PH_i computes $M_2 = E_{\text{sk}_2}(\text{Data}_i)$ and $\text{MAC}_2 = H(\text{sk}_2 \parallel R_2 \parallel M_2)$. Here, Data_i is the electronic health report composed by PH_i.

Step K4. When $\text{GW}_{i,j}$ is authenticated by SV, he/she can receive the report of the patient PA_i from SV. $\text{GW}_{i,j}$ computes the fresh session key sk_2' as follows:

$$
\text{sk}_2' = \hat{e}\left(s_1 \text{AD}_{\text{SV}}, \text{AD}_{\text{SV}}\right) \cdot \hat{e}\left(s_2 \text{AD}_{\text{PH}_i}, \text{AD}_{\text{PH}_i}\right)
$$
$$
\cdot \hat{e}\left(s_3 \text{AD}_{\text{GW}_{i,j}}, \text{AD}_{\text{PH}_i}\right) \cdot \hat{e}\left(\text{AD}_{\text{GW}_{i,j}}, R_2\right)^{s_4}. \tag{6}
$$

Then, $\text{GW}_{i,j}$ computes $\text{MAC}_2' = H(\text{sk}_2' \parallel R_2 \parallel M_2)$. Only if MAC_2' is equal to MAC_2 does $\text{GW}_{i,j}$ assure the correctness of sk_2'. Then, $\text{GW}_{i,j}$ decrypts M_2 to get Data_i by using the key sk_2'.

4. Security Analysis of Kim's Scheme

The author of [8] proposed a noninteractive key agreement scheme for freshness-preserving in WHMS. Under our security model, there is a weakness in the scheme as explained in the following section.

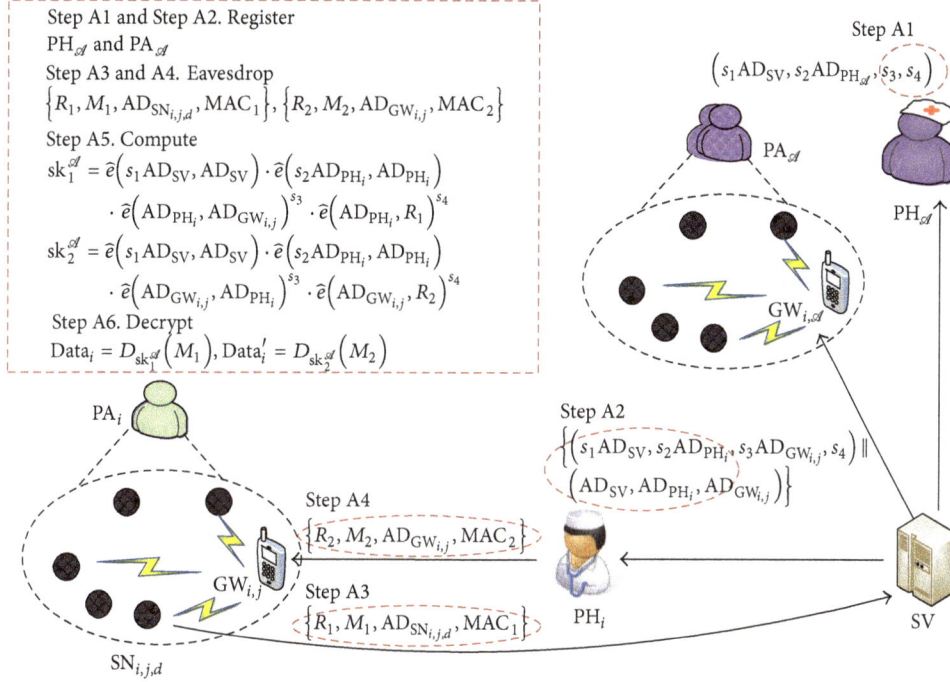

FIGURE 4: Security against collusion attack.

4.1. Security against Collusion Attack.

We now demonstrate that Kim's scheme is vulnerable to the collusion attack as claimed. One adversary $\mathscr{A}_1$ has registered as a legal physician $\text{PH}_{\mathscr{A}}$, and the other adversary $\mathscr{A}_2$ has registered as a normal patient $\text{PA}_{\mathscr{A}}$, as shown in Figure 4. The adversaries can obtain the electronic health data of any patient who is diagnosed by the same physician PH_i with the adversary $\mathscr{A}_2$. The adversaries attack a patient $\text{PA}_{i,j}$ as follows.

Step A1. Assume that $\mathscr{A}_1$ is an attacker who has registered as a physician $\text{PH}_{\mathscr{A}}$ in SV, and then he/she can legally receive a private key set $(s_1\text{AD}_{\text{SV}}, s_2\text{AD}_{\text{PH}_{\mathscr{A}}}, s_3, s_4)$ from SV (Step R1). Then, $\mathscr{A}_1$ sends a part of private key set $(s_1\text{AD}_{\text{SV}}, s_2\text{AD}_{\text{PH}_{\mathscr{A}}}, s_3, s_4)$ to $\mathscr{A}_2$.

Step A2. $\mathscr{A}_2$ is an adversary who has registered as a patient of the physician PH_i. He/she can legally receive a secure data set $(\text{AD}_{\text{SV}}, \text{AD}_{\text{PH}_i}, \text{AD}_{\text{GW}_{i,\mathscr{A}}})$ and a private key set $(s_1\text{AD}_{\text{SV}}, s_2\text{AD}_{\text{PH}_i}, s_3\text{AD}_{\text{GW}_{i,\mathscr{A}}}, s_4)$ of his/her gateway $\text{GW}_{i,\mathscr{A}}$ from SV (Step R2).

Step A3. Suppose $\text{SN}_{i,j,d}$ is a victim PA's smart node that sends information through the gateway $\text{GW}_{i,j}$. PA is diagnosed by the same physician PH_i with $\mathscr{A}_2$. When $\text{SN}_{i,j,d}$ runs the Step K1, an adversary can intercept the data $\{R_1, M_1, \text{AD}_{\text{SN}_{i,j,d}}, \text{MAC}_1\}$ because the communications are unsecure between $\text{SN}_{i,j,d}$ and SV.

Step A4. When PH_i sends the electronic health report to PA_i at the Step K3, an adversary can intercept data $\{R_2, M_2, \text{AD}_{\text{GW}_{i,j}}, \text{MAC}_2\}$ because the communications are also unsecure between PH_i and SV.

Step A5. $\mathscr{A}_2$ can compute the session key after the above steps. $\mathscr{A}_2$ receives (s_3, s_4) from $\mathscr{A}_1$ at Step A1. Then, he/she gets $(s_1\text{AD}_{\text{SV}}, s_2\text{AD}_{\text{PH}_i})$ and $(\text{AD}_{\text{SV}}, \text{AD}_{\text{PH}_i})$ at Step A2. Moreover, the information $\{R_1, M_1, \text{AD}_{\text{SN}_{i,j,d}}, \text{MAC}_1\}$ and $\{R_2, M_2, \text{AD}_{\text{GW}_{i,j}}, \text{MAC}_2\}$ is intercepted at Steps A3 and A4, separately. Therefore, $\mathscr{A}_2$ can compute the same session keys sk_1 and sk_2 as follows:

$$
\begin{aligned}
\text{sk}_1^{\mathscr{A}} &= \widehat{e}\left(s_1\text{AD}_{\text{SV}}, \text{AD}_{\text{SV}}\right) \cdot \widehat{e}\left(s_2\text{AD}_{\text{PH}_i}, \text{AD}_{\text{PH}_i}\right) \\
&\quad \cdot \widehat{e}\left(\text{AD}_{\text{GW}_{i,j}}, \text{AD}_{\text{PH}_i}\right)^{s_3} \cdot \widehat{e}\left(\text{AD}_{\text{PH}_i}, R_1\right)^{s_4} \\
&= \widehat{e}\left(s_1\text{AD}_{\text{SV}}, \text{AD}_{\text{SV}}\right) \cdot \widehat{e}\left(s_2\text{AD}_{\text{PH}_i}, \text{AD}_{\text{PH}_i}\right) \\
&\quad \cdot \widehat{e}\left(s_3\text{AD}_{\text{GW}_{i,j}}, \text{AD}_{\text{PH}_i}\right) \cdot \widehat{e}\left(s_4\text{AD}_{\text{SN}_{i,j,d}}, \text{AD}_{\text{PH}_i}\right)^{r_1} \\
&= \text{sk}_1, \\
\text{sk}_2^{\mathscr{A}} &= \widehat{e}\left(s_1\text{AD}_{\text{SV}}, \text{AD}_{\text{SV}}\right) \cdot \widehat{e}\left(s_2\text{AD}_{\text{PH}_i}, \text{AD}_{\text{PH}_i}\right) \\
&\quad \cdot \widehat{e}\left(\text{AD}_{\text{PH}_i}, \text{AD}_{\text{GW}_{i,j}}\right)^{s_3} \cdot \widehat{e}\left(\text{AD}_{\text{GW}_{i,j}}, R_2\right)^{s_4} \\
&= \widehat{e}\left(s_1\text{AD}_{\text{SV}}, \text{AD}_{\text{SV}}\right) \cdot \widehat{e}\left(s_2\text{AD}_{\text{PH}_i}, \text{AD}_{\text{PH}_i}\right) \\
&\quad \cdot \widehat{e}\left(\text{AD}_{\text{PH}_i}, \text{AD}_{\text{GW}_{i,j}}\right)^{s_3} \cdot \widehat{e}\left(\text{AD}_{\text{GW}_{i,j}}, \text{AD}_{\text{PH}_i}\right)^{s_4 r_2} \\
&= \text{sk}_2.
\end{aligned}
\tag{7}
$$

Step A6. $\mathscr{A}_2$ decrypts M_1 and M_2 to obtain the victim's medical information using the session keys $\text{sk}_1^{\mathscr{A}}$ and $\text{sk}_2^{\mathscr{A}}$, respectively.

5. Our Proposed Scheme

In this section, we propose an improved scheme that can overcome the flaw of Kim's scheme in Section 4. Our scheme construction is inspired by the practical noninteractive key distribution scheme in [12] and Kim's paper [8]. Our scheme consists of four operational phases: Setup Phase, Key Generation Phase, Key Agreement from SN to PH Phase, and Key Agreement from PH to GW Phase. The details of our scheme are described as follows.

5.1. Setup Phase. In this phase, the u-Health Server SV, as the Private Key Generator (PKG), takes as inputs a security parameter k and the maximal number of the physicians N. Then, SV outputs the system public parameters params and the master private key sets sk. SV publishes params and keeps sk private.

Similar to the identity-based cryptography scheme, SV generates two groups G_1 and G_2 of prime order p with a bilinear map $\hat{e} : G_1 \times G_1 \rightarrow G_2$. However, it chooses three cryptographic hash functions $H_1 : \{0,1\}^* \rightarrow G_1$, $H_2 : G_2 \times G_1^3 \rightarrow \{0,1\}^k$, and $H_3 : \{0,1\}^* \times G_1^3 \rightarrow \{0,1\}^k$. After that, SV generates $3 + N$ random numbers $\{s_1, s_2, s_3, sr_i \leftarrow Z_q^* \mid i \in (1,2,\ldots,N)\}$ and selects a random generator $P_0 \in G_1$. Finally, SV keeps the master key $\text{sk} = \{s_1, s_2, s_3, sr_i \mid i \in (1,2,\ldots,N)\}$ secret and publishes params $= (q, G_1, G_2, P_0, \hat{e}, H_1, H_2, H_3, s_1P_0, s_2P_0, s_3P_0, s_1sr_iP_0, s_2sr_iP_0, s_3sr_iP_0 \mid i \in (1,2,\ldots,N))$. Here, P_0 is used to verify the correctness of the secret key sets.

It is important to note that although our proposal increase the storage space because of the values sr_i, $i \in [1,N]$, there is a one-to-one mapping between a physician PH_i and a value sr_i. In addition, the list of physicians must be stored in SV. Thus, we can use the mapping to reduce the storage space. For instance, SV gets the list of the registration physicians. Then, SV chooses a secret hash function $H^* : \{0,1\}^q \rightarrow Z_q^*$ and a random value sr_0. Finally, SV can compute the i times hash function $H^*(\cdot)$ to get the secret value $sr_i = H^*(\cdots(H^*(sr_0)))$. In this way, SV only needs to store the selected hash function and initial value sr_0, secretly. On one hand, the proposal can save the storage resources by using the hash function. On the other hand, it increases the consumption of the computing resources. In order to balance the computing cost and the storage space, SV can store not only the initial value sr_0, but also some intermediate random values sr_j. We introduce the scheme by using the secret values sr_i, $i \in [1,N]$ to help the analysis.

5.2. Key Generation Phase. In this phase, SV takes the identity ID_{PH_i} as an input and outputs a secret key set sk_{PH_i}. Moreover, PH_i takes his/her secret key set sk_{PH_i} and the identities $\text{ID}_{\text{GW}_{i,j}}$ and $\text{ID}_{\text{SN}_{i,j,d}}$ as inputs and outputs two secret key sets $\text{sk}_{\text{GW}_{i,j}}$ and $\text{sk}_{\text{SN}_{i,j,d}}$, separately.

Step 1. A physician PH_i, $i \in (1,2,\ldots,N)$ submits his/her identity ID_{PH_i} to SV for registration. If the identity of PH_i is validated, Then, SV computes $\text{AD}_{\text{SV}_i} = H_1(\text{ID}_{\text{SV}} \parallel sr_i)$, $\text{AD}_{\text{PH}_i} = H_1(\text{ID}_{\text{PH}_i} \parallel sr_i)$, and a private key set of PH_i, $i \in (1,2,\ldots,N)$ as follows:

$$\text{sk}_{\text{PH}_i} = \left(s_1 sr_i \text{AD}_{\text{PH}_i}, s_2 sr_i, s_3 sr_i\right). \tag{8}$$

In addition, SV packs a data package containing a private key set and two amplified identities $\{\text{sk}_{\text{PH}_i} \parallel (\text{AD}_{\text{SV}_i}, \text{AD}_{\text{PH}_i})\}$ and delivers the data package to PH_i via a secure channel. Here, the secure channel could be a smart card passed by a trusted person. Finally, PH_i keeps the received information, securely.

Step 2. When a patient $\text{PA}_{i,j}$ goes to see a doctor in a real clinic, they decide to use the WHMS to monitor his/her health directed by a physician PH_i. The patient submits his/her identity $\text{ID}_{\text{PA}_{i,j}}$ and the identity of gateway $\text{ID}_{\text{GW}_{i,j}}$ and sensor nodes $\text{ID}_{\text{SN}_{i,j,d}}$ to PH_i for registration. If the patients' identity is validated, PH_i generates a random value $a_{i,j} \in Z_q^*$ and computes the amplified identities $\text{AD}_{\text{GW}_{i,j}} = H_1(\text{ID}_{\text{GW}_{i,j}} \parallel a_{i,j})$ and $\text{AD}_{\text{SN}_{i,j,d}} = H_1(\text{ID}_{\text{SN}_{i,j,d}} \parallel a_{i,j})$. PH_i computes the private key sets of $\text{GW}_{i,j}$ and $\text{SN}_{i,j,d}$ as follows:

$$\text{sk}_{\text{GW}_{i,j}} = \left(s_1 sr_i \text{AD}_{\text{PH}_i}, s_2 sr_i \text{AD}_{\text{GW}_{i,j}}\right),$$
$$\text{sk}_{\text{SN}_{i,j,d}} = \left(s_1 sr_i \text{AD}_{\text{PH}_i}, s_3 sr_i \text{AD}_{\text{SN}_{i,j,d}}\right). \tag{9}$$

Next, PH_i packs a data package containing a private key set and three amplified identities $\{\text{sk}_{\text{GW}_{i,j}} \parallel (\text{AD}_{\text{SV}_i}, \text{AD}_{\text{PH}_i}, \text{AD}_{\text{GW}_{i,j}})\}$ and delivers the data package to $\text{GW}_{i,j}$ via a secure channel. Furthermore, SV packs a data package containing a private key set $\{\text{sk}_{\text{SN}_{i,j,d}} \parallel (\text{AD}_{\text{SV}_i}, \text{AD}_{\text{PH}_i}, \text{AD}_{\text{GW}_{i,j}}, \text{AD}_{\text{SN}_{i,j,d}})\}$ and delivers it to $\text{SN}_{i,j,d}$ via a secure channel. Finally, $\text{GW}_{i,j}$ and $\text{SN}_{i,j,d}$ store their received information in a secure area, respectively.

5.3. Key Agreement from SN to PH Phase. In this phase, a sensor node $\text{SN}_{i,j,d}$ of the patient $\text{PA}_{i,j}$ makes a connection with the physician PH_i. The sensor node $\text{SN}_{i,j,d}$ and the physician PH_i achieve a key agreement.

Step 1. When a sensor node $\text{SN}_{i,j,d}$ wants to upload the patient's medical data, $\text{SN}_{i,j,d}$ chooses a random number r_1 and computes $R_1 = r_1 \text{AD}_{\text{SN}_{i,j,d}}$ using its amplified identity $\text{AD}_{\text{SN}_{i,j,d}}$. The session key sk_1 is calculated as follows:

$$K_1 = \hat{e}\left(\text{AD}_{\text{SN}_{i,j,d}}, s_1 sr_i \text{AD}_{\text{PH}_i}\right)$$
$$\cdot \hat{e}\left(s_3 sr_i \text{AD}_{\text{SN}_{i,j,d}}, \text{AD}_{\text{PH}_i}\right)^{r_1}, \tag{10}$$
$$\text{sk}_1 = H_2\left(K_1 \parallel R_1 \parallel \text{AD}_{\text{SN}_{i,j,d}} \parallel \text{AD}_{\text{PH}_i}\right).$$

Then, $\text{SN}_{i,j,d}$ computes $M_1 = E_{\text{sk}_1}(\text{Data}_{i,j,d})$ and $V_1 = H_3(\text{sk}_1 \parallel M_1 \parallel T_1 \parallel R_1 \parallel \text{AD}_{\text{SN}_{i,j,d}} \parallel \text{AD}_{\text{PH}_i})$. Here, $\text{Data}_{i,j,d}$ is the data collected by $\text{SN}_{i,j,d}$ and T_1 is a current timestamp. Finally, $SN_{i,j,d}$ sends a message package $D_1 = \{R_1, AD_{SN_{i,j,d}}, AD_{PH_i}, T_1, M_1, V_1\}$ to SV.

Step 2. After receiving the data package D_1, SV verifies the timestamp T_1 whether it is within the valid time for communication. If it is invalid, the key agreement terminates. Otherwise, it can assure the package by judging $V_1^* = V_1$ as follows:

$$K_1^* = \hat{e}\left(AD_{SN_{i,j,d}}, AD_{PH_i}\right)^{s_1 sr_i} \cdot \hat{e}\left(R_1, AD_{PH_i}\right)^{s_3 sr_i},$$

$$sk_1^* = H_2\left(K_1^* \parallel R_1 \parallel AD_{SN_{i,j,d}} \parallel AD_{PH_i}\right), \tag{11}$$

$$V_1^* = H_3\left(sk_1^* \parallel M_1 \parallel T_1 \parallel R_1 \parallel AD_{SN_{i,j,d}} \parallel AD_{PH_i}\right).$$

Only if V_1^* is equal to V_1 included in D_1 does SV assure the source of package from a sensor node $AD_{SN_{i,j,d}}$ and send a notice to PH_i. Finally, SV store the package D_1 in its database.

Step 3. When PH_i is authenticated by SV, he/she can check the data of a sensor node $SN_{i,j,d}$. PH_i computes the fresh session key sk_1' as follows:

$$K_1' = \hat{e}\left(AD_{SN_{i,j,d}}, s_1 sr_i AD_{PH_i}\right) \cdot \hat{e}\left(R_1, AD_{PH_i}\right)^{s_3 sr_i},$$

$$sk_1' = H_2\left(K_1' \parallel R_1 \parallel AD_{SN_{i,j,d}} \parallel AD_{PH_i}\right). \tag{12}$$

In addition, PH_i computes $V_1' = H_3(sk_1' \parallel M_1 \parallel T_1 \parallel R_1 \parallel AD_{SN_{i,j,d}} \parallel AD_{PH_i})$ by using the information of D_1. Only if V_1' is equal to V_1 does PH_i assure the correctness of sk_1' and decrypt M_1 to get $Data_{i,j,d}$ by using the key sk_1'.

5.4. Key Agreement from PH to GW Phase. In this phase, the physician PH_i makes a connection with a patient's gateway $GW_{i,j}$, and they agree on a fresh session key for communication.

Step 1. When PH_i wants to communicate with $PA_{i,j}$ such as sending the electronic health report, he/she chooses a random number r_2 and computes $R_2 = r_2 AD_{PH_i}$. PH_i computes the fresh session key sk_2 as follows:

$$K_2 = \hat{e}\left(AD_{GW_{i,j}}, s_1 sr_i AD_{PH_i}\right) \cdot \hat{e}\left(AD_{GW_{i,j}}, AD_{PH_i}\right)^{s_2 sr_i r_2}, \tag{13}$$

$$sk_2 = H_2\left(K_2 \parallel R_2 \parallel AD_{PH_i} \parallel AD_{GW_{i,j}}\right).$$

In addition, PH_i computes $M_2 = E_{sk_2}(Data_{i,j})$ and $V_2 = H_3(sk_2 \parallel M_2 \parallel T_2 \parallel R_2 \parallel AD_{PH_i} \parallel AD_{GW_{i,j}})$. Here, $Data_{i,j}$ is the electronic health report composed by PH_i, and T_2 is a current timestamp. Finally, PH_i sends a message package $D_2 = \{R_2, AD_{PH_i}, AD_{GW_{i,j}}, T_2, M_2, V_2\}$ to SV.

Step 2. After receiving the data package D_2, SV checks the validity of the timestamp T_2. If it has grown stale, SV quits

the session. Otherwise, SV can assure the package by judging $V_2^* = V_2$ as follows:

$$K_2^* = \hat{e}\left(AD_{GW_{i,j}}, AD_{PH_i}\right)^{s_1 sr_i} \cdot \hat{e}\left(AD_{GW_{i,j}}, R_2\right)^{s_2 sr_i},$$

$$sk_2^* = H_2\left(K_2^* \parallel R_2 \parallel AD_{PH_i} \parallel AD_{GW_{i,j}}\right), \tag{14}$$

$$V_2^* = H_3\left(sk_2^* \parallel M_2 \parallel T_2 \parallel R_2 \parallel AD_{PH_i} \parallel AD_{GW_{i,j}}\right).$$

Only if V_2^* is equal to V_2 included in D_2 does SV assure the source of package from a physician AD_{PH_i} and send a notice to $PA_{i,j}$. Finally, SV stores the package D_2 in its database.

Step 3. When $GW_{i,j}$ is authenticated by SV, he/she can get the report of the patient $PA_{i,j}$ from SV. $GW_{i,j}$ computes the fresh session key sk_2' as follows:

$$K_2' = \hat{e}\left(AD_{GW_{i,j}}, s_1 sr_i AD_{PH_i}\right) \cdot \hat{e}\left(s_2 sr_i AD_{GW_{i,j}}, R_2\right), \tag{15}$$

$$sk_2' = H_2\left(K_2' \parallel R_2 \parallel AD_{PH_i} \parallel AD_{GW_{i,j}}\right).$$

Then, $GW_{i,j}$ computes $V_2' = H_3(sk_2' \parallel M_2 \parallel T_2 \parallel R_2 \parallel AD_{PH_i} \parallel AD_{GW_{i,j}})$. Only if V_2' is equal to V_2 does $GW_{i,j}$ assure the correctness of sk_2' and decrypt M_2 to get $Data_{i,j}$ by using the key sk_2'.

6. Correctness and Security

In this section, we present the correctness of our improved scheme. Then, we illustrate that our enhanced key agreement scheme can overcome the two security weaknesses of security analysis of FNKAP by security analysis.

6.1. Correctness. We verify the correctness of key agreement in our scheme as follows:

$$K_1 = \hat{e}\left(AD_{SN_{i,j,d}}, s_1 sr_i AD_{PH_i}\right)$$

$$\cdot \hat{e}\left(s_3 sr_i AD_{SN_{i,j,d}}, AD_{PH_i}\right)^{r_1}$$

$$= \hat{e}\left(AD_{SN_{i,j,d}}, AD_{PH_i}\right)^{s_1 sr_i}$$

$$\cdot \hat{e}\left(r_1 AD_{SN_{i,j,d}}, AD_{PH_i}\right)^{s_3 sr_i}$$

$$= \hat{e}\left(AD_{SN_{i,j,d}}, AD_{PH_i}\right)^{s_1 sr_i} \cdot \hat{e}\left(R_1, AD_{PH_i}\right)^{s_3 sr_i} \tag{16}$$

$$= K_1^*$$

$$= \hat{e}\left(AD_{SN_{i,j,d}}, s_1 sr_i AD_{PH_i}\right) \cdot \hat{e}\left(R_1, AD_{PH_i}\right)^{s_3 sr_i}$$

$$= K_1'.$$

Thus, the agreed session keys sk_1, sk_1^*, and sk_1' computed by PH_i, SV, and $SN_{i,j,d}$ are equal. The same as above, we prove that sk_2 is equal to sk_2' because K_2 is equal to K_2^* and K_2':

$$
\begin{aligned}
K_2 &= \hat{e}\left(AD_{GW_{i,j}}, s_1 sr_i AD_{PH_i}\right) \\
&\quad \cdot \hat{e}\left(AD_{GW_{i,j}}, AD_{PH_i}\right)^{s_2 sr_i r_2} \\
&= \hat{e}\left(AD_{GW_{i,j}}, AD_{PH_i}\right)^{s_1 sr_i} \\
&\quad \cdot \hat{e}\left(AD_{GW_{i,j}}, r_2 AD_{PH_i}\right)^{s_2 sr_i} \\
&= \hat{e}\left(AD_{GW_{i,j}}, AD_{PH_i}\right)^{s_1 sr_i} \cdot \hat{e}\left(AD_{GW_{i,j}}, R_2\right)^{s_2 sr_i} \\
&= K_2^* \\
&= \hat{e}\left(AD_{GW_{i,j}}, s_1 sr_i AD_{PH_i}\right) \cdot \hat{e}\left(s_2 sr_i AD_{GW_{i,j}}, R_2\right) \\
&= K_2'.
\end{aligned}
\tag{17}
$$

6.2. Security Proof. In the following, we will show that our scheme is provably secure under DBDH assumption in the random oracle model. We treat H_1, H_2, and H_3 as three random oracles.

Theorem 6. *Let G_1 and G_2 be two groups of order q and $\hat{e}$ be a bilinear mapping that together satisfy the DBDH assumption. Let the hash functions H_1, H_2, and H_3 used in the scheme be modeled as the random oracles. Suppose that the DBDH assumption holds; the proposed scheme is a secure key agreement in our security model.*

Proof. Suppose an adversary $\mathscr{A}$ is an attack algorithm that breaks our scheme in the probability ϵ; we will show how to use the ability of $\mathscr{A}$ to build an algorithm $\mathscr{B}$ that solves the DBDH assumption with probability of at least ϵ'. Thus, $\mathscr{A}$'s advantage must be negligible because the DBDH assumption holds. □

We refer to $\mathscr{B}$ as "the simulator" because it simulates a real attacking environment for $\mathscr{A}$. $\mathscr{B}$ is initialized with the DBDH parameters $\{G_1, G_2, \hat{e}, q\}$ and the points $\{P, aP, bP, cP \in G_1, D, R \in G_2\}$, $D = \hat{e}(P, P)^{abc}$, and $R = \hat{e}(P, P)^r$. The idea of the proof is that $\mathscr{B}$ will embed the DBDH problem into the queries issued by $\mathscr{A}$. Since the hash function H_2 is modeled as random oracle, after the adversary issues the test query, it has only two unneglected cases to distinguish the tested session key sk_1 or sk_2 from a random string.

Case 1 (key-replication attack). The adversary $\mathscr{A}$ forces a nonmatching session to have the same session key with the Test($\prod_{A,B}^s$). In this case, the adversary $\mathscr{A}$ can get the session key by querying the nonmatching session. However, the input of hash function H_2 includes the entities' identities and the random nonce. Furthermore, they and a timestamp are integrally protected by H_3. For example, in Step 1, the session key $sk_1 = H_2(K_1 \parallel R_1 \parallel AD_{SN_{i,j,d}} \parallel AD_{PH_i})$ includes

the identities AD_{PH_i} and $AD_{SN_{i,j,d}}$ and the random nonce R_1. The certification value $V_1 = H_3(sk_1 \parallel M_1 \parallel T_1 \parallel R_1 \parallel AD_{SN_{i,j,d}} \parallel AD_{PH_i})$ includes them and the timestamp T_1. Therefore, two nonmatching sessions cannot have the same values and when H_2 and H_3 are modeled as a random oracle, the success probability of key-replication attack is negligible.

Case 2 (forging attack). The adversary $\mathscr{A}$ queries H_2 on the value $H_2(K_1 \parallel R_1 \parallel AD_{SN_{i,j,d}} \parallel AD_{PH_i})$ or $H_2(K_2 \parallel R_2 \parallel AD_{PH_i} \parallel AD_{GW_{i,j}})$ in the test query. Obviously, in this case the adversary $\mathscr{A}$ can compute the value K_1 or K_2 by itself.

In the following, we mainly analyze the Case 2 forging attack. A simulator $\mathscr{B}$ is interested in using the $\mathscr{A}$ to turn $\mathscr{A}$'s advantage in distinguishing the tested session key from a random string into an advantage in solving the DBDH problem. During the game, $\mathscr{B}$ has to answer all queries of the $\mathscr{A}$.

Setup. $\mathscr{B}$ simulates the *Setup* algorithm as follows. $\mathscr{B}$ starts by choosing security and public parameters for our scheme using its input DBDH parameters $\{G_1, G_2, \hat{e}, q\}$ and $\{P, aP, bP, cP \in G_1, Q = D \text{ or } R, D = \hat{e}(P, P)^{abc}, R = \hat{e}(P, P)^r\}$. $\mathscr{B}$ also chooses a random master key set $\{s_1, s_2, s_3, sr_i \mid i \in (1, \ldots, N)\}$ from Z_q^* as the PKG would do. Using these keys, $\mathscr{B}$ sets the random generator $P_0 = P$, and then SV's public parameters are params $= \{q, G_1, G_2, P, \hat{e}, H_1, H_2, H_3, cP, s_2P, s_3P, sr_icP, s_2sr_iP, s_3sr_iP \mid i \in (1, \ldots, N)\}$. $\mathscr{B}$ invokes the adversary $\mathscr{A}$, providing it with the public parameters params. Note that the DBDH parameters cP have been embedded in the game and the simulator $\mathscr{B}$ has no idea about the value c. With the probability at least $1/n(k)^2$, $\mathscr{B}$ guesses the adversary $\mathscr{A}$ will select one patient $ID_{PA_{i,j}}$ and his/her physician ID_{PH_i}. With the probability at least $1/s(k)$, $\mathscr{B}$ guesses the adversary $\mathscr{A}$ will select the session s as test session.

Queries. When the adversary $\mathscr{A}$ makes his/her queries, the simulator $\mathscr{B}$ answers the queries in arbitrary order as follows. Note that ID_{PH_i}, $ID_{GW_{i,j}}$, and $ID_{SN_{i,j,d}}$ are the guessed victims physician and devices.

$H_1(\cdot)$. In order to enhance simulation's fidelity, $\mathscr{B}$ maintains an initially empty list H_1^{list} of tuples $(ID_A, R_A, u_A, h_{A,R_A}) \in \{0, 1\}^* \times G_1 \times Z_q^* \times \{0, 1\}^k$. When $\mathscr{A}$ queries the oracle H_1 as an input $(ID_A \parallel R_A)$, $\mathscr{B}$ responds to the query in the following way.

(i) $\mathscr{B}$ checks the list H_1^{list}; if (ID_A and R_A) are already there, then $\mathscr{B}$ responds with stored value h_{A,R_A}.

(ii) Otherwise, if $ID_A = ID_{PH_i}$ and $R_A = sr_i$, $\mathscr{B}$ randomly chooses $u_{PH_i} \in Z_q^*$, and it computes $h_{PH_i,sr_i} = u_{PH_i}aP$. Then, it inserts $(ID_{PH_i}, sr_i, u_{PH_i}, h_{PH_i,sr_i})$ into the H_1^{list}. Finally, it responds with $H_1(ID_{PH_i} \parallel sr_i) = h_{PH_i,sr_i}$.

(iii) Otherwise, if $ID_A = ID_{GW_{i,j}}$, $\mathscr{B}$ randomly chooses $u_{GW_{i,j}} \in Z_q^*$ and computes the value $h_{GW_{i,j},a_{i,j}} = u_{GW_{i,j}}bP$. Then, it inserts $(ID_{GW_{i,j}}, a_{i,j}, u_{GW_{i,j}}, h_{GW_{i,j},a_{i,j}})$

into the H_1^{list}. Here, $a_{i,j} = R_A$. Finally, it responds with $H_1(\text{ID}_{\text{GW}_{i,j}} \| a_{i,j}) = h_{\text{GW}_{i,j},a_{i,j}} = u_{\text{GW}_{i,j}}bP$.

(iv) Otherwise, if $\text{ID}_A = \text{ID}_{\text{SN}_{i,j,d}}$, $\mathscr{B}$ randomly chooses $u_{\text{SN}_{i,j,d}} \in Z_q^*$ and computes the value $h_{\text{SN}_{i,j,d},a_{i,j}} = u_{\text{SN}_{i,j,d}}bP$. Then, it inserts $(\text{ID}_{\text{SN}_{i,j,d}}, a_{i,j}, u_{\text{SN}_{i,j,d}}, h_{\text{SN}_{i,j,d},a_{i,j}})$ into the H_1^{list}. Here, $a_{i,j} = R_A$. Finally, it responds with $H_1(\text{ID}_{\text{GW}_{i,j}} \| a_{i,j}) = h_{\text{GW}_{i,j},a_{i,j}} = u_{\text{SN}_{i,j,d}}bP$.

(v) Otherwise, $\mathscr{B}$ randomly chooses $u_A \in Z_q^*$, computes $h_{A,R_A} = u_A P$, and inserts $(\text{ID}_A, R_A, u_A, h_{A,R_A})$ in the list. Finally, it responds with $H_1(\text{ID}_A \| R_A) = h_{A,R_A} = u_A P$.

$H_2(\cdot)$. The simulator $\mathscr{B}$ maintains an initially empty list H_2^{list} with entries of the form $(K_{AB}, R_{AB}, \text{AD}_A, \text{AD}_B, h_{AB,R_{AB}}) \in G_2 \times G_1^3 \times \{0,1\}^k$. When $\mathscr{A}$ queries the oracle H_2 as a input $(K_{AB}, R_{AB}, \text{AD}_A, \text{AD}_B)$, the simulator $\mathscr{B}$ responds to the query in the following way.

(i) $\mathscr{B}$ checks the list H_2^{list}; if $(K_{AB}, R_{AB}, \text{AD}_A, \text{AD}_B)$ is already there, $\mathscr{B}$ responds with the value $h_{AB,R_{AB}}$.

(ii) Otherwise, $\mathscr{B}$ randomly chooses $h_{AB,R_{AB}} \in \{0,1\}^k$ and sends back the value to $\mathscr{A}$. Finally, $\mathscr{B}$ stores the new tuple $(K_{AB}, R_{AB}, \text{AD}_A, \text{AD}_B, h_{AB,R_{AB}})$ in the list H_2^{list}.

$H_3(\cdot)$. The simulator $\mathscr{B}$ maintains an initially empty list H_3^{list} with entries of the form $(K_{AB}, M_{AB}, T_{AB}, R_{AB}, \text{AD}_A, \text{AD}_B, h_{AB,T_{AB}}) \in \{0,1\}^k \times \{0,1\}^* \times G_1^3 \times \{0,1\}^k$. The simulator $\mathscr{B}$ responds to these queries in the following ways.

(i) $\mathscr{B}$ checks the list H_3^{list}; if $(K_{AB}, M_{AB}, T_{AB}, R_{AB}, \text{AD}_A, \text{AD}_B)$ is already there, $\mathscr{B}$ responds with the value $h_{AB,R_{AB}}$.

(ii) Otherwise, $\mathscr{B}$ randomly chooses $h_{AB,R_{AB}} \in \{0,1\}^k$ and sends back $h_{AB,R_{AB}}$ to $\mathscr{A}$. Finally, $\mathscr{B}$ stores the new tuple $(K_{AB}, M_{AB}, T_{AB}, R_{AB}, \text{AD}_A, \text{AD}_B, h_{AB,T_{AB}})$ in the list H_3^{list}.

$Establish(\text{ID}_A)$. When receiving this query, $\mathscr{B}$ responds to the query in the following way.

(i) If ID_A is the target physician or the target patient's gateway or sensor nodes, $\mathscr{B}$ aborts the game.

(ii) Otherwise, if ID_A is a physician PH_j, $\mathscr{B}$ looks in H_1^{list} for the entries $(\text{ID}_{\text{SV}_j}, sr_j, u_{\text{SV}_j}, h_{\text{SV}_j,sr_j})$ and $(\text{ID}_{\text{PH}_j}, sr_j, u_{\text{PH}_j}, h_{\text{PH}_j,sr_j})$. Then, $\mathscr{B}$ returns $\{(sr_j u_{\text{PH}_j}cP, s_2 sr_j, s_3 sr_j) \| u_{\text{SV}_j}P, u_{\text{PH}_j}P\}$.

(iii) Otherwise, if ID_A is a patient's $\text{GW}_{j,k}$, $\mathscr{B}$ looks in H_1^{list} for the entries $(\text{ID}_{\text{SV}_j}, sr_j, u_{\text{SV}_j}, h_{\text{SV}_j,sr_j})$, $(\text{ID}_{\text{PH}_j}, sr_j, u_{\text{PH}_j}, h_{\text{PH}_j,sr_j})$, and $(\text{ID}_{\text{GW}_{j,k}}, sr_j, u_{\text{GW}_{j,k}}, h_{\text{GW}_{j,k},sr_j})$. Then, $\mathscr{B}$ returns $\{(sr_j u_{\text{PH}_j}cP, s_2 sr_j u_{\text{GW}_{j,k}}P) \| u_{\text{SV}_j}P, u_{\text{PH}_j}P, u_{\text{GW}_{j,k}}P\}$.

(iv) Otherwise, if ID_A is a physician $\text{SN}_{j,k,d}$, the simulator $\mathscr{B}$ looks in H_1^{list} for the entries $(\text{ID}_{\text{SV}_j}, sr_j, u_{\text{SV}_j}, h_{\text{SV}_j,sr_j})$, $(\text{ID}_{\text{PH}_j}, sr_j, u_{\text{PH}_j}, h_{\text{PH}_j,sr_j})$, $(\text{ID}_{\text{GW}_{j,k}}, sr_j, u_{\text{GW}_{j,k}}, h_{\text{GW}_{j,k},sr_j})$, and $(\text{ID}_{\text{SN}_{j,k,d}}, sr_j, u_{\text{SN}_{j,k,d}}, h_{\text{SN}_{j,k,d},sr_j})$. Then, $\mathscr{B}$ returns $\{(sr_j u_{\text{PH}_j}cP, s_3 sr_j u_{\text{SN}_{j,k,d}}P) \| u_{\text{SV}_j}P, u_{\text{PH}_j}P, u_{\text{GW}_{j,k}}P, u_{\text{SN}_{j,k,d}}P\}$.

$Send(\prod_{A,B}^s, D)$. Since the H_2 and H_3 are the random oracles, the adversary cannot change the communication message. the simulator $\mathscr{B}$ needs only to store the values according to the scheme. Moreover, the parameters are included in the data D, which can be found in the lists H_3^{list} and H_2^{list}.

$Reveal(\prod_{A,B}^s)$. $\mathscr{B}$ maintains a list sk^{list} with tuples of the form $(\text{ID}_A, \text{ID}_B, R_{AB}, \prod_{A,B}^s)$. The simulator $\mathscr{B}$ responds to the query in the following way.

(i) If ID_A and ID_B are the target physician and the target patient's gateway or sensor nodes, $\mathscr{B}$ aborts the game.

(ii) Otherwise, if ID_A is a target physician PH_i and ID_B is not target patients' facilities, $\mathscr{B}$ proceeds in the following way to respond:

(a) If ID_B is an identity of gateway, $\mathscr{B}$ computes $K_{AB} = \hat{e}(u_{\text{GW}_{i,j^*}}cP, sr_i u_{\text{PH}_i}aP) \cdot \hat{e}(s_2 sr_i u_{\text{GW}_{i,j^*}}P, R_2)$. Then, $\mathscr{B}$ finds the value $h_{AB,R_{AB}}$ from H_2^{list} and returns $h_{AB,R_{AB}}$ as the response.

(b) Otherwise ID_B should be an identity of sensor node; the simulator $\mathscr{B}$ computes $K_{AB} = \hat{e}(u_{\text{SN}_{i,j^*,d^*}}cP, sr_i u_{\text{PH}_i}aP) \cdot \hat{e}(R_1, s_3 sr_i u_{\text{PH}_i}aP)$. Then, $\mathscr{B}$ finds the value $h_{AB,R_{AB}}$ from H_2^{list} and returns $h_{AB,R_{AB}}$ as the response.

(iii) Otherwise, if ID_A is another physician PH_j and ID_B is his/her patients facilities, $\mathscr{B}$ proceeds in the following way to respond:

(a) If ID_B is an identity of gateway, $\mathscr{B}$ computes $K_{AB} = \hat{e}(u_{\text{GW}_{j,j^*}}cP, sr_j u_{\text{PH}_j}P) \cdot \hat{e}(s_2 sr_j u_{\text{GW}_{j,j^*}}P, R_2)$. Then, $\mathscr{B}$ finds the value $h_{AB,R_{AB}}$ from H_2^{list} and returns $h_{AB,R_{AB}}$ as the response.

(b) Otherwise ID_B should be an identity of sensor node; $\mathscr{B}$ computes $K_{AB} = \hat{e}(u_{\text{SN}_{j,j^*,d^*}}cP, sr_j u_{\text{PH}_j}P) \cdot \hat{e}(R_1, s_3 sr_j u_{\text{PH}_j}P)$. Then, $\mathscr{B}$ finds the value $h_{AB,R_{AB}}$ from H_2^{list} and returns $h_{AB,R_{AB}}$ as the response.

$Test(\prod_{A,B}^s)$. $\mathscr{A}$ issues a test query. Suppose the identity tuple of the first node A is ID_{PH_i} and the second target node B is $\text{ID}_{\text{GW}_{i,j}}$ or $\text{ID}_{\text{SN}_{i,j,d}}$.

(i) If A and B do not belong to our guessed victims PH_i and $\text{PA}_{i,j}$, $\mathscr{B}$ aborts the game.

(ii) Otherwise, $\mathscr{B}$ queries AD_{SV_i}, AD_{PH_i}, $\text{AD}_{\text{GW}_{i,j}}$, and $\text{AD}_{\text{SN}_{i,j,d}}$.

(a) If $B = \text{GW}_{i,j}$, $\mathscr{B}$ computes $K_2 = Q^{u_{\text{GW}_{i,j}} u_{\text{PH}_i} sr_i} \cdot \hat{e}(\text{AD}_{\text{GW}_{i,j}}, R_2)^{s_2 sr_i}$.

(b) Otherwise, $\mathscr{B}$ computes $K_1 = Q^{u_{\text{SN}_{i,j,d}} u_{\text{PH}_i} sr_i} \cdot \hat{e}(R_1, \text{AD}_{\text{PH}_i})^{s_3 sr_i}$.

$\mathscr{B}$ looks in the list H_2^{list} and returns the value sk_1 or sk_2 to the adversary $\mathscr{A}$.

The test query is answered by $\mathscr{B}$ with its DBDH input D or R. Consider the following two cases:

(i) If $Q = D$, since $D = \hat{e}(P,P)^{abc}$ in the DBDH instance, then

$$
\begin{aligned}
K_1 &= D^{u_{\text{SN}_{i,j,d}} u_{\text{PH}_i} sr_i} \cdot \hat{e}\left(R_1, \text{AD}_{\text{PH}_i}\right)^{s_3 sr_i} \\
&= \hat{e}\left(u_{\text{SN}_{i,j,d}} bP, u_{\text{PH}_i} aP\right)^{sr_i c} \cdot \hat{e}\left(R_1, \text{AD}_{\text{PH}_i}\right)^{s_3 sr_i} \\
&= \hat{e}\left(\text{AD}_{\text{SN}_{i,j,d}}, \text{AD}_{\text{PH}_i}\right)^{s_1 sr_i} \cdot \hat{e}\left(R_1, \text{AD}_{\text{PH}_i}\right)^{s_3 sr_i}, \\
K_2 &= D^{u_{\text{GW}_{i,j}} u_{\text{PH}_i} sr_i} \cdot \hat{e}\left(\text{AD}_{\text{GW}_{i,j}}, R_2\right)^{s_2 sr_i} \\
&= \hat{e}\left(u_{\text{GW}_{i,j}} bP, u_{\text{PH}_i} aP\right)^{sr_i c} \cdot \hat{e}\left(\text{AD}_{\text{GW}_{i,j}}, R_2\right)^{s_2 sr_i} \\
&= \hat{e}\left(\text{AD}_{\text{GW}_{i,j}}, \text{AD}_{\text{PH}_i}\right)^{s_1 sr_i} \cdot \hat{e}\left(\text{AD}_{\text{GW}_{i,j}}, R_2\right)^{s_2 sr_i}.
\end{aligned}
\tag{18}
$$

Thus, the response by $\mathscr{B}$ corresponds to the real values sk_1 and sk_2.

(ii) If $Q = R$, since $R = \hat{e}(P,P)^r$ is random, then the response by $\mathscr{B}$ to the test query of $\mathscr{A}$ is a random element in G_2.

If the adversary $\mathscr{A}$ succeeds in getting the session key sk_1 or sk_2, it shall distinguish between the value sk_1 or sk_2 and a random value; then, it outputs the correct bit $b = 1$ or $b = 0$. $\mathscr{B}$ can give the correct answer to the DBDH problem by using $\mathscr{A}$'s output.

The success probability of $\mathscr{B}$ is

$$
\epsilon' \geq \frac{1}{s(k) n(k)^2 t(k)} \epsilon.
\tag{19}
$$

Here, ϵ is the probability that the adversary $\mathscr{A}$ succeeds in launching the attack. $t(k)$ is the polynomial bound on the number of the adversary $\mathscr{A}$'s queries.

If the adversary $\mathscr{A}$ succeeds with nonnegligible probability to attack our scheme, we can also solve the DBDH problem with a nonnegligible probability. Thus, our scheme is based on the DBDH problem.

6.3. Security Analysis. In the following, we will directly analyze how our proposed scheme achieves entity anonymity and untraceability and resists collusion attack and whether the security requirements have been satisfied.

Proposition 7. *The proposed scheme can resist the replay attack.*

Proof. It should be noted that our proposed scheme inherits the structure of FNKAP. We also use the random numbers r_1 and r_2 to achieve the freshness key agreement. The adversary cannot compute the r_i, $i \in \{1,2\}$ from $R_1 = r_1 \text{AD}_{\text{SN}_{i,j,d}}$ and $R_2 = r_2 \text{AD}_{\text{PH}_i}$ because of the difficulty of the ECDL problem. Moreover, the proposed scheme can efficiently resist the replay attack by considering the following scenarios. (1) An adversary cannot replay the data package D_1 to cheat SV and PH_i. During the Key Agreement from SN to PH Phase, when SV receives a data package D_1, it verifies the timestamp T_1 with the current time. If the data package is a replay attack, SV will detect it. Moreover, if the adversary changes the timestamp T_1 in D_1, SV will find the behavior by checking the equation $V_1^* = H_3(\text{sk}_1^* \parallel M_1 \parallel T_1 \parallel R_1 \parallel \text{AD}_{\text{SN}_{i,j,d}} \parallel \text{AD}_{\text{PH}_i})$ because it cannot obtain the session key sk_1^*. (2) An adversary cannot replay the data package D_2 to cheat SV and PH_i. Similar to the above, an adversary cannot replay the data package D_2 to cheat SV and $\text{PA}_{i,j}$. During the key agreement from PH to GW Phase, when SV receives a data package D_2, it verifies the timestamp T_2 with the current time. If the data package is a replay attack, then SV will detect it. Moreover, if the adversary changes the timestamp T_2 in D_2, SV will find the behavior by checking the equation $V_2^* = H_3(\text{sk}_2^* \parallel M_2 \parallel T_2 \parallel R_2 \parallel \text{AD}_{\text{PH}_i} \parallel \text{AD}_{\text{GW}_{i,j}})$ because it cannot know the session key sk_2^*.

Proposition 8. *The proposed scheme can provide basic forward secrecy.*

Proof. To establish session key between SN and PH, $\text{SN}_{i,j,d}$ and PH_i use various $r_1 \text{AD}_{\text{SN}_{i,j,d}}$ for each session. Thus, the current session key $\text{sk}_1 = H_2(K_1 \parallel R_1 \parallel \text{AD}_{\text{SN}_{i,j,d}} \parallel \text{AD}_{\text{PH}_i})$ is disclosed, and an adversary cannot obtain the information about $K_1 = \hat{e}(\text{AD}_{\text{SN}_{i,j,d}}, s_1 sr_i \text{AD}_{\text{PH}_i}) \cdot \hat{e}(s_3 sr_i \text{AD}_{\text{SN}_{i,j,d}}, \text{AD}_{\text{PH}_i})^{r_1}$. In other words, the adversary cannot get more opportunities to guess previous key $\text{sk}_1^* = H_2(K_1^* \parallel R_1 \parallel \text{AD}_{\text{SN}_{i,j,d}} \parallel \text{AD}_{\text{PH}_i})$ than before, even if he/she knows the current key sk_1. Similarly, because R_2 is equal to $r_2 \text{AD}_{\text{PH}_i}$, the adversary cannot gain any benefits to guess previous key $\text{sk}_2^* = H_2(K_2^* \parallel R_2 \parallel \text{AD}_{\text{PH}_i} \parallel \text{AD}_{\text{GW}_{i,j}})$ between PH and GW compared to before, even if he/she knows the current key sk_2. Thus, our proposal can provide basic forward secrecy.

Proposition 9. *The proposed scheme can prevent fraud attack.*

Proof. Our proposal provides mutual authentication between PH_i and $\text{GW}_{i,j}$ or PH_i and $\text{SN}_{i,j,d}$. The proposed scheme can prevent fraud attack by considering the following scenarios. (1) An adversary cannot impersonate $\text{SN}_{i,j,d}$ to cheat PH_i. PH_i can authenticate $\text{SN}_{i,j,d}$ by verifying V_1' in Step 3. Since the adversary cannot obtain $s_3 sr_i$ or $s_3 sr_i r_1 \text{AD}_{\text{SN}_{i,j,d}}$, he/she cannot compute $K_1' = \hat{e}(\text{AD}_{\text{SN}_{i,j,d}}, s_1 sr_i \text{AD}_{\text{PH}_i}) \cdot \hat{e}(R_1, \text{AD}_{\text{PH}_i})^{s_3 sr_i}$, $K_1^* = \hat{e}(\text{AD}_{\text{SN}_{i,j,d}}, \text{AD}_{\text{PH}_i})^{s_1 sr_i} \cdot \hat{e}(R_1, \text{AD}_{\text{PH}_i})^{s_3 sr_i}$, or $K_1 = \hat{e}(\text{AD}_{\text{SN}_{i,j,d}}, s_1 sr_i \text{AD}_{\text{PH}_i}) \cdot \hat{e}(s_3 sr_i \text{AD}_{\text{SN}_{i,j,d}}, \text{AD}_{\text{PH}_i})^{r_1}$. Thus, the adversary cannot get $\text{sk}_1' = H_2(K_1' \parallel R_1 \parallel \text{AD}_{\text{SN}_{i,j,d}} \parallel \text{AD}_{\text{PH}_i})$ and $V_1' = H_3(\text{sk}_1' \parallel R_1 \parallel M_1 \parallel \text{AD}_{\text{SN}_{i,j,d}} \parallel \text{AD}_{\text{PH}_i} \parallel T_1)$, sequentially. Thus, the adversary cannot generate the valid verifier to PH_i. (2) An adversary cannot impersonate PH_i to cheat $GW_{i,j}$. Similar to the above, $GW_{i,j}$ can authenticate PH_i by verifying V_2' in Step 3. Since the adversary cannot obtain $s_2 sr_i$ or $s_2 sr_i AD_{GW_{i,j}}$, he/she cannot compute K_2', K_2^*, or K_2. Thus, the adversary cannot get sk_2' and V_2', sequentially. Thus, the adversary cannot generate the valid verifier to $GW_{i,j}$.

Proposition 10. *The proposed scheme can provide entity anonymity and untraceability.*

Proof. In the proposed scheme, the adversary can obtain the amplified identities $H_1(ID_{SV} \parallel sr_i)$, $H_1(ID_{PH_i} \parallel sr_i)$, $H_1(ID_{GW_{i,j}} \parallel a_{i,j})$, and $H_1(ID_{SN_{i,j,d}} \parallel a_{i,j})$ instead of $H(ID_{SV})$, $H(ID_{PH_i})$, $H(ID_{GW_{i,j}})$, and $H(ID_{SN_{i,j,d}})$ in Steps K1 and K3. Here, sr_i and $a_{i,j}$ are big random numbers in Z_q^*. Therefore, the adversary cannot verify whether the guessed identity is correct or incorrect by testing all possible identities without the secret sr_i and $a_{i,j}$. For example, to guess $H_1(ID_{PH_i} \parallel sr_i)$, the adversary should input the guess values of ID_{PH_i} and sr_i at the same time. Suppose the identity ID_{PH_i} is composed of m bits; it is infeasible for adversary to launch an exhausted search for 2^{m+q} possible solutions. Here, q is the group order of Z_q^*, and it is a big random number. In particular, if the physicians reregister on a period, sr_i would be fresh regularly. Thus, this risk of corruption will be lower to ID_{PH_i}. Moreover, $a_{i,j}$ is also a big random number in Z_q^*, and each patient has a different value. Even if it is the same patient, there are different values on the various diagnoses. Based on the similar reason, the adversary cannot know the identities of $ID_{GW_{i,j}}$ and $ID_{SN_{i,j,d}}$ or trace them. Furthermore, it is also intractable to derive the identity from $H_1(ID_{SV} \parallel sr_i)$, $H_1(ID_{PH_i} \parallel sr_i)$, $H_1(ID_{GW_{i,j}} \parallel a_{i,j})$, and $H_1(ID_{SN_{i,j,d}} \parallel a_{i,j})$ because H_1 is a secure one-way cryptography hash function. Thus, our proposal can achieve anonymity and untraceability.

Proposition 11. *The proposed scheme can withstand the collusion attack.*

Proof. In our proposal, SV distributes different secret values $(s_1 sr_i AD_{PH_i}, s_2 sr_i, s_3 sr_i)$ for various physicians PH_i. Thus, the adversary physician $PH_{\mathscr{A}}$ and his/her patients $PA_{\mathscr{A},j}$ cannot get the victim's information $s_2 sr_i$, $s_3 sr_i$, directly. Furthermore, the adversary $\mathscr{A}$ who has registered as a normal patient of the physician PH_i can legally obtain $(s_1 sr_i AD_{PH_i}, s_2 sr_i AD_{GW_{i,j}})$ from SV. However, he/she cannot obtain $s_1 sr_i$ or $s_2 sr_i$ from $(s_1 sr_i AD_{PH_i}, s_2 sr_i AD_{GW_{i,j}})$ except when he/she can solve the ECDL problem. Similarly, the adversary $\mathscr{A}$ cannot obtain $s_3 sr_i$ from $s_3 sr_i AD_{SN_{i,j,d}}$ because of the difficulty of the ECDL problem. Obviously, our scheme destructs the attack conditions at Steps A1 and A2 in Section 4. As a result, the scheme can resist the collusion attack and prevent the adversary from generating the session keys sk_1 and sk_2. Furthermore, if an insider adversary wants to attack the key agreement from SN to PH, he/she should get the secure information about $s_3 sr_i AD_{SN_{i,j,d}}$. The adversary receives up to $s_3 sr_i AD_{SN_{i,j,d}}$;

he/she cannot get the information except whens he/she can solve the ECDL problem. Similarly, an insider adversary cannot launch an attack to the key agreement from PH to GW phase, because he/she cannot get the value $s_2 sr_i AD_{GW_{i,j}}$. Thus, our proposal resists the collusion attack, effectively.

7. Functionality and Performance Comparison

In this section, security and functionality are compared between our scheme and FNKAP. Then, we illustrate a comparison of the communication and computation costing performances.

7.1. Functionality Comparison. As shown in Table 1, our scheme not only provides the functionality in [8] but also resists the collusion attack. Therefore, we can conclude that the proposed scheme achieves a higher security level than FNKAP.

7.2. Performance Comparison. To compare the actual computational costs, we have implemented our scheme and Kim's scheme with JPBC Library (Java Pairing-Based Cryptography Library [19]) in an ARM platform and a desktop platform. The detailed parameters of the platform are listed in Table 2. To provide a similar environment in WHMS, the weak processing ability is simulated on an android smartphone (HTC M7) running Android 4.1 with Snapdragon APQ8064 1.7 GHz, and the powerful processing ability is simulated on a desktop computer running Windows 7 with Intel Core i5-3470.

Table 3 summarizes the detailed parameters about the elliptic curve and pairing parameters for JPBC. We use a 512 bits elliptic curve $y^3 = x^3 + x$ to evaluate our scheme in the platforms. In Table 4, top row is the results in the ARM platform, and the second row is the result in the desktop platform. Here, all the experiment results are averaged over 10 independent runs.

In order to provide detailed comparison, we test the basic operation in Z_q^*, G_1, and G_2, separately. The time of a pairing computation is indicated by T_p. The time of a hash operation is indicated by T_h. The time complexity of computing multiplication in Z_q^*, G_1, and G_2 is indicated by T_{m_q}, T_{m_1}, and T_{m_2}, respectively. The time of the addition in Z_q^* and G_1 is indicated by T_{a_q} and T_{a_1}, independently. The time of the exponentiation in G_2 is indicated by T_e. Note that the time of hash operation T_h is the smallest because it needs very limited computation. On the contrary, the time of pairing operation T_p is the highest consumption.

Tables 5 and 6 illustrate the performance comparison with Kim's scheme. In Tables 5 and 6, the notation id is a unit length of identity; the notation pr is a unit length of private key. First, in order to achieve the session key freshness, we maintain one-round communication to exchange a random value R_1 or R_2 in FNKAP. Second, our scheme increases the amplified identity randomness against the passive offline attack. However, the amplified identity space is equal to that of FNKAP because the amplified identity is still a hash value. Third, the private key space of GW and SN decreases because

TABLE 1: Security and functionality comparison with Kim's scheme.

	Replay and fraud attack	Basic forward secrecy	Anonymity and untraceability	Collusion attack
Kim's	Resistance	Yes	Yes	No
Ours	Resistance	Yes	Yes	Resistance

TABLE 2: Detailed platform parameters.

Testbed	CPU	RAM	Operating system
ARM	Snapdragon APQ8064 1.7 GHz	2 GB	Android 4.1
Desktop	Intel Core i5 3.2 × 4 GHz	4 GB	Windows 7–32 bits and Open JDK 1.8.0

TABLE 3: Detailed elliptic curve and pairing parameters.

Elliptic curve	Group order of Z_q^*	Element size in G_1
Type A: $y^3 = x^3 + x$	Around 160 bits	Around 512 bits

TABLE 4: Detailed performance parameters.

	T_p	T_h	T_{m_q}	T_{a_q}
ARM	442 ms	≪1 ms	<1 ms	<1 ms
Desktop	25 ms	≪1 ms	≪1 ms	=1 ms
	T_e	T_{m_2}	T_{m_1}	T_{a_1}
ARM	450 ms	1 ms	124 ms	<1 ms
Desktop	4 ms	≪1 ms	22 ms	<1 ms

TABLE 5: Basic performance comparison with Kim's scheme [8].

	Round	Private key space			
		SV	PH	GW	SN
Kim's	One	4pr	4pr	4pr	4pr
Ours	One	$(3 + N)$pr	3pr	2pr	2pr

we reduce the redundancy of private key information to GW and SN. Moreover, it shrinks the risk of insider attack because only SV knows total secure information. Fourth, the computation time of our scheme is near half of FNKAP because we decrease half of the pairing operations. Finally, we should point out that our scheme computation and store cost for the SV are higher than those of Kim's work. More precisely, we should choose and store N random numbers more than FNKAP, and $4N$ multiplications in Z_p^* should be added in Initial Section. Commonly, the above propositions only increase the computation cost and the storage requirement in SV. SV has enough computing and storing power to hold the operations because the u-Health Server is usually a server cluster. Furthermore, the computing operations are only increased in Initial Phase. For the resources limited entities GW and SN, the computation and storage requirements do not increase instead of decreasing. Thus, the scheme is feasible to key agreement in WHMS.

Our proposed scheme inherits the advantage of Kim's hierarchical scheme in WHMS. At the same time, our scheme provides security enhancement against collusion attack in our security model. Furthermore, it preserves the low computation and private key space in SN and GW compared to FNKAP. Therefore, it is an enhanced security hierarchical key agreement scheme with the noninteractive property that is suitable for the application in WHMS.

8. Conclusions

In this paper, we have illustrated that there is a security weakness in Kim's work [8] under a practical security model with the physicians corruption. The security flaw is due to the fact that the physicians' parts of the private key are the same. Therefore, the adversary, as a legal physician, can acquire the entire patient's private information. To enhance the scheme, we proposed an authenticated key agreement scheme which randomizes each physician's private key. Moreover, we have reduced the numbers of the private keys and the operations of the bilinear pairing. Thus, the performance of our scheme is more suitable for the WHMS environment than Kim's work. We also prove the security of our scheme. The proof shows that the proposed scheme is secure under the DBDH assumption in the random oracle model.

Notations

PH_i:	The ith physician
$PA_{i,j}$:	The ith physician's jth patient
SV:	The u-Health Server
$GW_{i,j}$:	The $PA_{i,j}$'s gateway
$SN_{i,j,d}$:	The $PA_{i,j}$'s dth sensor node
ID_A:	The identity of an entity A
AD_A:	The amplified identity of ID_A
sk_1 and sk_2:	The session key established between two entities
$H(\cdot), H_1(\cdot), H_2(\cdot),$ and $H_3(\cdot)$:	The cryptographic hash functions
$E_K(M)$:	Encryption of a message M using an symmetric key K
$\cdot$:	Multiplication operator
$\|$:	Concatenation operator.

TABLE 6: Computation performance comparison with Kim's scheme [8].

	ID space	Computation		
		PH	SN/GW	Sum
Kim's SN to PH	4id	$T_h + 2T_{a_1} + 4T_p + 3T_{m_2} \approx 102$ ms	$T_h + 2T_{a_1} + 4T_p + 3T_{m_2} \approx 1773$ ms	1875 ms
Our SN to PH	4id	$2T_h + T_{a_1} + 2T_p + T_{m_2} \approx 51$ ms	$2T_h + 2T_{a_1} + 2T_p + T_{m_2} \approx 887$ ms	938 ms
Kim's PH to GW	4id	$T_h + 3T_{a_1} + 4T_p + 3T_{m_2} \approx 103$ ms	$T_h + T_{a_1} + 4T_p + 3T_{m_2} \approx 1772$ ms	1875 ms
Our PH to GW	4id	$2T_h + 2T_{a_1} + 2T_p + T_{m_2} \approx 52$ ms	$2T_h + 2T_p + T_{m_2} \approx 885$ ms	937 ms

Competing Interests

The authors declare that they have no competing interests.

Acknowledgments

The authors thank Dr. Jianghong Wei, Dr. Haosu Cheng, Dr. Hui Lu, and Dr. Honghao Zhao for their support. This paper is supported by the National Key Basic Research Program (973 Program) through Project 2012CB315905, by the Natural Science Foundation through Projects 61370190, 61272501, 61173154, 61402029, and 61003214, and by the Beijing Natural Science Foundation through Projects 4132056 and 4122041.

References

[1] V. Custodio, F. J. Herrera, G. López, and J. I. Moreno, "A review on architectures and communications technologies for wearable health-monitoring systems," *Sensors*, vol. 12, no. 10, pp. 13907–13946, 2012.

[2] F. Touati and R. Tabish, "U-healthcare system: state-of-the-art review and challenges," *Journal of Medical Systems*, vol. 37, no. 3, pp. 9949–9969, 2013.

[3] H. F. Rashvand, V. Traver Salcedo, E. Montón Sánchez, and D. Iliescu, "Ubiquitous wireless telemedicine," *IET Communications*, vol. 2, no. 2, pp. 237–254, 2008.

[4] R. Lu, X. Lin, X. Liang, and X. Shen, "A secure handshake scheme with symptoms-matching for mHealthcare social network," *Mobile Networks and Applications*, vol. 16, no. 6, pp. 683–694, 2011.

[5] W. Liu, J. Liu, Q. Wu, W. Susilo, H. Deng, and B. Qin, "SAKE: scalable authenticated key exchange for mobile e-health networks," *Security and Communication Networks*, 2015.

[6] X. Lin, R. Lu, X. S. Shen, Y. Nemoto, and N. Kato, "SAGE: a strong privacy-preserving scheme against global eavesdropping for ehealth systems," *IEEE Journal on Selected Areas in Communications*, vol. 27, no. 4, pp. 365–378, 2009.

[7] Q. Huang, X. Yang, and S. Li, "Identity authentication and context privacy preservation in wireless health monitoring," *International Journal of Computer Network and Information Security*, vol. 4, pp. 53–60, 2011.

[8] H. Kim, "Freshness-preserving non-interactive hierarchical key agreement protocol over WHMS," *Sensors*, vol. 14, no. 12, pp. 23742–23757, 2014.

[9] D.-C. Lou, T.-F. Lee, and T.-H. Lin, "Efficient biometric authenticated key agreements based on extended chaotic maps for telecare medicine information systems," *Journal of Medical Systems*, vol. 39, no. 5, pp. 58–68, 2015.

[10] S. H. Erfani, H. H. S. Javadi, and A. M. Rahmani, "A dynamic key management scheme for dynamic wireless sensor networks," *Security and Communication Networks*, vol. 8, no. 6, pp. 1040–1049, 2015.

[11] R. Gennaro, S. Halevi, H. Krawczyk, T. Rabin, S. Reidt, and S. D. Wolthusen, "Strongly-resilient and non-interactive hierarchical key-agreement in MANETs," in *Proceedings of the 13th European Symposium on Research in Computer Security (ESORICS '08)*, pp. 49–65, Málaga, Spain, October 2008.

[12] R. Dupont and A. Enge, "Provably secure non-interactive key distribution based on pairings," *Discrete Applied Mathematics*, vol. 154, no. 2, pp. 270–276, 2006.

[13] Y. Yang, "Broadcast encryption based non-interactive key distribution in MANETs," *Journal of Computer and System Sciences*, vol. 80, no. 3, pp. 533–545, 2014.

[14] H. Guo, Y. Mu, Z. Li, and X. Zhang, "An efficient and non-interactive hierarchical key agreement protocol," *Computers and Security*, vol. 30, no. 1, pp. 28–34, 2011.

[15] Chad Garland. Cedars-Sinai reports possible breach of patients' medical data, June 2015, http://www.latimes.com/business/la-fi-cedars-breach-20140823-story.html.

[16] M. Bellare and P. Rogaway, "Entity authentication and key distribution," in *Advances in Cryptology— CRYPTO '93*, D. R. Stinson, Ed., vol. 773 of *Lecture Notes in Computer Science*, pp. 232–249, Springer, New York, NY, USA, 1994.

[17] D. Boneh and M. Franklin, "Identity-based encryption from the Weil pairing," *SIAM Journal on Computing*, vol. 32, no. 3, pp. 586–615, 2003.

[18] D. Hankerson and A. Menezes, "Elliptic curve discrete logarithm problem," in *Encyclopedia of Cryptography and Security*, C. A. Van Tilborg and S. Jajodia, Eds., pp. 397–400, Springer US, 2011.

[19] A. De Caro and V. Iovino, "jPBC: Java pairing based cryptography," in *Proceedings of the 16th IEEE Symposium on Computers and Communications (ISCC '11)*, pp. 850–855, IEEE, Kerkyra, Greece, July 2011.

14

The Routing Algorithm based on Fuzzy Logic Applied to the Individual Physiological Monitoring Wearable Wireless Sensor Network

Jie Jiang, Yun Liu, Fuxing Song, Ronghao Du, and Mengsen Huang

Department of Electronic and Information Engineering, Beijing Jiaotong University, Beijing 100044, China

Correspondence should be addressed to Yun Liu; liuyun@bjtu.edu.cn

Academic Editor: Chin-Feng Lai

In recent years, the research of individual wearable physiological monitoring wireless sensor network is in the primary stage. The monitor of physiology and geographical position used in wearable wireless sensor network requires performances such as real time, reliability, and energy balance. According to these requirements, this paper introduces a design of individual wearable wireless sensor network monitoring system; what is more important, based on this background, this paper improves the classical Collection Tree Protocol and puts forward the improved routing protocol F-CTP based on the fuzzy logic routing algorithm. Simulation results illustrate that, with the F-CTP protocol, the sensor node can transmit data to the sink node in real time with higher reliability and the energy of the nodes consumes balance. The sensor node can make full use of network resources reasonably and prolong the network life.

1. Introduction

The wearable wireless sensor network [1] is applied to monitor the physiological information (heart rate, blood oxygen, breath, blood pressure, body temperature, and so on) and movement information (the speed, gait, trajectory, and the consumption of energy in the sport) of human body and the external environment (such as temperature, humidity, gas composition, and location) dynamically and continuously for a long time.

Focus on the wearable wireless sensor network among bodies, the wearable sensor node which is placed on the mobile human bodies constitutes the sensor network, the network topology of which changes fiercely. The wearable wireless sensor network which is applied in the individual physiological information monitoring is currently in the early research. And because the mobility of the human body is bigger and the topology changes dramatically, these require that sensor nodes transmit data in real time with high reliability and the energy of the nodes consumes balance which can prolong the network life. Therefore, new requirements for traditional wireless sensor network routing protocol are put forward.

The typical protocol of wireless sensor network mostly takes into account single performance. Many papers make improvement for these protocols: some take into account reducing the time delay performance and some take into account the energy utilization performance to make the network balance. All of them rarely involve the multiobjective optimization problem. Which are important for the application of the wireless sensor network in the wearable field.

The main contributions of this paper and the future outlook are summarized as follows.

We give a design of hardware system which is used for individual physiological monitoring. It uses wireless sensor network technology to sense the individual physiological information and integrates PC and PAD to achieve long-distance medical care purposes.

Based on the system we design, we put forward F-CTP protocol taking into account multiobjective optimization problem which is based on the fuzzy logic algorithm. It uses the distributed dynamic optimization to ensure that the sensor node can transmit data to the sink node in real time with higher reliability and the energy of the node consumes balance. The sensor node can make full use of network resources reasonably and prolong the network life.

It makes the wireless sensor network better applied to wearable physiological monitoring area. The achievements in this paper will be the basis for development of wearable wireless sensor network applied in hospitals, the army, community-oriented and so on.

The rest of this paper is organized as follows. The related work is presented in Section 2. Section 3 gives the hardware system designed for individual wearable wireless sensor network. In Section 4, we briefly explain the CTP protocol; then, we illustrate the F-CTP mainly including three parts. First, we describe several definitions for fuzzy algorithm. Then, we give the membership function of input and output parameters and give the fuzzy rules. At last, we give the defuzzification method to rank the candidate neighbor node. In Section 5, the performance of F-CTP is evaluated. At last, we summarized this paper in Section 6.

2. Related Work

The wearable sensor network is a new discipline in recent years, there are a lot of researches about wearable sensor network development platform at home and abroad; however, there are less researches on the wearable sensor network communication protocols.

Firstly, some research results about the wearable sensing development platform are listed below. BSN node [2], which is developed by Imperial College London, is suitable for the field of medical care. The hardware system is composed of CC2420 and MSP430F14 of TI Company and the software is running on TinyOS. CodeBlue system [3] developed by Harvard University, which is used in an emergency situation to provide wireless medical monitoring. It uses wireless sensor network technology to sense the relevant physiological information and integrate PC, PAD, and other monitoring medical facilities to achieve long-distance medical care purposes. In paper [4], J. P. Carmo and his companions developed the smart electronic clothing, which can collect the physiological data of human body and transmit it to the remote terminal and then carry on further processing and analysis on the remote terminal. In paper [5], S. Choi and his companions developed the new wearable heart electrical signal monitoring equipment which can monitor body information during the sleep; the device uses the conduction tissue and PVDF, which is harmless to the human body for long time use. In paper [6], Otto et al. developed the product that not only can collect and save the human body's ECG signal but also can obtain the human body's motion state through the sensor. Therefore, it is a kind of system which can simultaneously monitor the ECG and motion state information. Po et al.'s [7] invention utilizes a smart sensor to detect physiological signals such as ECG and heart rate. And it can transmit these signals to the terminal through Bluetooth. These systems use a new type of sensor to obtain the physiological signals of the human body; some of the systems also have the function of wireless transmission. The current research is to study the communication of wearable wireless sensor networks from the aspects of sensor nodes, remote communication, sensors, and wearable embedding method, and most of the theoretical

model of communication is still unpractical and cannot get a good promotion. This paper gives a design of hardware system used for individual physiological monitoring which is realized. The communication protocol of the system is based on wireless sensor network; on the other hand, it integrates positioning module using the Big Dipper on the basis of physiological monitoring such as ECG status and gesture information.

At present, the research on the communication of wearable wireless sensor networks is mainly carried out from the aspects of sensor nodes, remote communication, sensors, and wearable embedding method; the research on communication protocol of routing layer is rare. Basically all the wearable wireless sensor network communication protocols are based on wireless sensor network. There are specific needs for a wireless sensor network in the field of wearable wireless sensor which is introduced in Section 1. But the existing literature has not yet been targeted at multiobjective optimization about delay, energy balance, and delivery reliability for the wearable wireless sensor network. In the case of [8], a kind of CTP routing algorithm based on energy consumption for wireless sensor network is proposed, which takes into account the remaining energy of the next hop neighbor nodes in the data packet transmission process to achieve dynamic equilibrium network energy consumption. Paper [9] proposes a new clustering algorithm based on genetic algorithm for load balancing in wireless sensor networks. In the literature [10], Ronghui et al. proposed a kind of approximate algorithm based on the delay and the cost of two targets for routing. Paper [11] proposes a kind of multiobjective route based on ant colony algorithm, which integrates energy consumption, time delay, and robustness and transmission efficiency. In paper [12], the FLCHE clustering routing protocol based on fuzzy logic method is proposed, which considers the residual energy, energy consumption rate, and the distribution density of nodes. In paper [13], an energy-saving routing architecture with a uniform clustering algorithm is proposed in this paper to reduce the energy consumption in wireless body sensor networks. Paper [1] introduces that Pareto multiobjective optimization method and ant colony algorithm are used to improve the reliability and real time routing protocol. This paper proposed a multiobjective routing algorithms based on fuzzy logic algorithm, which integrates energy consumption, time delay, and reliability for wearable wireless network which makes the wireless sensor network better applied to wearable physiological monitoring area.

3. Individual Wearable Wireless Sensor Network Monitoring System

This section describes the design of an individual wearable wireless sensor network monitoring system [7]. The system is composed of three parts: extension, relay, and host which, respectively, are the soldier wearable device, relay terminal, and central data receiving control platform. For example, for a team composed of several members, the team leader equipped with relay terminal is regarded as the sink node

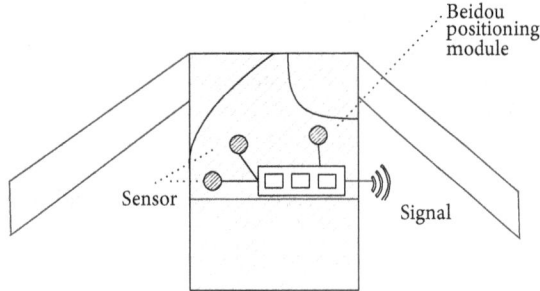

FIGURE 1: Appearance of the individual wearable device.

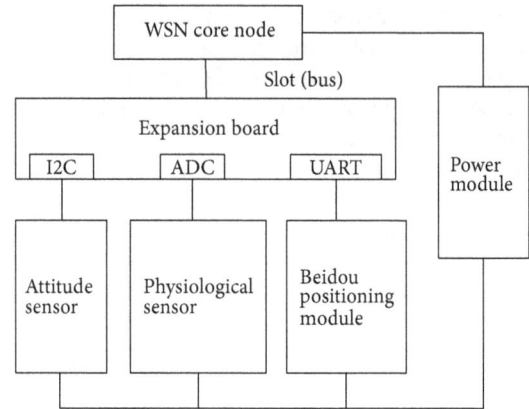

FIGURE 2: The structure diagram of the individual wearable device.

(Sink), which can display the monitoring data. Ordinary team members are equipped with wearable devices which are responsible for the collection of individual physiological information, attitude information (acceleration), and location information.

We build the wireless sensor network using Zigbee technology and use the CTP routing protocol (Collection Tree Protocol) to communicate between the sink node and sensor nodes. In order to guarantee the safety of communication, communication between the sensor nodes can be encrypted. The sensor node transmits its data or others nodes' data to the sink node by multihops transmission. The sink node collects all the data and then uploads them to a central data receiving control platform.

The individual wearable devices are placed on a wearable vest. The purpose is to monitor the individual position, gesture, and ECG status information in real time. The design principle is compact and easy to wear in the greatest extent to avoid the inconvenience of daily behavior. Appearance of the individual wearable device is shown in Figure 1.

The main module of individual wearable device includes WSN core module, BD positioning module, attitude sensor, physical sensor module, and battery module. The WSN core module consists of MSP430 microprocessor and CC2420 wireless transmission chip. The positioning part uses the BD2 positioning Module. The power management module includes lithium battery protection board, power indicator board, and boost regulator board to ensure the normal power supply. The structure diagram of the individual wearable device is shown in Figure 2.

4. The Design of F-CTP

4.1. The Classic CTP Routing Algorithm. CTP [14] is an aggregation protocol based on tree structure and sensor node delivery data to the sink node by anycast multihop. CTP uses the ETX value as the routing gradient, and the ETX value of sink node is 0, the ETX of the other nodes is the ETX of its parent node plus the ETX of its parent link. The calculation of ETX is formulas (1) and (2). In formula (1), data_total represents the total packet delivered between two sensor nodes, and data_success represents the packet delivered successfully between two sensor nodes. In

formula (2), ETX_{parent} presents the ETX value of parent node, and $ETX_{linkparent}$ presents the ETX value of parent link:

$$ETX = \left[\frac{data\_total}{data\_success} - 1 \right] \times 10, \qquad (1)$$

$$ETX\,(node) = ETX_{parent} + ETX_{linkparent}, \qquad (2)$$

where $ETX(node)$ presents the ETX value of sensor node in wireless sensor network, when the nodes choose the path, they choose the path with the smallest ETX as the routing path. From formulas above, we can see that when the sensor node chooses the next hop the classic CTP only considers the packet transmission success rate, which is the reliability performance that we are concerned with.

4.2. A Routing Algorithm Based on Fuzzy Logic. For the requirements of the wearable wireless sensor network that is applied in the field of individual physiological monitoring, we propose a routing selection algorithm based on the fuzzy logic to improve the CTP protocol. The basic idea is calculating a reasonable value by fuzzy logic taking into account three parameters: reliability, time delay, and energy to replace the original ETX value. The routing algorithm can be divided into three phases. The first stage is defining the input and output parameters, respectively, and give the membership function which uses the language set to express the parameters. Then, fuse the language information using fuzzy rules and get the evaluation results of candidate parent node. At last, defuse the evaluation results by center of gravity method and choose the best path.

4.2.1. The Description of the Input and Output Parameters. The transmission model of the wireless networks is shown in Figure 3. We give some definitions as following.

Definition 1 (neighbor node set). The effective coverage range of the wireless signal from *A* becomes its transmission region. The neighbors of *A* are the nodes which fall into the transmission region of *A*. *B*, *C*, and *D* are *A*'s neighbors shown in Figure 3.

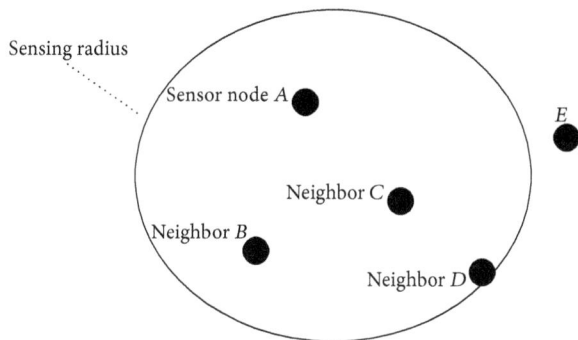

FIGURE 3: The transmission model for the sensor networks.

TABLE 1: Fuzzy rules.

Rule	Reliability	Delay	Energy	Output
1	High	Small	Enough	Perfect
2	High	Small	General	Good
3	High	Small	Few	Acceptable
4	High	Medium	Enough	Good
5	High	Medium	General	Acceptable
6	High	Medium	Few	Unperfected
7	High	Large	Enough	Acceptable
8	High	Large	General	Unperfected
9	High	Large	Few	Bad

Definition 2 (the reliability parameter R). Calculate the probability of packets that are successfully transmitted between two sensor nodes. The formula is (2). Assume that A sent m packets to B with a certain time, and the number of packets B successfully received is n. Without considering the difference of two-way communication link, we assume that $R(A, B) = R(B, A)$:

$$R(A, B) = \frac{n}{m}. \tag{3}$$

Definition 3 (time delay parameter T). Calculate the average time delay that B successfully receives packets from A. The formula is (3). Assume that A sent n packets to B continuously in a certain time; the number of the packet that B successfully received is n. The send time series is $\{T_1, T_2, \ldots, T_{m-1}, T_m\}$ (the send time of m packet) and the receiving time series is $\{T_1^*, T_2^*, \ldots, T_m^*\}$ (the receiving time of n packet), ignoring the lost package:

$$T(A, B) = \sum_{i \in n} \left(T_i^* - T_i\right). \tag{4}$$

Definition 4 (energy parameter E). Calculate the proportion of residual energy of one node accounted for the largest energy. The formula is (4). Assume that E_r represents the residual energy of B and $E_{\max}$ represents the largest energy among the neighbor nodes:

$$E = \frac{E_r}{E_{\max}}. \tag{5}$$

Definition 5 (one hop link quality table). A piece of entry in the routing table maintained by node A is (A, parent, $R(A, S)$, and $T(A, S)$). A represents the current node, and parent represents a random candidate parent node among neighbor nodes. Parameter E is not recorded in the routing entry, which is used for the calculation for routing selection.

4.2.2. Fuzzification. The membership functions and fuzzy rules are defined in Figure 4 and Table 1. The membership functions for the three input parameters R, T, and E are defined according to our experimental data as (a), (b), and (c), based on output member function defined in (d).

Establish the appropriate fuzzy rules to fuse the three parameters and try to make the routing algorithm reach the optimal evaluation. The fuzzy rules are composed of a series of fuzzy conditional statements in "IF-THEN" type, as in Table 1. We use $3 * 3 * 3$ fuzzy rules. Part of the rules are shown in Table 1.

From the table, we can see the nodes with high reliability are more likely to be selected as the parent node. Nodes with lower delay also have a greater chance of being selected as the parent node. In addition, we also consider the energy index. The sensor node with more energy has more opportunities to be selected as the parent node. Comprehending three factors, we can find the F-CTP can select the optimal path.

4.2.3. Defuzzification. Center of gravity (COG) method is used to defuzzify the fuzzy result. Since the fuzzy logic can reconcile conflicting objectives, this step can provide a quick ranking of multiple candidates (neighbor nodes).

Each node maintains a routing table using the "IF-THEN" criterion. Each routing entry will output a representation with natural language, which is the fuzzy quantity. However, we really want to evaluate which routing entry is most perfect; we should change the fuzzy quantity as a clear number, which is called defuzzification. Here we select the center of gravity (COG) method for defuzzification, to get the accurate value of the probability of a candidate parent node becoming a parent node. So we can sort the probability of each routing entry; then, we can get the maximum value that is the optimal path.

5. Performance Evaluation

We assume there are only five nodes in the wireless sensor network and node 1 acts as a sink node. Besides, (R, T) are parameters between each pair of nodes. The whole topology of the network is shown in Figure 5.

Parameters (R, T) between nodes are recorded in routing table and the initial energy parameters are assumed to be the same. In addition, we further assume each node that sends and forwards a package consumes the same energy. After initial phase, each node maintains a routing table and the routing table of node 5 is shown in Table 2.

According to the classic CTP, we should choose entry 2 (red line) with highest reliability, which ignores the real time. With the routing algorithm based on fuzzy logic, we should choose entry 4 (blue line). We perform a further analysis

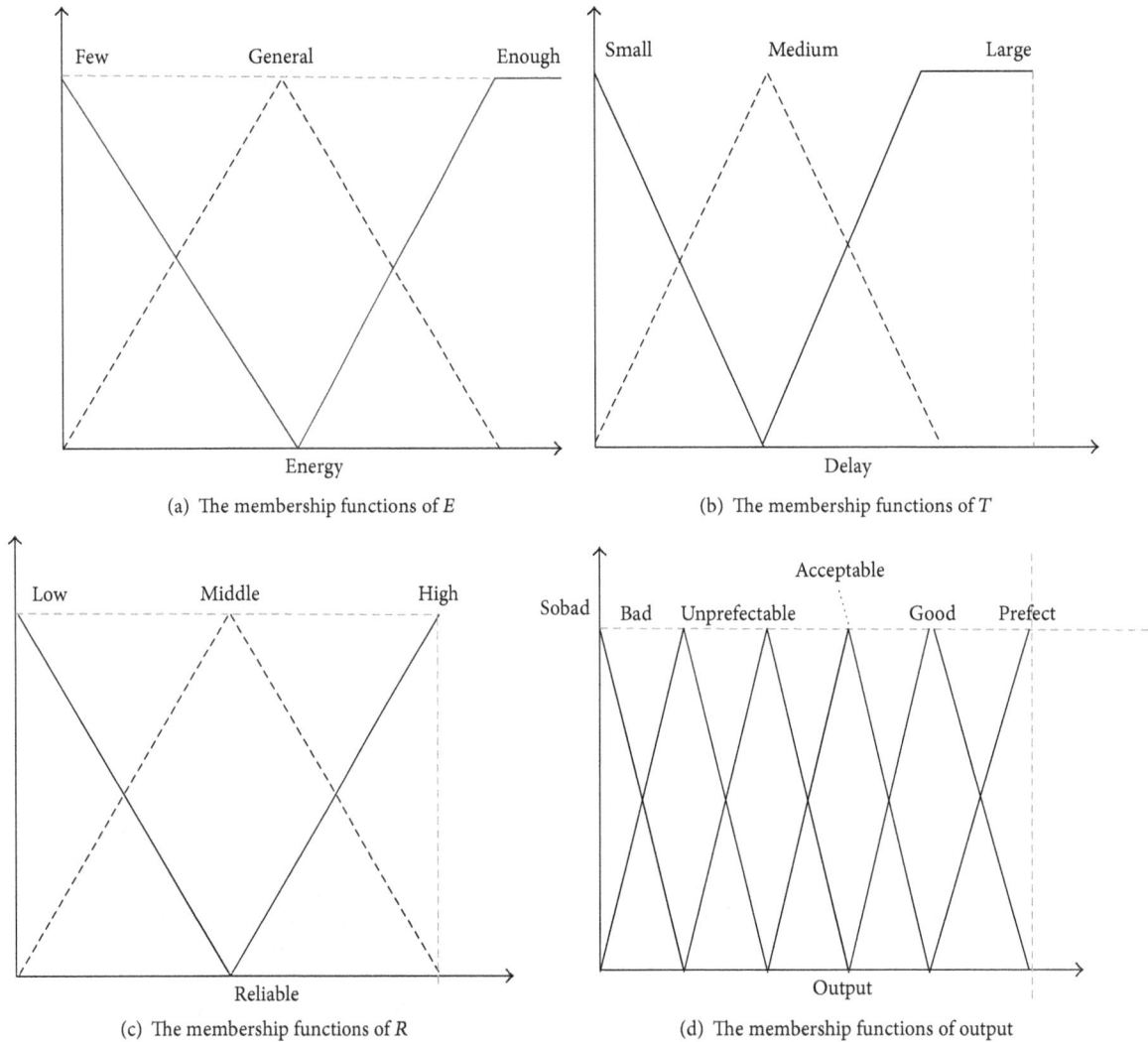

(a) The membership functions of E

(b) The membership functions of T

(c) The membership functions of R

(d) The membership functions of output

FIGURE 4: The membership function of the input and output.

TABLE 2: Routing table of node 5.

Current	Parent	(R, T)
5	2	(0.36, 50)
5	3	(0.48, 90)
5	3	(0.288, 100)
5	4	(0.432, 20)
5	4	(0.2592, 30)

of performance of the algorithm in case the network load increases. In the case, the number of data packets is 50 and 100, we compare entries 2 and 4 of the performance of time delay and reliability.

With the increase of load, node consumes a certain amount of energy, under the condition of invariability of R and T. When the 31st packet is transferred, because of the loss of node energy, the transmission path change to entry 1. It proves that selection of transmission path will change with the increase of network load in the network environment,

which reasonably uses the network resources, as is shown by the red line in the figure. The time delay and success ratio about classic CTP and F-CTP are shown in Figure 6 with the load of 50 and 100.

6. Conclusion

The fast growth of the biological physiological sensors, integrated circuit with low consumption, and the wireless communication makes the wearable wireless sensor network develop towards the individual physiological monitoring field. Routing Algorithm Research on wearable wireless network is the key to ensure the communication of nodes in the network. Therefore, the research of this paper will provide important evidence for the application of WSN in the individual physiological monitoring field.

This paper gives a design of hardware system used for individual physiological monitoring which integrates positioning module using the Big Dipper on the basis of physiological monitoring such as ECG status and gesture

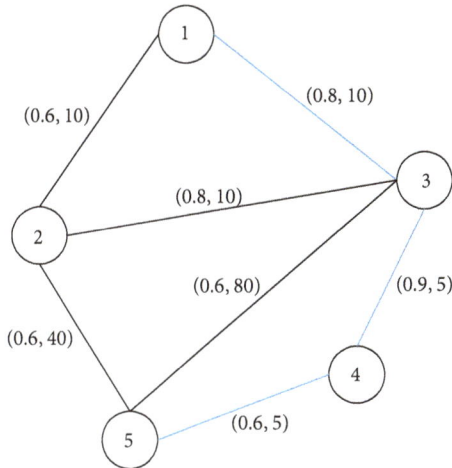

FIGURE 5: The topology of wireless sensor network for simulation.

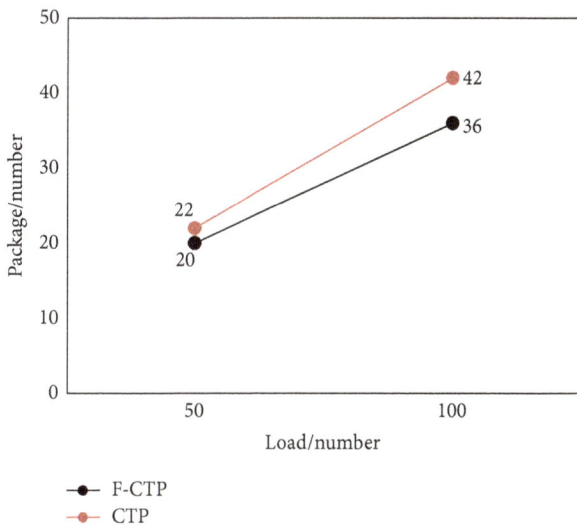

(a) Transmission quality comparison of different load cases

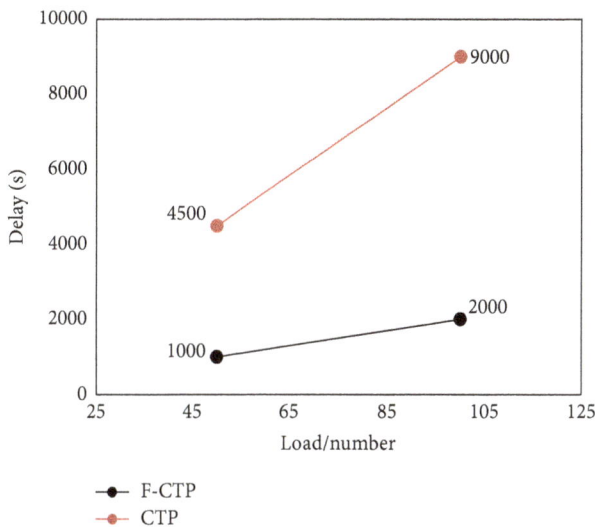

(b) Delay comparison of different load cases

FIGURE 6: The result of simulation.

information. Multiobjective optimization of real time, reliability, and energy balance is one of the important research directions in the field of wearable wireless sensor. In this paper, we develop a routing algorithm using the fuzzy logic based on the CTP protocol, which is applied in the individual wearable wireless sensor network monitoring system. The improved F-CTP optimizes the selection of parent node, fully considering the time delay, packet success receiving rate, and residual energy, so that the whole network is more optimized and the life cycle of the whole network is effectively prolonged. But the problem is that our simulation is just a simple verification; the F-CTP put forward in this paper has not been applied in our hardware system and some membership functions used in fuzzy logic algorithms are also needed to be adjusted in more practice in the future research. Otherwise, a new heuristic algorithm can be given, which can search for a suitable routing path faster and better. We hope that wireless sensor network can be better applied to wearable physiological monitoring area.

Conflict of Interests

The authors declare that there is no conflict of interests regarding the publication of this paper.

Acknowledgments

This research is supported by Fundamental Research Funds for the Central Universities (2015YJS027) and National Natural Science Foundation under Grant 61371071.

References

[1] D. P. Quan, *Wireless Sensor Network QoS Routing Studied for the Wearable Physiological Testing*, Donghua University, 2012.

[2] B. P. L. Lo, S. Thiemjarus, R. King, and G.-Z. Yang, "Body sensor network—a wireless sensor platform for pervasive healthcare monitoring," in *Proceedings of the 3rd International Conference on Pervasive Computing*, Munich, Germany, 2005.

[3] M. Yu, S. J. Xiahou, and X. Y. Li, "A survey of studying on task scheduling mechanism for TinyOS," in *Proceedings of the International Conference on Wireless Communications, Networking and Mobile Computing (WiCOM '08)*, pp. 1–4, IEEE, Dalian, China, October 2008.

[4] J. P. Carmo, P. M. Mendes, C. Couto, and J. H. Correia, "2.4 GHz wireless sensor network for smart electronic shirts," in *Smart Sensors, Actuators, and MEMS II*, vol. 5836 of *Proceedings of SPIE*, pp. 579–586, May 2005.

[5] S. Choi and Z. Jiang, "A novel wearable sensor device with conductive fabric and PVDF film for monitoring cardiorespiratory signals," *Sensors and Actuators A: Physical*, vol. 128, no. 2, pp. 317–326, 2006.

[6] C. Otto, A. Milenković, C. Sanders, and E. Jovanov, "System architecture of a wireless body area sensor network for ubiquitous health monitoring," *Journal of Mobile Multimedia*, vol. 1, no. 4, pp. 307–326, 2006.

[7] S. N. C. Po, G. Dagang, M. D. B. M. Hapipi et al., "Overview of MEMSWear II—incorporating MEMS technology into smart shirt for geriatric care," *Journal of Physics: Conference Series*, vol. 34, no. 1, pp. 1079–1085, 2006.

[8] W. S. Zhu and W. Li, "The CTP algorithm based on energy awareness in wireless sensor network routing," *Application Research of Computers*, no. 11, 2014.

[9] P. Kuila, S. K. Gupta, and P. K. Jana, "A novel evolutionary approach for load balanced clustering problem for wireless sensor networks," *Swarm & Evolutionary Computation*, vol. 12, pp. 48–56, 2013.

[10] H. Ronghui, K.-S. Lui, K.-C. Leung, and F. Baker, "An approximation algorithm for QoS routing with two additive constraints," in *Proceedings of the IEEE International Conference on Network Protocols (ICNP '08)*, pp. 328–337, IEEE, Orlando, Fla, USA, October 2008.

[11] W. Wei and L. Zhi, "The multi-objective routing optimization of WSNs based on an improved ant colony algorithm," in *Proceedings of the 6th International Conference on Wireless Communications, Networking & Mobile Computing (WiCOM '10)*, pp. 1–4, IEEE, Chengdu, China, September 2010.

[12] G. D. Su and X. Wang, "The clustering routing protocol based on the fuzzy logical of wireless sensor network," *Computer Application Research*, no. 7, pp. 117–121, 2013.

[13] J.-Y. Chang and P.-H. Ju, "An energy-saving routing architecture with a uniform clustering algorithm for wireless body sensor networks," *Future Generation Computer Systems*, vol. 35, no. 3, pp. 128–140, 2014.

[14] The Collection Tree Protocol (CTP) [OL], 2010, http://www.tinyos.net/tinyos-2.1.0/doc/html/tep123.html.

15

The High Security Mechanisms Algorithm of Similarity Metrics for Wireless and Mobile Networking

Xingwang Wang[1,2]

[1]*School of Computer Engineering and Science, Shanghai University, Shanghai 200444, China*
[2]*Shanghai Vocational Technical College of Agriculture & Forestry, Shanghai 201699, China*

Correspondence should be addressed to Xingwang Wang; wangx_w@shu.edu.cn

Academic Editor: Arun K. Sangaiah

With the development of human society and the development of Internet of things, wireless and mobile networking have been applied to every field of scientific research and social production. In this scenario, security and privacy have become the decisive factors. The traditional safety mechanisms give criminals an opportunity to exploit. Association rules are an important topic in data mining, and they have a broad application prospect in wireless and mobile networking as they can discover interesting correlations between items hidden in a large number of data. Apriori, the most influential algorithm of association rules mining, needs to scan a database many times, and the efficiency is low when the database is huge. To solve the security mechanisms problem and improve the efficiency, this paper proposes a new algorithm. The new algorithm scans the database only one time and the scale of data to deal with is getting smaller and smaller with the algorithm running. Experiment results show that the new algorithm can efficiently discover useful association rules when applied to data.

1. Introduction

With the rapid development of web technology, the number of choices is becoming overwhelming. It takes a long time to filter, prioritize, and efficiently deliver relevant information so as to alleviate the problem of information overload. Recommender systems [1] have grown so fast that they can meet the needs of users' ambiguous requirements. They utilize statistic method and knowledge discovery technology, providing users with personalized content and services by searching through large volume of dynamically generated information. Recently, various approaches for building recommender systems have been developed, which can utilize collaborative filtering, content-based filtering, or hybrid filtering [2–4]. Among the above filtering techniques, collaborative filtering recommendation is the most mature and the most commonly implemented. Collaborative filtering technique can be divided into two classifications; they are model-based filtering and memory-based filtering. The model-based filtering learns a model from the user-item ratings which can be computed offline. Once the model is generated, the process of prediction will be easy and fast. Lots of model-based filtering techniques have been proposed by researchers such as Latent Semantic Indexing (LSI) [5], decision tree [6], Bayesian network approach models [7], and cluster models [8, 9]. Usually, model-based algorithm has better scalability but lower accuracy than memory-based algorithm. Although collaborative filtering technique is commonly used, it still encounters one crucial issue remaining to be solved, namely, data sparsity problem [10–13], thus leading to the nonoptimal nearest neighbors because the core of the collaborative filtering algorithm is to find the k-nearest neighbors [14–18]. For lack of reference rating values, this step of searching neighbors causes big inaccuracy. In the traditional collaborative filtering algorithm, such similarity metrics are used to calculate the similarity between users or items as cosine, Pearson-correlation, and modified cosine [19–22]. All of them present poor performance when they are applied to big data with high sparsity. This paper proposes a new algorithm, considering both user similarity and item ones. Matrix prefilling, a method of preprocessing, is based on association rules which are not proposed by others before when measuring similarity. Experimental results of the proposed model on a real dataset: the dataset proves to generate more accurate prediction results compared to the traditional ones. The remainder of this paper is organized as follows. Section 2 is a brief introduction of association rule whose concept and algorithm will be used in Section 3 to propose a new algorithm. Section 3 focuses on the algorithm for wireless and mobile networking which is the highlight of this paper. Experimental results and analyses are displayed in Section 4. Section 5 is the final part of this paper in which conclusion is reached.

2. Related Work

2.1. Related Concepts of Association Rules. Transaction database $D = \{t_1, t_2, \ldots, t_k, \ldots, t_n\}$ is the set of all the transactions. $I = \{i_1, i_2, \ldots, i_m\}$ is the set of all the items in D [23–25]. Every transaction contains a set of items which is the subset of I. Item set is a collection that contains 0 or more items. If the number of items an item set contains is k, then the item set is called k-item set. *Support* count is an important property of an item set. It indicates the number of a particular item set contained in the transactions. $\sigma(X)$, the *support* count of item set X, is defined as follows:

$$\sigma X = \left| \{t_i \mid X \subseteq t_i,\ t_i \in D\} \right|. \tag{1}$$

$|\cdots|$ represents the number of elements in the collection.

A rule is defined as an implication form $X \rightarrow Y$, where $X \subset I, Y \subset I$, and $X \cap Y = \phi$. *Support* and *confidence* are two important measures of association rules. *Support* indicates the frequency of the rule in a dataset. It is defined as

$$support\ (X \longrightarrow Y) = \frac{\sigma X \cup Y}{N}. \tag{2}$$

N is the total number of transactions.

The *confidence* of a rule $X \rightarrow Y$ is the proportion of transactions that contains X which also contains Y. It is defined as

$$confidence\ (X \longrightarrow Y) = \frac{\sigma X \cup Y}{\sigma X}. \tag{3}$$

Support and *confidence* are two important measures to evaluate association rules. Rules with low *support* may occur only occasionally which are meaningless in most cases. Therefore, *support* is often used to delete those meaningless rules. *Confidence* is a measure of accuracy of association rules. If the *confidence* of the rule $X \rightarrow Y$ is high, the possibility of Y appearing in the transactions which contain X is larger.

2.2. Apriori Algorithm. Apriori is a typical algorithm with candidate set generated. It uses the *support* based pruning method and a level-wise and breadth-first search to discover the frequent item sets. Apriori uses two properties below to compress search space.

Lemma 1. *If the item set X is frequent, then all nonempty subsets are frequent too.*

Lemma 2. *If the item set X is nonfrequent, then all supersets are nonfrequent too.*

Candidate item set generation is a very critical step. It should ensure that the candidate item sets are complete while avoiding too many unnecessary candidates. This step consists of two parts.

(1) In the join step, this paper joins two frequent $(k\text{-}1)$-item sets L1 and L2 to generate candidate k-item sets. This paper should make sure that the first $k\text{-}2$ items of L1 and L2 are the same. Then, the first $k\text{-}2$ items and the last item of L1 as well as the last item of L2 compose the candidate k-item set.

(2) In the pruning step, this paper uses a strategy to delete some unnecessary candidates. According to Lemmas 1 and 2, for each k-item set generated, this paper examines whether all the $k\text{-}1$ subsets are frequent. If not, this paper removes it from the candidate k-item sets.

Apriori algorithm effectively filters the unnecessary candidates. It will get a good data mining result, especially for short pattern data. However, the weakness is that the database needs to be scanned many times. It will produce tremendous I/O cost. Another weakness is that a lot of candidate item sets may be generated. It will cost a lot of time and memory space.

2.3. FP-Growth Algorithm. FP-Growth is a classic algorithm without candidate item sets generated. It compresses the data into a structure called FP-tree. The frequent item sets are discovered by doing a recursive search of the FP-tree.

The process of FP-Growth mainly consists of two steps.

(1) Constructing the FP-Tree. When the database is scanned for the first time, this paper selects the items which satisfy the minimum *support* and puts these items to a header table with a descending sort order according to *support*. When the database is scanned for the second time, the items contained in a transaction are sorted according to their order in the header table and are inserted in the FP-tree. Then combine the same paths in the tree.

(2) Discovering Frequent Item Sets by Searching the FP-Tree. If the FP-tree contains only one path, enumerate all the possible item sets. If not, for each item in the header table, this paper creates its conditional pattern base so as to construct the conditional pattern tree. The recursive process will not stop until the tree is empty.

FP-growth algorithm scans the database only two times and avoids the generation of candidate item sets, but the weakness is that when the database is huge, the FP-tree is too large and even cannot be constructed in memory because all the records in database are compressed into the FP-tree.

3. The Improved Apriori Algorithm Based on Matrix

To avoid the weakness of apriori algorithm, this paper proposes an improved algorithm on the basis of apriori algorithm. This paper converts the transaction database to

TABLE 1: A transaction database.

TID	Items
1	I_2, I_5
2	I_1, I_2, I_4
3	I_1, I_3, I_4
4	I_2, I_3, I_4, I_5

TABLE 2: A Boolean matrix.

R	I_1	I_2	I_3	I_4	I_5
T_1	0	1	0	0	1
T_2	1	1	0	1	0
T_3	1	0	1	1	0
T_4	0	1	1	1	1

a Boolean matrix and deletes the unnecessary rows and columns of the matrix to reduce the scale of the data.

3.1. Related Concept. Association rules usually focus on transaction databases. If this paper converts the transaction database to a Boolean matrix, on the one hand, the database can be scanned only one time so as to reduce the cost of I/O and, on the other hand, it may reduce the memory consumption when the data is in the form of 0 and 1.

Definition 3. Let $I = \{I_1, I_2, \ldots, I_n\}$ be an item set and $T = \{T_1, T_2, \ldots, T_m\}$ be a set of transactions in the database and each transaction in T has a unique transaction id called TID. The method by which transactions are converted into a Boolean matrix is as follows: let R be the binary relation from I to T. $r_{ij} = R(T_i, I_j)$, $R = (r_{ij})_{m \times n}$. Then

$$r_{ij} = \begin{cases} 1, & I_j \in T_i \\ 0, & I_j \notin T_i \end{cases} \quad i = 1, 2, \ldots, m; \; j = 1, 2, \ldots, n. \quad (4)$$

An example of a transaction database is in Table 1. The Boolean matrix of the database is in Table 2.

The column vector I_j of the Boolean matrix is defined as $I_j = \{r_{1j}, r_{2j}, \ldots, r_{mj}\}$. The *support* count of I_j is

$$support\_count \{I_j\} = \sum_{i=1}^{n} (r_{ij}). \quad (5)$$

For k-item set $\{I_1, I_2, \ldots, I_k\}$, its *support* count is

$$support\_count \{I_1, I_2, \ldots, I_k\}$$
$$= \sum_{i=1}^{n} (r_{i1} \wedge r_{i2} \wedge \cdots \wedge r_{ik}). \quad (6)$$

$\wedge$ is "and" operation. When $I_1, I_2, \ldots, I_k$ are simultaneously 1, the *support* count is incremented by 1.

Lemma 4. *If the number of "1" instances contained in a row of Boolean matrix is less than k, then when this paper counts the support of k-item set, this row can be deleted from the matrix.*

According to the definition of support count,

$$support\_count \{I_1, I_2, \ldots, I_k\}$$
$$= \sum_{i=1}^{n} (r_{i1} \wedge r_{i2} \wedge \cdots \wedge r_{ik}). \quad (7)$$

If the number of "1" instances contained in a row is less than k, there will exist j which makes $r_{ij} = 0$; then $r_{i1} \wedge r_{i2} \wedge \cdots \wedge r_{ik} = 0$. Therefore this row makes no contribution to the support count of k-item set.

Lemma 5. *If there is an item I_j, the number of I_j instances that appear in frequent k-item sets L_k is less than k; the column of I_j can be deleted in the process of frequent $(k + 1)$-item set generation.*

Let Y be a frequent $(k + 1)$-item set; then all its k-subsets are frequent. For each $I_j \in Y$, the number of I_j instances that appear in frequent k-item sets should be k. if the number is less than k, then I_j will not be the element of the frequent $(k + 1)$-item set.

3.2. The Searching of k-Nearest Neighbors. After the process above, this paper takes user similarity into account. The similarity of user i and user j is computed as (8). I denotes the set of all the items.

$$sim(i, j) = \frac{\sum_{c \in I} \left(R_{i,c} - \overline{M_i}\right)\left(R_{j,c} - \overline{R_j}\right)}{\sqrt{\sum_{c \in I} \left(R_{i,c} - \overline{R_i}\right)^2} \sqrt{\sum_{c \in I} \left(R_{j,c} - \overline{R_j}\right)^2}}. \quad (8)$$

For each user u, the aim to find the k-nearest neighbor is to find a user set $U = \{U_1, U_2, \ldots, U_k\}$, $u \notin U$, $sim(u, U_1)$ has the highest value, and $sim(u, U_2)$ has the second highest value, and so on.

3.3. The Generation of Recommendation. After the step of finding the k-nearest neighbors, the next step is to generate recommendations. Let the set of k-nearest neighbors of user u be NN_u and the rating that user u give to the item i be $R_{u,i}$; the calculation is as follows:

$$R_{u,i} = \begin{cases} \overline{R_u} + \dfrac{\sum_{n \in NN_u} sim(u, n) \times \left(R_{n,i} - \overline{R_n}\right)}{\sum_{n \in NN_u} \left(|sim(u, n)|\right)}, & \text{if } r_{u,i} \cdot flag = 2 \\ r_{u,i}, & \text{if } r_{u,i} \cdot flag = 0. \end{cases} \quad (9)$$

3.4. Description of the Improved Algorithm. The process of the improved algorithm is described in Algorithm 1. First this paper converts a database to a Boolean matrix. Then according to Lemmas 4 and 5 the unnecessary rows and columns of the matrix are deleted with the algorithm running.

The improved algorithm based on matrix is shown in Algorithm 1.

3.5. Evaluation Criteria. Not all association rules are useful, so it is necessary to select the association rules in which we

Input: dataset D, minimum *support* minsup
Output: all the itemsets satisfied with minsup
(1) Scan transaction database D and convert it to a Boolean matrix.
(2) Calculate the *support* of every 1-itemset. Item sets whose *support* are not less than min sup compose frequent 1-item sets L_1. Delete the columns of the infrequent items. Delete the rows in which the number of "1" contained is less than 2.
(3) for ($k = 2$; $L_{k-1} \neq \phi$; k++) do begin
(4) Combine the items of each column and generate candidate k- item sets C_k.
(5) Calculate the *support* of C_k.
(6) Item sets whose *support* is not less than min sup compose frequent k-item sets L_k.
(7) Delete the columns of the Items which are contained in infrequent item sets and the number Of which appear in L_k is less than k.
(8) Delete the rows in which the number of "1" contained is less than $k + 1$
(9) End

ALGORITHM 1: The procedure of the improved algorithm based on matrix.

are interested. *Support* and *confidence* are two basic criteria to evaluate if an association rule is useful. However in some case the two criteria may give us an unexpected suggestion. So this paper uses the criterion called *lift* to evaluate the association rules in addition to *support* and *confidence*. The *lift* of a rule $X \rightarrow Y$ is defined as

$$\text{lift}(X \longrightarrow Y) = \frac{confidence(X \longrightarrow Y)}{support(Y)}. \tag{10}$$

Lift is the radio of a rule's *confidence* and the consequent's *support*. If the value of *lift* is 1, X and Y are independent. If the value is above 1, X and Y are positively correlated. If the value is below 1, X and Y are negatively correlated.

3.6. Performance Analysis. Compared with apriori algorithm, the improved algorithm scans the database only one time. It converts the transaction database to a matrix. The remaining steps are operated on the matrix without scanning the database again. This will reduce the I/O cost. The other advantage of the improved algorithm is that the scale of data to be dealt with is getting smaller and smaller with the algorithm running. In the process of frequent item sets generation, the columns of items which will not be contained in frequent item sets and the rows which make no contribution to the *support* count will be deleted. Therefore the scale of the matrix will be smaller and smaller and the efficiency will be improved a lot. On the other hand, when a transaction contains many items, compared with transaction list, a Boolean matrix occupies less memory space.

4. Results and Analysis

To access the performance of the improved algorithm, this paper uses apriori algorithm and the improved algorithm proposed in this paper to mine frequent item sets from different agricultural databases. The experiments were performed on an Intel i5-2450 processor 2.5 GHz with 4G memory, running Windows 8. This paper used R language to code the algorithms.

TABLE 3: Runtime of the two algorithms on mushroom dataset.

Support	0.60	0.65	0.70	0.75	0.80
Apriori algorithm	3.25	2.12	1.95	1.85	1.72
Improved algorithm	2.72	1.73	1.63	1.54	1.42

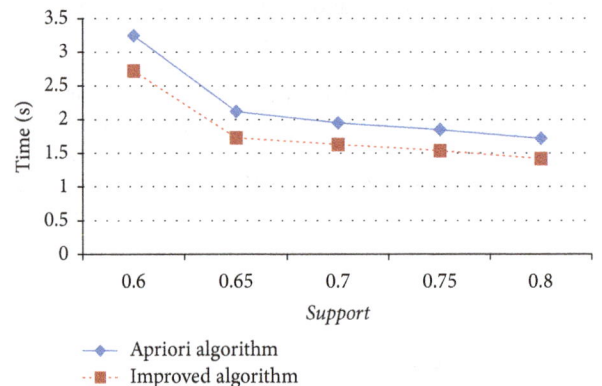

FIGURE 1: Runtime of the two algorithms on mushroom dataset.

Table 3 and Figure 1 show the performance of the two algorithms in the UCI dataset named mushroom. The dataset contains 7847 records and 118 items. The minimum *confidence* is set to be 0.5 and the minimum *support* is set, respectively, to be 0.60, 0.65, 0.70, 0.75, and 0.80.

Table 4 and Figure 2 show the performance of the two algorithms in the UCI dataset named soybean. The dataset contains 5264 records and 655 items. The minimum *confidence* is set to be 0.5 and the minimum *support* is set, respectively, to be 0.75, 0.76, 0.77, 0.78, 0.79, and 0.80.

The results show that the runtime of the improved algorithm is less than apriori algorithm. The improved algorithm is more effective than apriori algorithm.

The evaluation method *lift* is used to optimize the mining result. A subset of the mining result of mushroom dataset is shown in Algorithm 2. Algorithm 2 shows the association rules whose *support* and *confidence* are high but whose *lift* is 1. It means that the antecedent and consequent are

TABLE 4: Runtime of the two algorithms on soybean dataset.

Support	0.75	0.76	0.77	0.78	0.79	0.80
Apriori algorithm	0.29	0.20	0.17	0.12	0.10	0.07
Improved algorithm	0.21	0.135	0.082	0.075	0.065	0.05

(1) Gill-attachment = free → veil-type = partial *support* = 0.9732382, *confidence* = 1, *lift* = 1

(2) Veil-color = white → veil-type = partial *support* = 0.9745126, *confidence* = 1, *lift* = 1

(3) Ring-number = one → veil-type = partial *support* = 0.9336052, *confidence* = 1, *lift* = 1

(4) Gill-attachment = free, Veil-color = white → veil-type = partial *support* = 0.9722187, *confidence* = 1, *lift* = 1

(5) Gill-attachment = free, Ring-number = one → veil-type = partial *support* = 0.9091372, *confidence* = 1, *lift* = 1

(6) Veil-color = white, Ring-number = one → veil-type = partial *support* = 0.9081178, *confidence* = 1, *lift* = 1

(7) Gill-attachment = free, Veil-color = white, Ring-number = one → veil-type = partial *support* = 0.9081178, *confidence* = 1, *lift* = 1

ALGORITHM 2: A subset of the mining result of mushroom dataset.

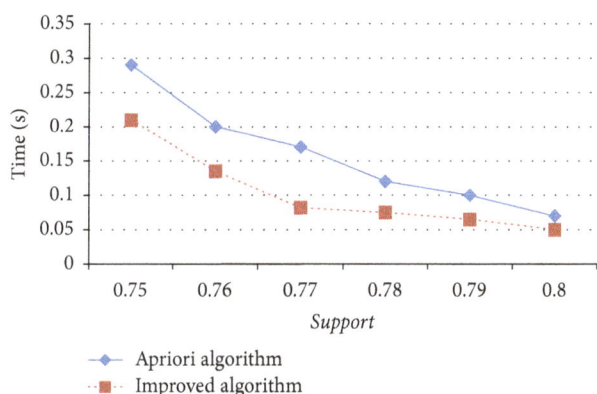

FIGURE 2: Runtime of the two algorithms on soybean dataset.

independent, and these association rules are not the rules this paper expect even though they have high *support* and *confidence*.

5. Conclusion

To avoid the weakness of apriori algorithm, this paper proposes an improved algorithm based on matrix and applies the improved algorithm to agricultural datasets. Experimental results show that the improved algorithm can efficiently discover useful association rules for the reason that database will be scanned only one time and that the data to deal with is getting smaller and smaller with the algorithm running. The improved algorithm is more applicable when the database is huge. But it is not that efficient compared with apriori when the database is not that large due to the fact that the scale data to deal with is small but the improved algorithm has an extra operation to covert the database to a matrix. Further research should be focused on the optimization of the proposed algorithm so as to further improve the efficiency

when applied to big data. Algorithm parallelization can be taken into account. Therefore our future work is to improve our algorithm so as to be applicable for more kinds of database. Besides, new evaluation criteria can be used to optimize our mining result.

Conflicts of Interest

There are no conflicts of interest.

References

[1] Y. Pang, Y. Jin, Y. Zhang, and T. Zhu, "Collaborative filtering recommendation for MOOC application," *Computer Applications in Engineering Education*, vol. 25, no. 1, pp. 120–128, 2017.

[2] J. Wei, J. He, K. Chen, Y. Zhou, and Z. Tang, "Collaborative filtering and deep learning based recommendation system for cold start items," *Expert Systems with Applications*, vol. 69, pp. 29–39, 2017.

[3] L. Qi, W. Dou, and X. Zhang, "Service recommendation based on social balance theory and collaborative filtering," in *Proceedings of the Intelligence and Lecture Notes in Bioinformatics, 14th International Conference*, vol. 9936 of *Lecture Notes in Computer Science*, pp. 637–645, Springer International Publishing, Basel, Switzerland, 2016.

[4] B. Kyoungsoo and S.-G. Cheongju, "Social group recommendation based on dynamic profiles and collaborative filtering," *Neurocomputing*, vol. 209, pp. 3–13, 2016.

[5] Y. Gao, "Collaborative filtering recommendation model based on normalization method," *International Journal of Grid and Distributed Computing*, vol. 9, no. 10, pp. 291–300, 2016.

[6] M. Sun, F. Li, J. Lee et al., "Learning multiple-question decision trees for cold-start recommendation," in *Proceedings of the 6th ACM International Conference on Web Search and Data Mining (WSDM '13)*, pp. 445–454, ACM, Rome, Italy, February 2013.

[7] T.-H. Ma, L.-M. Guo, M. Li, M.-L. Tang, Y. Tian, and A. Mznah, "A collaborative filtering recommendation algorithm based on hierarchical structure and time awareness," *IEICE Transactions on Information and Systems*, no. 6, pp. 1512–1520, 2016.

[8] H. L. dos Santos, C. Cechinel, R. M. Araujo, and M.-Á. Sicilia, "Clustering learning objects for improving their recommendation via collaborative filtering algorithms," *Communications in Computer and Information Science*, vol. 544, pp. 183–194, 2015.

[9] P. Krupa, A. Thakkar, C. Shah, and K. Makvana, "A state of art survey on shilling attack in collaborative filtering based recommendation system," *Smart Innovation, Systems and Technologies*, vol. 50, pp. 377–385, 2016.

[10] P. Mirko and A. Fabio, "Kernel based collaborative filtering for very large scale top-N item recommendation," in *Proceedings of the 24th European Symposiumon Artificial Neural Networks*, pp. 11–16, 2016.

[11] Y. Shen, T.-G. Lv, X. Chen, and Y.-D. Wang, "A collaborative filtering based social recommender system for E-commerce," *International Journal of Simulation: Systems, Science and Technology*, vol. 17, no. 22, pp. 91–96, 2016.

[12] S. Rossi, F. Barile, D. Improta, and L. Russo, "Towards a collaborative filtering framework for recommendation in museums: from preference elicitation to group's visits," in *Proceedings of the 7th International Conference on Emerging Ubiquitous Systems and Pervasive Networks, EUSPN 2016 / The 6th International Conference on Current and Future Trends of Information and Communication Technologies in Healthcare, ICTH '16*, pp. 431–436, Elsevier, London, UK, September 2016.

[13] B. Urszula, "Differential evolution in a recommendation system based on collaborative filtering," in *Proceedings of The Lecture Notes in Computer Science Including Subseries Lecture Notes in Artificial Intelligence and Lecture Notes in Bioinformatics*, Lecture Notes in Computer Science, pp. 113–122, Springer International Publishing.

[14] W. Ebisa and W. Vitor, "User-based collaborative filtering recommender systems approach in industrial engineering curriculum design and review process," in *Proceedings of the ASEE Annual Conference and Exposition*, 2016.

[15] J. Jinhyun, B. Sangwon, and P. Geunduk, "Implementation of a recommendation system using association rules and collaborative filtering," in *Proceedings of the Proceedings of the 4th International Conference on Information Technology and Quantitative Management, ITQM '16*, pp. 944–952, 2016.

[16] Y. K. Ng, "Recommending books for children based on the collaborative and content-based filtering approaches," in *Proceedings of the Computational Science and Its Applications-ICCSA '16*, Lecture Notes in Computer Science, pp. 302–317, Springer International Publishing.

[17] H. H. Qiu, Y. Liu, Z. J. Zhang, and G. X. Luo, "An improved collaborative filtering recommendation algorithm for microblog based on community detection," in *Proceedings of the 2014 Tenth International Conference on Intelligent Information Hiding and Multimedia Signal Processing (IIH-MSP)*, pp. 876–879, IEEE, Kitakyushu, Japan, August 2014.

[18] K. Kim and H. Ahn, "Recommender systems using cluster-indexing collaborative filtering and social data analytics," *International Journal of Production Research*, vol. 55, no. 17, pp. 5037–5049, 2017.

[19] C.-Y. Li and K.-J. He, "An optimized map reduce for item-based collaborative filtering recommendation algorithm with empirical analysis," *Concurrency Computation*, 2017.

[20] N. Polatidis and C. K. Georgiadis, "A dynamic multi-level collaborative filtering method for improved recommendations," *Computer Standards & Interfaces*, vol. 51, pp. 14–21, 2017.

[21] R. L. Palak, "An effective collaborative filtering based method for movie recommendation," *Advances in Intelligent Systems and Computing*, vol. 506, pp. 149–159, 2017.

[22] M. Liu, Z. Zeng, W. Pan, X. Peng, Z. Shan, and Z. Ming, "Hybrid One-Class Collaborative Filtering for Job Recommendation," in *Smart Computing and Communication*, Lecture Notes in Computer Science, pp. 267–276, Springer International Publishing, 2017.

[23] M. Sridevi and R. R. Rao, "An enhanced personalized recommendation utilizing expert's opinion via collaborative filtering and clustering techniques," in *Proceedings of the 2016 International Conference on Inventive Computation Technologies (ICICT)*, pp. 1–4, IEEE, Coimbatore, India, August 2016.

[24] H. Zhang, I. Ganchev, N. S. Nikolov, and M. O'droma, "A trust-enriched approach for item-based collaborative filtering recommendations," in *Proceedings of the 12th IEEE International Conference on Intelligent Computer Communication and Processing, ICCP '16*, pp. 65–68, September 2016.

[25] L.-Y. Dong, G.-L. Zhu, Q. Zhu, and Y.-L. Li, "Research on collaborative filtering recommendation based on k-means clustering," *ICIC Express Letters*, pp. 2493–2498, 2016.

16

Optimal Contract Design for Cooperative Relay Incentive Mechanism under Moral Hazard

Nan Zhao, Minghu Wu, Wei Xiong, and Cong Liu

Hubei Collaborative Innovation Center for High-Efficiency Utilization of Solar Energy, Hubei University of Technology, Wuhan 430068, China

Correspondence should be addressed to Nan Zhao; nzhao@mail.hbut.edu.cn

Academic Editor: George S. Tombras

Cooperative relay can effectively improve spectrum efficiency by exploiting the spatial diversity in the wireless networks. However, wireless nodes may acquire different network information with various users' location and mobility, channels' conditions, and other factors, which results in asymmetric information between the source and the relay nodes (RNs). In this paper, the relay incentive mechanism between relay nodes and the source is investigated under the asymmetric information. By modelling multiuser cooperative relay as a labour market, a contract model with moral hazard for relay incentive is proposed. To effectively incentivize the potential RNs to participate in cooperative relay, the optimization problems are formulated to maximize the source's utility while meeting the feasible conditions under both symmetric and asymmetric information scenarios. Numerical simulation results demonstrate the effectiveness of the proposed contract design scheme for cooperative relay.

1. Introduction

With the explosive development of wireless services and devices, wireless spectrum has become more congested and scarce [1]. Cooperative relay technology [2, 3] exploits spatial diversity and weakens unfavourable effects of wireless channels, which can improve spectrum efficiency effectively.

However, to design an efficient cooperative relay mechanism is considered to be challenging. As spectrum and energy are the natural resources in all the wireless communication systems [4], the relay nodes (RNs) may compete for the limited spectrum resources [5] considering their own benefit because of their selfishness [6]. Thus, the RNs may be unwilling to offer their relay help without any extra incentives. Moreover, due to the mobility of wireless nodes and the shadowing and fading effects of wireless channels, certain network information (i.e., node locations, relay actions) may not be available to all the users [7], which causes the network information to be asymmetric between the RNs and the source. Therefore, the objective of this work is to address these challenging issues by proposing a contract-theoretic cooperative relay mechanism under asymmetric information scenario.

Contract theory [8] is an economic concept to investigate the mutually agreeable contract among economic players in the presence of asymmetric or incomplete information scenarios [9]. Recently, contract-based methods have been suggested for cooperative systems with various strategies, such as resource exchange [7], integrated contraction [10], and profit incentive [11]. Our prior work developed an efficient contract model for adverse selection in the presence of the RNs' hidden relay information [12]. However, unlike the existing works, in this paper, we intentionally concentrate on the *moral hazard* problem caused by the RNs' hidden relay actions. To the best of our knowledge, this relay incentive mechanism had not been studied.

In this paper, the relay incentive mechanism between the RNs and the source is investigated under asymmetric information scenario. The main contributions of this paper are as follows:

(i) The contract-theoretic model for relay incentive is proposed to achieve the *effort-incentive* objective. The *bonus ratio* related to the RNs' relay performance is introduced to motivate the RNs to carry out relay effectively. The RNs' basic wage paid by the source is various with their different relay efforts and relay abilities.

(ii) The optimal contract designs are investigated under both symmetric and asymmetric information scenarios. In the presence of the symmetric information, each RN can get a retained utility by choosing the optimal contract. And a principal-agent model with *moral hazard* is proposed to combat detrimental effects of the RNs' hidden relay actions in the presence of the asymmetric information.

(iii) The optimization problems are formulated to maximize the source's utility while meeting the feasible conditions of the potential relay RNs under the above two information scenarios. Simulations demonstrate the performance of optimal contract-based cooperative communication mechanism. The optimal contract can achieve the same source's utility as under both symmetric information and asymmetric information scenarios.

The rest of the paper is organized as follows. Section 2 introduces the system model. The optimal contract designs under both symmetric information and asymmetric information scenarios are discussed in Sections 3 and 4, respectively. Section 5 demonstrates the performance evaluation results, and the conclusion is given in Section 6.

2. System Model

In this paper, a typical wireless cooperative network with one source and multiple RNs is considered, as shown in Figure 1. The source has high priority to access the licensed spectrum with poor channel condition between its transmitter and its receiver; thus, it needs the RNs' relay help. Due to the selfish nature of wireless nodes, the source wants to get the RNs' relay help as much as possible, which is against the RNs' interests. And the RNs want to offer a little help with large reward. Thus, to deal with the conflicting objectives between the source and the RNs, in this cooperative communication scheme, relay incentive is modelled as a labour market. The source, as the employer, offers the contract to recruit the certain RNs for the relay help. And RN, as the employee, chooses one of the contract items to participate in cooperative relay. The set of the involved RNs is denoted as κ. The contract with several different items related to the different combinations of *relay effort*, *basic wage*, and *performance bonus* is utilized.

In Sections 2.1 and 2.2, we will describe in detail how the source and the RNs evaluate the trade-off among *relay effort*, *basic wage*, and *performance bonus*. Further, the contract-theoretic model to balance the interests of both sides is proposed in Section 2.3.

2.1. Source Modelling. In this subsection, the source's model with the RNs' relay help and reward is considered. The source's increased profit due to cooperative relaying of the *i*th RN can be expressed as

$$\pi_i = \rho \log \left(1 + \frac{p_i}{n_0}\right), \tag{1}$$

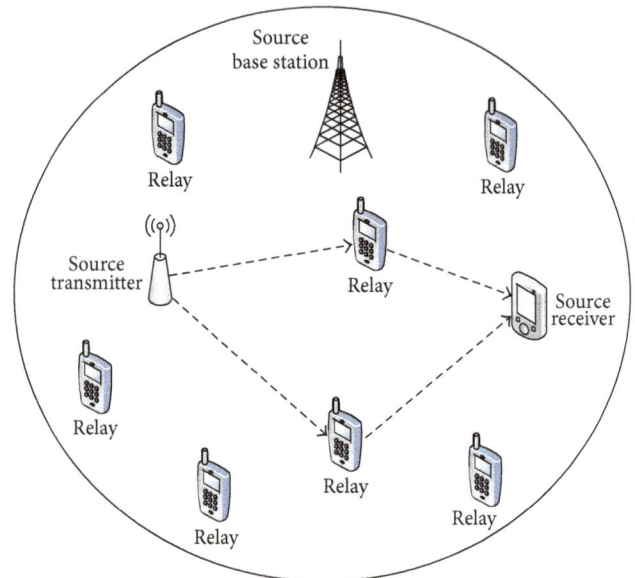

FIGURE 1: Wireless cooperative network.

where p_i is the *i*th RN's transmitting power at the source's receiver, n_0 is the noise power, and $\rho > 0$ is the equivalent profit of per-unit channel capacity, which is assumed to be identical for all the RNs.

Based on the linear sharing scheme [13], the payment w_i to the *i*th RN is defined as

$$w_i = \alpha_i + \beta_i p_i, \tag{2}$$

where α_i is the basic wage of the *i*th RN and $\beta_i \in [0, 1]$ is the performance-based bonus coefficient. Due to the different relay abilities and actions, the RNs may obtain the different basic wages and bonuses.

Then, the source's utility achieved by cooperative communication is defined as

$$U_S = \sum_{i \in \kappa} (\pi_i - w_i) = \sum_{i \in \kappa} \left[\rho \log \left(1 + \frac{p_i}{n_0}\right) - \alpha_i - \beta_i p_i\right]. \tag{3}$$

In the cooperative networks, a RN transmitter can transmit simultaneously with the other RN transmitter. And the interference among the users can be mitigated by utilizing the spectrum sensing [14], for example, through asymmetric transmitter cooperation (i.e., asymmetric cognitive behaviour) or symmetric transmitter cooperation (i.e., cooperative behaviour) [15]. Then, there could be no competition or interference either among the RNs or between the RNs and the source. Note that the source only chooses to offer a cooperative relay contract when its cooperative utility is larger than zero. In that case, the number of the relay RNs is at least one in cooperative communication. Without loss of generality, the value of n_0 is normalized to be 1 for simplification in the following analysis.

2.2. Relay Node Modelling. Next, the relay communication cost $C_i(p_i)$ of the *i*th RN with the relay powers p_i is considered. Since the more relay powers the RNs consume, the more relay cost the RNs pay, $C_i(p_i)$ is nonnegative and monotone increasing on p_i. Moreover, $C_i(p_i)$ grows more rapidly in the large relay power than it does in the small relay power [16]. Thus, we have $C_i'(p_i) > 0$ and $C_i''(p_i) > 0$. Then, the relay cost $C_i(p_i)$ of the ith RN can be simply represented as

$$C_i(p_i) = \theta_i p_i^2, \tag{4}$$

where θ_i is the relay coefficient of the ith RN, which describes the ith RN's transmission information (i.e., relay channel's condition or battery technology). Assume that the relay coefficient remains constant during the cooperative communication. A lower θ_i means that the ith RN has a lower relay cost and a higher relay ability.

Then, the ith RN's utility can be written as

$$U_{\text{RN}_i} = w_i - C_i(p_i) = \alpha_i + \beta_i p_i - \theta_i p_i^2. \tag{5}$$

Notice that the RN's utility U_{RN_i} is increasing in the basic wage α_i and the performance-based bonus β_i.

2.3. Contract Formulation. In this subsection, the contract-based incentive mechanism is presented to motivate the RNs to carry out relay efficiently in the presence of the hidden action. After contracting between the source and the RNs, the source may not monitor the RNs' exact relay actions because of the asymmetric information, which would influence the performance of cooperative relay. Therefore, the source needs to design a contract to incentivize the RNs to participate in relay communications efficiently and credibly.

Considering that the contracts with a finite number of values can be easily and efficiently broadcast in the cooperative networks, in this work, we focus on contract design with N discrete RN types denoted by a set $\Theta = \{\theta_1, \theta_2, \ldots, \theta_N\}$. Without loss of generality, assume that $0 < \theta_1 < \theta_2 < \cdots < \theta_N$. And the type-$\theta_i$ RNs' number is N_i.

In this work, we consider optimal contract design for the following two information scenarios. Symmetric information scenario in Section 3 is a benchmark case, where the source knows each RN's relay information. In this case, the source will achieve the maximum relay utility, which is considered as an upper bound of the source's achievable utility under any network information scenario. And, under asymmetric information scenario in Section 4, the source does not know each RN's relay action. In this case, the source will achieve the same maximum utility with the optimal contract design as in the symmetric information scenario, which will be demonstrated in the following illustrations.

Once the optimal contract is designed, the complete representation of relay incentive mechanism can be described as follows.

Step 1. The source broadcasts a set of contracts to all the geographically potential RNs.

Step 2. After receiving the contract, each RN that is willing to accept the certain contract informs the source of its choice.

Step 3. After receiving all the RNs' confirmations, the source informs the involved RNs of cooperative instructions and messages (i.e., the space-time codes to use), and the RNs help to transmit the source's traffic.

3. Optimal Contract Design under Symmetric Information Scenario

In the presence of the symmetric information, relay information of each RN is opened to the source clearly, and only one issue is for confirmation by the source that each RN accepts only the contract item $\{p_i, \alpha_i\}$ designed with the utility at least as much as it would obtain by declining the contract. And this minimum utility is called the *retained utility* $\overline{U}$. Thus, the following individually rational (IR) constraint has to be satisfied for the contract:

$$\text{(IR)} \quad \alpha_i + \beta_i p_i - \theta_i p_i^2 \geq \overline{U}, \quad \forall i \in \Omega, \tag{6}$$

where $\Omega = \{1, 2, \ldots, N\}$.

To achieve the sources maximum utility under symmetric information scenario, an optimal contract has to be fulfilled and it could be designed as follows:

$$\max_{\{\{p_i, \alpha_i\} \geq 0, \forall i \in \Omega\}} \sum_{i \in \Omega} N_i \left[\rho \log(1 + p_i) - \alpha_i - \beta_i p_i \right],$$
$$\text{s.t.} \quad \text{(IR)} \quad \alpha_i + \beta_i p_i - \theta_i p_i^2 \geq \overline{U}, \quad \forall i \in \Omega. \tag{7}$$

Lemma 1. *To maximize the source's utility, each RN obtains the retained utility; that is, $\alpha_i^* = \overline{U} + \theta_i p_i^2 - \beta_i p_i$, $\forall i \in \Omega$.*

Proof. We provide proof by contradiction. An optimal contract item (p_i, α_i) with $\alpha_i + \beta_i p_i - \theta_i p_i^2 > \overline{U}$ is supposed. It is known that the source's utility described in (3) decreases with α_i, and the sources maximum utility could be obtained by decreasing α_i until $\alpha_i + \beta_i p_i - \theta_i p_i^2 = \overline{U}$. It contradicts the above assumption, and thus the lemma is proved. □

According to Lemma 1, problem (7) could be simplified as

$$\max_{\{p_i \geq 0, \forall i \in \Omega\}} \sum_{i \in \Omega} N_i \left[\rho \log(1 + p_i) - \theta_i p_i^2 - \overline{U} \right]. \tag{8}$$

Lemma 2. *To achieve the maximum utility of the source, the contract item for the lowest type θ_1 is positive while all the others are zero; that is, $(p_1, \alpha_1) > 0$ and $(p_i, \alpha_i) = 0, \forall i \in \Omega$.*

Proof. The lemma could be proved by contradiction. It is supposed that an optimal contract item with $p_i > 0$, for $i > 1$ (for type-θ_i RNs), exists. By making $P' = \sum_{i \in \Omega} N_i [\rho \log(1 + p_i) - \overline{U}]$, the source's utility is

$$U_1 = P' - \sum_{i \in \Omega} N_i \theta_i p_i^2. \tag{9}$$

Next, given fixed P', a larger utility for the source can be achieved by allocating positive relay power only to the lowest type RNs:

$$U_2 = P' - N_1 \theta_1 p_1^2. \tag{10}$$

This is because $\sum_{i\in\Omega} N_i\theta_i p_i^2 > N_1\theta_1 p_1^2$ in (9) and $\sum_{i\in\Omega} N_i\theta_i p_i^2 = N_1\theta_1 p_1^2$ in (10); thus, (10) is larger than (9). It contradicts the assumption and thus finishes the proof.

Based on Lemma 2, the optimization problem in (8) is further simplified as

$$\max_{p_1\geq 0} N_1\left[\rho\log\left(1+p_1\right)-\theta_1 p_1^2-\overline{U}\right]. \quad (11)$$

Note that the optimization problem for the source from involving $2N$ variables $\{p_i, \alpha_i, \forall i \in \Omega\}$ in (7) could be simplified to a single variable p_1 in (11). And the optimal solution $\hat{p}_1$ to problem (11) should satisfy

$$\left.\frac{dU_S(p_1)}{dp_1}\right|_{p_1=\hat{p}_1} = N_1\left(\frac{\rho}{1+\hat{p}_1}-2\theta_1\hat{p}_1\right)=0. \quad (12)$$

Then, the second-order derivative of (11) can be expressed as

$$\left.\frac{\partial^2 U_S(p_1)}{\partial p_1^2}\right|_{p_1=\hat{p}_1} = -N_1\left[\frac{\rho}{(1+\hat{p}_1)^2}+2\theta_1\right]<0, \quad (13)$$

which means that problem (11) has the unique and globally optimal solution. Then, $p_1^* = \max(\sqrt{(\theta_1+2\rho)/4\theta_1}-1/2, 0)$. Therefore, the source offers the RNs the optimal contract items with (p_1^*, α_1^*) to ensure that each RN receives a nonnegative payoff by accepting the contract item and the source obtains the maximum utility with the RNs' relay service.

4. Optimal Contract Design under Asymmetric Information Scenario

In this section, we present the optimal contract design in the presence of the asymmetric information. Due to the selfishness of relay nodes and the finiteness of spectrum resources, the RNs may deviate from the contract to maximize their own benefit. Because of the asymmetry of network information, the relay actions of the RNs are unobservable to the principal (the source). Thus, the RNs' hidden relay actions give rise to the *moral hazard* problem.

As the source knows little information about RN's relay effort or action, it needs to ensure that each RN accepts only the contract item $\{\alpha_i, \beta_i, \forall i \in \Omega\}$ to maximize their own utility; that is, the contract needs to satisfy the following incentive compatibility (IC) constraint:

$$\text{(IC)}\ \max_{p_i\geq 0} U_{RN_i} = \alpha_i + \beta_i p_i - \theta_i p_i^2, \quad \forall i \in \Omega. \quad (14)$$

Then, to maximize the source's utility, an optimal contract under symmetric information scenario can be designed as follows:

$$\max_{\{\{\alpha_i,\beta_i\}\geq 0, \forall i\in\Omega\}} \sum_{i\in\Omega} N_i\left[\rho\log\left(1+p_i\right)-\alpha_i-\beta_i p_i\right],$$

$$\text{s.t.}\quad \text{(IC)}\ \max_{p_i\geq 0} U_{SU_i} = \alpha_i + \beta_i p_i - \theta_i p_i^2, \quad (15)$$

$$\text{(IR)}\ \alpha_i + \beta_i p_i - \theta_i p_i^2 \geq \overline{U}, \quad \forall i \in \Omega.$$

The two constraints in (15) correspond to IC and IR constraints, respectively. Specifically, the IC constraint ensures that each RN will get the maximum payoff by choosing the optimal relay power p_i^*. The IR constraint ensures that each RN can get a retained utility by choosing the optimal relay power p_i^*.

From the IC constraint, we obtain the optimal relay power p_i^*:

$$p_i^* = \frac{\beta_i}{2\theta_i}. \quad (16)$$

Similar to Lemma 1, we can obtain the optimal basic wage α_i^* from the IR constraint:

$$\alpha_i^* = \overline{U} + \theta_i p_i^{*2} - \beta_i p_i^*. \quad (17)$$

Then, the optimization problem in (15) can be simplified as

$$\max_{\{0\leq\beta_i\leq 1, \forall i\in\Omega\}} \sum_{i\in\Omega} N_i\left[\rho\log\left(1+\frac{\beta_i}{2\theta_i}\right)-\overline{U}-\frac{\beta_i^2}{4\theta_i}\right]. \quad (18)$$

Similar to Lemma 2, to obtain the source's maximum utility, only the contract item for the lowest type θ_1 is positive and all the other contract items are zero; that is, $(\alpha_1, \beta_1) > 0$ and $(\alpha_1, \beta_1) = 0, \forall i \in \Omega$.

Thus, the optimization problem in (18) can be further simplified as

$$\max_{0\leq\beta_1\leq 1} N_1\left[\rho\log\left(1+\frac{\beta_1}{2\theta_1}\right)-\overline{U}-\frac{\beta_1^2}{4\theta_1}\right]. \quad (19)$$

Following a similar analysis to that under the symmetric information scenario, the optimal solution to problem (19) is given by

$$\beta_1^* = \max\left(\sqrt{2\theta_1\rho+\theta_1^2}-\theta_1, 0\right). \quad (20)$$

By combining (16), (19), and (20), we have

$$U_S^* = N_1\left[\rho\log\left(1+p_1^*\right)-\theta_1\left(p_1^*\right)^2-\overline{U}\right],$$

$$p_1^* = \frac{\beta_1}{2\theta_1} = \max\left(\frac{\sqrt{2\theta_1\rho+\theta_1^2}-\theta_1}{2\theta_1}, 0\right) \quad (21)$$

$$= \max\left(\sqrt{\frac{\theta_1+2\rho}{4\theta_1}}-\frac{1}{2}, 0\right).$$

It is easy to see that the source's maximum utility under asymmetric information scenario is the same as that in symmetric information scenario. Therefore, the source offers the RNs the optimal contract items with (α_1^*, β_1^*), which ensures that each RN gets the maximum payoff by accepting the contract item to avoid the moral hazard problem and the source obtains the maximum utility with the RNs' relay service.

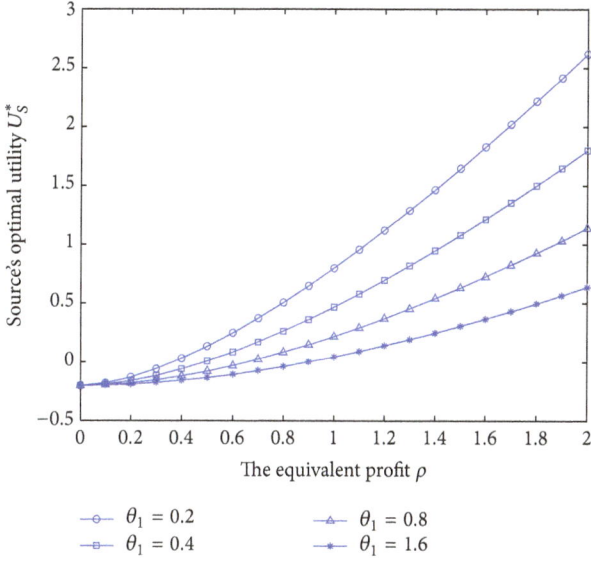

FIGURE 2: Source's optimal utility U_S^* versus the equivalent profit ρ for fixed $\beta_1 = 0.1$, $N_1 = 2$, and $\overline{U} = 0.1$.

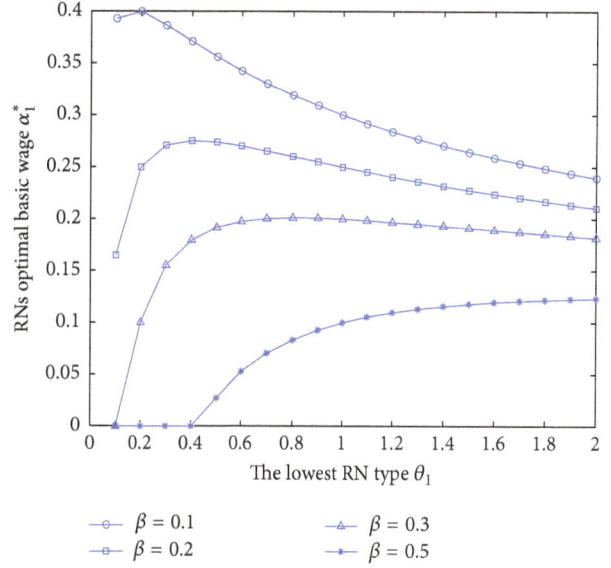

FIGURE 3: RNs' optimal basic wage α_1^* versus the lowest RN type θ_1 for fixed $\overline{U} = 0.1$ and $\rho = 1.5$.

5. Results and Discussions

To assess the performance of our proposed contract-based relay incentive mechanism, numerical simulation results under both symmetric and asymmetric information scenarios are presented in this section.

5.1. Symmetric Information Scenario. Figure 2 demonstrates that the source's optimal utility is increasing in the equivalent profit ρ and decreasing in the lowest RNs' type θ_1. As ρ increases, the RNs have more incentives to provide the relay communication with the source, and the source obtains more utility from the RNs' cooperative relay. As θ_1 increases, the RNs of that type have a higher relay cost for relay communication; thus, the source obtains less utility by hiring the higher RNs' type θ_1. When θ_1 becomes very large, the source's optimal utility is reduced to zero by hiring the RNs with very high relay cost. And, in this case, the source may choose not to hire any RN to participate in cooperative communication.

Figure 3 shows that the RNs' optimal basic wage of the lowest type θ_1, α_1^*, decreases in the performance-based bonus coefficient β. As β becomes large, the source only needs to allocate a little amount of reward to the RNs for enough relay help; thus, the RNs' optimal basic wage α_1^* is strictly decreasing in β. As β becomes very large, the lowest type θ_1 RNs can get enough bonus payoff without any basic wage; thus, there are some zero points in certain θ_1 cases.

5.2. Asymmetric Information Scenario. Next, cooperative communication strategy under asymmetric information scenario with the RNs' hidden relay action is considered. Figures 4, 5, and 6 illustrate the RNs' optimal bonus coefficient β_1^*, the optimal basic wage α_1^*, and the source's optimal utility

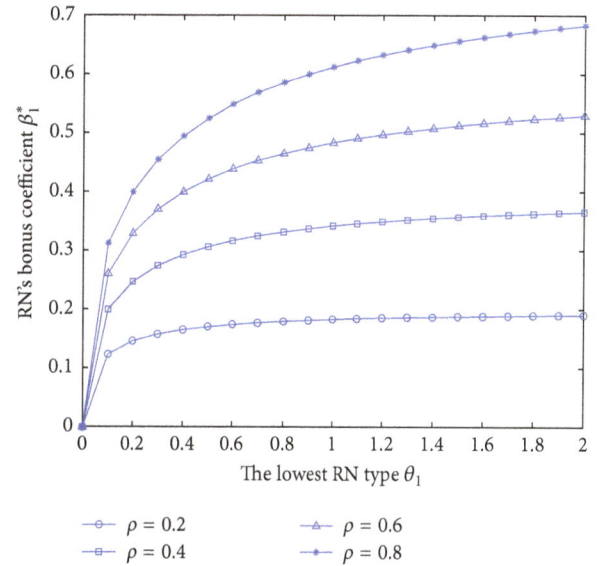

FIGURE 4: RNs' optimal basic wage β_1^* versus the lowest RN type θ_1 for fixed $N_1 = 2$ and $\overline{U} = 0.1$.

U_S^* with the lowest RN type θ_1 and the equivalent profit ρ, respectively.

Figure 4 illustrates that the bonus coefficient β_1^* increases in the lowest RN type θ_1 and the equivalent profit ρ. Figure 5 shows that the RNs' optimal basic wage α_1^* increases in the lowest RN type θ_1 and decreases in the equivalent profit ρ. And Figure 6 presents that the source's optimal utility U_S^* decreases in the lowest RN type θ_1 and increases in the equivalent profit ρ. On the one hand, as θ_1 increases, the RNs' relay cost of that type is increased; thus, the source may allocate much more wage and bonus to attract the RNs to participate in cooperative relay, which will reduce the source's

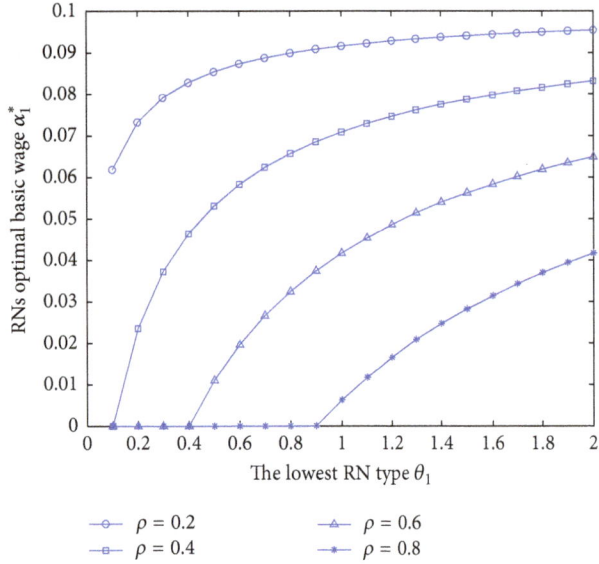

FIGURE 5: RNs' bonus coefficient α_1^* versus the lowest RN type θ_1 for fixed $N_1 = 2$ and $\overline{U} = 0.1$.

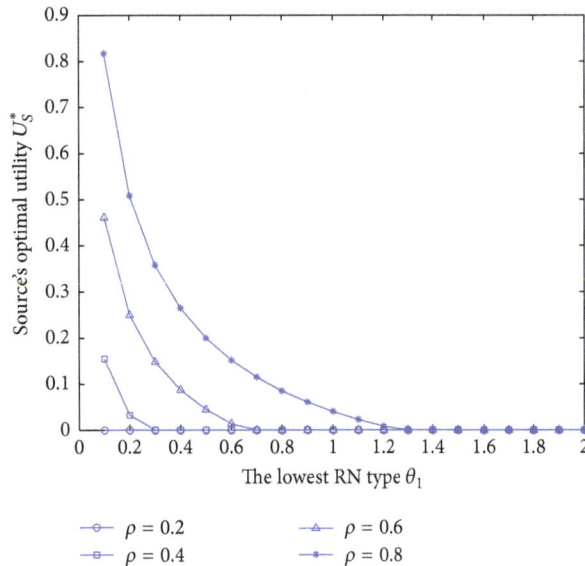

FIGURE 6: Source's optimal utility U_S^* versus the lowest RN type θ_1 for fixed $N_1 = 2$ and $\overline{U} = 0.1$.

final utility. On the other hand, the increased equivalent profit *rho* will tend to enhance the RNs' interest to participate in relay communication with the source and result in higher performance-based bonus and more source's utility. As the source offers higher performance-based bonus β_1^*, the source only needs to allocate less amount of wage α_1^* to the RNs for enough relay help.

6. Conclusion

In this paper, the relay incentive mechanism between the source and the RNs is investigated in the wireless cooperative

networks. Due to the selfish nature of the source and the RNs, their objectives are conflicted to compete for the limited spectrum resources. In order to incentivize the RNs to participate in cooperative relay efficiently and credibly, the cooperative communication is modelled as a labour market. Moreover, according to the characteristics of network information, the optimal contract designs for the two information scenarios are investigated in this work. In the presence of the symmetric information, the optimal contract can be obtained if and only if it is IR for each RN. And, in the presence of the asymmetric information, a *moral hazard* contract model is discussed to combat detrimental effects of the RNs' hidden relay actions. Simulation results show that the optimal contract can achieve the same source's utility as under both symmetric information and asymmetric information scenarios. And our proposed contract-based relay incentive mechanism can effectively improve the performance of cooperative relay.

Conflict of Interests

The authors declare that there is no conflict of interests regarding the publication of this paper.

Acknowledgments

This work was supported by the National Natural Science Foundation of China (61501178, 61471162), Natural Science Foundation of Hubei Province (2015CFB646), Open Foundation of Hubei Collaborative Innovation Center for High-Efficiency Utilization of Solar Energy (HBSKFMS2014033), and Ph.D. Research Startup Foundation of Hubei University of Technology (BSQD13029). The authors would like to acknowledge the anonymous reviewers whose constructive criticism, comments, and suggestions led to a greatly improved paper.

References

[1] N. Zhao, F. Pu, X. Xu, and N. Chen, "Cognitive wideband spectrum sensing using cosine-modulated filter banks," *International Journal of Electronics*, vol. 102, no. 11, pp. 1890–1901, 2015.

[2] A. Nosratinia, T. E. Hunter, and A. Hedayat, "Cooperative communication in wireless networks," *IEEE Communications Magazine*, vol. 42, no. 10, pp. 74–80, 2004.

[3] Z. Zhou, S. Zhou, J.-H. Cui, and S. Cui, "Energy-efficient cooperative communication based on power control and selective single-relay in wireless sensor networks," *IEEE Transactions on Wireless Communications*, vol. 7, no. 8, pp. 3066–3078, 2008.

[4] X. Hong, J. Wang, C.-X. Wang, and J. Shi, "Cognitive radio in 5G: a perspective on energy-spectral efficiency trade-off," *IEEE Communications Magazine*, vol. 52, no. 7, pp. 46–53, 2014.

[5] S. A. Astaneh and S. Gazor, "Resource allocation and relay selection for collaborative communications," *IEEE Transactions on Wireless Communications*, vol. 8, no. 12, pp. 6126–6133, 2009.

[6] Z. Hasan and V. K. Bhargava, "Relay selection for OFDM wireless systems under asymmetric information: a contract-theory based approach," *IEEE Transactions on Wireless Communications*, vol. 12, no. 8, pp. 3824–3837, 2013.

[7] L. Duan, L. Gao, and J. Huang, "Cooperative spectrum sharing: a contract-based approach," *IEEE Transactions on Mobile Computing*, vol. 13, no. 1, pp. 174–187, 2014.

[8] P. Bolton and M. Dewatripont, *Contract Theory*, MIT Press, 2005.

[9] B. Salani, *The Economics of Contracts: A Primer*, MIT Press, 2005.

[10] L. Gao, J. Huang, Y.-J. Chen, and B. Shou, "An integrated contract and auction design for secondary spectrum trading," *IEEE Journal on Selected Areas in Communications*, vol. 31, no. 3, pp. 581–592, 2013.

[11] S.-P. Sheng and M. Liu, "Profit incentive in trading nonexclusive access on a secondary spectrum market through contract design," *IEEE/ACM Transactions on Networking*, vol. 22, no. 4, pp. 1190–1203, 2014.

[12] N. Zhao, M. Wu, W. Xiong, and C. Liu, "Cooperative communication in cognitive radio networks under asymmetric information: a contract-theory based approach," *International Journal of Distributed Sensor Networks*, In press.

[13] R. Gibbons, "Incentives between firms (and within)," *Management Science*, vol. 51, no. 1, pp. 2–17, 2005.

[14] N. Zhao, F. Pu, X. Xu, and N. Chen, "Optimisation of multichannel cooperative sensing in cognitive radio networks," *IET Communications*, vol. 7, no. 12, pp. 1177–1190, 2013.

[15] E. Hossain and V. K. Bhargava, *Cognitive Wireless Communication Networks*, Springer Science & Business Media, 2007.

[16] L. Gao, X. Wang, Y. Xu, and Q. Zhang, "Spectrum trading in cognitive radio networks: a contract-theoretic modeling approach," *IEEE Journal on Selected Areas in Communications*, vol. 29, no. 4, pp. 843–855, 2011.

17

Power Optimized Single Relay Selection with an Improved Link-Adaptive-Regenerative Protocol

Jie Li ⓘ, Jianrong Bao ⓘ, Shenji Luan ⓘ, Bin Jiang ⓘ, and Chao Liu ⓘ

Information Engineering School, Hangzhou Dianzi University, Hangzhou, Zhejiang, 310018, China

Correspondence should be addressed to Jianrong Bao; baojr@hdu.edu.cn

Academic Editor: Vinod Sharma

To improve the reliability and efficiency in cooperative communications, a power optimized single relay selection scheme is proposed by increasing the diversity effort with an improved link-adaptive-regenerative (ILAR) protocol. The protocol determines the forwarding power of a relay node by comparing the signal-to-noise ratio (SNR) at both sides of the node; thus it improves the power efficiency. Moreover, it also proposes a single relay selection strategy to maximize the instantaneous SNR product, which ensures the approximate best channel link quality for good relay forwarding. And the system adjusts the forwarding power in real time and also selects the best relay node participated in the cooperative forwarding. In addition, the cooperation in the protocol is analyzed and the approximate expression of the bit-error-rate (BER) and the outage probability at high SNRs are also derived. Simulation results indicate that the BER and outage probability of the relay selection scheme by the ILAR protocol outperform other contrast schemes of current existing protocols. At BER of 10^{-2}, the proposed scheme with ILAR protocol outperforms those of the decoded-and-forward (DF), the selected DF (SDF), and the amplify-and-forward (AF) protocols by 3.5, 3.5 and 7 dB, respectively. Moreover, the outage probability of the relay system decreases with the growth of the relay number. Therefore, the proposed relay selection scheme with ILAR strategies can be properly used in cooperative communications for good reliability and high power efficiency.

1. Introduction

In wireless communications, multipath fading deteriorated the transmission rate and quality of the communications system and the diversity was one of the effective ways to combat it [1]. As a typical space diversity technology, the multiple-input multiple-output (MIMO) system was then proposed to improve the achievable rate and the BER performance [2]. However, modern mobile communications, especially for the handset mobile phones, etc., were strictly confined to the physical size, power, locations, etc., in practice, which limited the wide applications of the technique. So, cooperative communications were proposed to exploit nearby relay nodes to forward the messages to achieve cooperative diversity, which actually formed a virtual MIMO system [3–5]. Nowadays, with the rapid development of the next generation mobile communications, cooperative communications have become one of the research hot-spots in wireless cooperative communications.

The relay selection is one of the key issues in cooperative communications. For the relay selection in literature, there had been several schemes based on the decision of the outage probability, the SNR threshold, the BER performance and the channel status information (CSI), and so on. Two relay selection schemes for cooperative diversity were discussed in [6]. And a threshold-based adaptive relay selection was suggested to minimize the forwarding relay number given a fixed outage requirement. So the computation complexity was reduced and the probability of the outage events was prevented as much as possible. In addition, power selection cooperation can be used to obtain better performance than those of the maximum ratio combination (MRC) based protocol [7]. In [8], performance analysis of a single relay selection in Rayleigh fading was discussed and analyzed. And the closed-form expressions of the outage probability and the bit error probability (BEP) of an uncoded threshold-based opportunistic relaying (OR) and a selection cooperation (SC) were provided for the optimized forwarding. Finally, relay selection based on statistical CSI was proposed in [9]. And the closed-form expressions for the outage and BEP were derived with the DF relays at the Rayleigh fading channel. So the outage probability, the SNR threshold, the BER performance, and the channel status information (CSI) can be referenced for the improvement of a cooperative relay communication system.

The performances of the cooperative communications are mainly determined by the forwarding, and the combining schemes, etc. For the forwarding schemes, there were several existing relay forwarding protocols, as well as their improved version, such as the amplify-and-forward (AF), the decode and-forward (DF), the selective DF (SDF), the hybrid of the above three schemes, and so on [10, 11]. In the AF, the destination node can achieve available diversity with maximum ratio combining (MRC) [12]. Nevertheless, it may be less pragmatic, because it required the relay nodes to store these analog information waveforms, which required huge storage resources [13]. The DF protocol was then proposed with decoding and forwarding for better performance and thus more practical. But the full diversity gain cannot be easily acquired in a poor source-relay-destination link [14]. The SDF protocol with cyclic redundancy check (CRC) codes can detect errors at relay node and selectively forwarded the received messages to the destination [15]. It enabled diversity at the cost of some decoding delay and inefficiency spectrum utilization due to the CRC codes. But it cannot get full diversity gains due to the error propagation phenomena. Actually, the DF protocol only utilized the simple optimization of the combing information at destinations [16]. Then, the maximum likelihood (ML) combining was put forward to solving the diversity problem in the early years [4]. However, it was too complicated to be implemented. Recently, a cooperative MRC (C-MRC) algorithm was proposed and investigated in [14] for full diversity. With C-MRC measure at the destination, the relay system can achieve full diversity gains by the DF at the cost of rather large signal processing overhead. Subsequently, a link-adaptive-regenerative (LAR) protocol was proposed with a C-MRC to overcome the above deficiency [17]. By the DF strategy in a LAR scheme, the decoded messages at the relay were firstly scaled in power before being forwarded to the destination. For the LAR protocol, the scale was firstly adopted at the relay node, which was closely correlated with the signal-to-noise ratio (SNR) of the source-relay and destination-relay links. After that, the C-MRC was adopted at the destination to guarantee the maximum diversity gains [17]. Therefore, it gave a new approach to solve the diversity problems by the DF like strategies.

For the LAR like scheme, the performance of the multibranch relay system can be globally optimized by selecting the proper cooperative relay nodes. In a cooperative diversity system, all relays participate in sending the source signals to the destination. So the destination can combine all received signals from the source-relay-destination or the source-destination link with the MRC mechanism. In the best relay selection scheme, the destination node combines the best source-relay-destination link with the source-destination link only. The main advantages of the best relay selection can

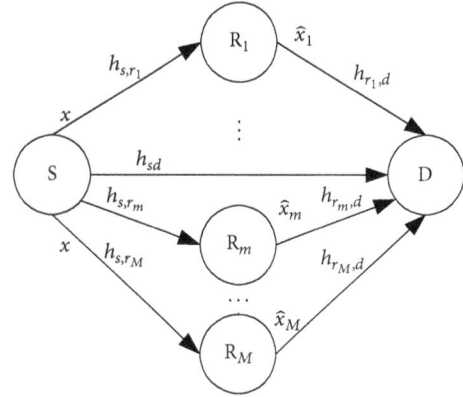

FIGURE 1: A multibranch cooperative system model.

be evaluated by low complexity and high resource utilization [18]. So in this paper, based on the single relay system, the end-to-end performance of the relay system with LAR protocol is introduced and analyzed. And the main contribution of this paper is listed as follows:

(1) The ILAR protocol is introduced to solve the diversity gains problem by introducing a new channel scaling factor to match the channel status information properly in the DF like transmission schemes.

(2) Based on the above ILAR protocol, the best single relay selection scheme is proposed, where the relay node with the largest equivalent channel SNR is optimally chosen to forward the messages for best cooperative transmission. And the equivalent channel SNR is generated with the above channel scaling factor and the instantaneous and average SNRs, which is much more suitable for the relay channel status.

(3) The BER performance and outage probability of the relay selection system with ILAR protocol are analyzed and explained for the good performance brought by the suggested power optimized single relay selection scheme with the ILAR protocol.

The remainder of the paper is organized as follows. In Section 2, the cooperative system model is described and the ILAR protocol is briefly introduced. In Section 3, by the above proposed ILAR protocol, the best single relay selection in a multibranch relay system is accomplished and analyzed, including the closed-form analytic expression of the BER and the outage probability in the ILAR relay selection system. In Section 4, the numerical simulations and the corresponding result analyses are performed to verify the effectiveness of the proposed scheme. Finally, the conclusion is drawn in Section 5.

2. Cooperative System Model and ILAR Protocol

2.1. Cooperative System Model. As a typical multibranch cooperative relay system depicted in Figure 1, there are one source node (S), one destination node (D), and M relay nodes (R_m) with $m = 1, 2, \ldots, M$. In this model, besides the message transmission link from the source node S direct to the destination node D, messages are also transmitted from the node S to the node D via one of the mth relay R_m node. The relay node R_m have no own data to be transmitted and they just help transmit the messages from the source node S to the destination node D indirectly. Even if the nodes of S, R_m, and D are all equipped with the only one antenna, the messages can arrive at the destination node D from link S-R_m-D and link S-D, respectively, and can be tackled by the MRC criterion efficiently to merge the whole messages for high space diversity gains and optimal power efficiency. And the transmission can be performed by a time-division half-duplex mode.

Because the time-division scheme is widely exploited, this paper mainly concerns the half-duplex relay channel model. It is easily extended to much more relay models, such as multibranch relay, etc. Suppose that the symbols x from the source node S and $\hat{x}_m$ from the relay node R_m are transmitted over the half-duplex relay channel by time-division mode. Each time period T is divided into two time slots for data transmission.

In the first time slot lT with $0 < l < 1$, the source node S broadcasts the symbol x with power P_s. Then the received information y_{sr_m} and y_{sd} at the relay and destination nodes of R_m and D are expressed, respectively, as

$$y_{sr_m} = \sqrt{P_s} h_{sr_m} x + z_{sr_m}, \tag{1}$$

$$y_{sd} = \sqrt{P_s} h_{sd} x + z_{sd}, \tag{2}$$

where h_{sr_m} and h_{sd} are the fading coefficients of the link S-R_m and the link S-D, respectively. z_{sr_m} and z_{sd} are the additive white Gaussian noise (AWGN) in the S-R_m and S-D links, respectively.

In the second time slot $(1-l)T$, the information received by the relay nodes R_m is detected and forwarded to the destination node D with power $\sqrt{P_{R_m} \alpha_m}$. So the received information $y_{r_m d}$ at the destination node D is

$$y_{r_m d} = h_{r_m d} \sqrt{P_{R_m} \alpha_m} \hat{x}_m + z_{r_m d}, \tag{3}$$

where $h_{r_m d}$ and $z_{r_m d}$ are the fading coefficients and the AWGN parameter in the link R_m-D. P_{R_m} is transmit power at R_m and α_m represents a power control coefficient.

2.2. ILAR Protocol. Based on the above model, the ILAR protocol can be operated on the following three conditions.

Firstly, assume that the wireless channels are Rayleigh fading channels and the fading coefficients are subject to the complex Gaussian random distribution, i.e., $h_{sr_m} \sim CN(0, \sigma^2_{sr_m})$, $h_{sd} \sim CN(0, \sigma^2_{sd})$, and $h_{r_m d} \sim CN(0, \sigma^2_{r_m d})$, with $\sigma^2_{sr_m} = E\{|h^2_{sr_m}|\}$, $\sigma^2_{sd} = E\{|h^2_{sd}|\}$, $\sigma^2_{r_m d} = E\{|h^2_{r_m d}|\}$, and z_{sr_m}, z_{sd}, and $z_{r_m d} \sim CN(0, N_0)$. $CN(0, \sigma^2)$ is defined as the circular symmetric complex Gaussian distribution with zero mean and variance σ^2. Hence, the corresponding instantaneous and

average SNR sets $(\gamma_{sr_m}, \gamma_{r_m d}, \gamma_{sd})$ and $(\bar{\gamma}_{sr_m}, \bar{\gamma}_{r_m d}, \bar{\gamma}_{sd})$ of S-R_m, R_m-D, and S-D links are expressed, respectively, as follows:

$$\gamma_{sr_m} = \left|h_{sr_m}\right|^2 \bar{\gamma},$$

$$\gamma_{r_m d} = \left|h_{r_m d}\right|^2 \bar{\gamma}, \tag{4}$$

$$\gamma_{sd} = \left|h_{sd}\right|^2 \bar{\gamma},$$

$$\bar{\gamma}_{sr_m} = \sigma^2_{sr_m} \bar{\gamma},$$

$$\bar{\gamma}_{r_m d} = \sigma^2_{r_m d} \bar{\gamma}, \tag{5}$$

$$\bar{\gamma}_{sd} = \sigma^2_{sd} \bar{\gamma},$$

where $\bar{\gamma} = P_S / N_0$ and P_S denotes the average transmit power at the source node S. And the transmission power at each relay node is supposed to be the same as that in the source node, i.e., $P_{R_m} = P_S$.

Secondly, symbols flows from the source node S are detected at the relay nodes R_m with maximum likelihood (ML) demodulation as follows:

$$\hat{x}_m = \arg \min_{x \in A_x} \left|y_{sr_m} - h_{sr_m} x\right|^2, \tag{6}$$

where $|Ax| = \theta$ is defined as the size of the θ-ary constellation. Then the detected symbol $\hat{x}_m$ is remodulated and forwarded to the destination node D, with power control coefficient of

$$\alpha^{inst}_m = \frac{\min\left(\gamma_{sr_m}, \gamma_{r_m d}\right)}{\gamma_{r_m d}} = \begin{cases} \dfrac{\gamma_{sr_m}}{\gamma_{r_m d}}, & \gamma_{sr_m} < \gamma_{r_m d} \\ 1, & \gamma_{sr_m} \geq \gamma_{r_m d}. \end{cases} \tag{7}$$

The instantaneous SNR of each link is difficult to be achieved. So γ_{sr_m} and $\gamma_{r_m d}$ are always used instead in practice, where $\bar{\gamma}_{r_m d}$ is the average SNR of link S-R_m. Then (7) is rewritten as

$$\alpha_m = \frac{\min\left(\gamma_{sr_m}, \bar{\gamma}_{r_m d}\right)}{\bar{\gamma}_{r_m d}} = \begin{cases} \dfrac{\gamma_{sr_m}}{\bar{\gamma}_{r_m d}}, & \gamma_{sr_m} < \bar{\gamma}_{r_m d} \\ 1, & \gamma_{sr_m} \geq \bar{\gamma}_{r_m d}. \end{cases} \tag{8}$$

In (7), $\gamma_{r_m d}$ is hard to be obtained. So α_m is clearly practical even on a fast Rayleigh fading channel, because $\bar{\gamma}_{r_m d}$ can be easily estimated under the given stationary channel.

Compared with other protocols, the main difference of the ILAR protocol is that the relay transmission power is related to the R_m-D link. For the DF protocol, $\alpha=1$ and the transmit power at relay is independent with the channel status information. According to (8), $\alpha \in [0, 1]$ is proper for the ILAR protocol. When the link S-R_m is more reliable than the link R_m-D, the relay node transmits information with full power, i.e., $\alpha=1$. Otherwise, the transmit power is scaled with the coefficient of $\alpha \in (0, 1]$. When $\alpha=0$, it means that the relay node is idle in the subslot and the source node S performs the retransmission.

Finally, the demodulation at the destination node D is analyzed. For the weight coefficients $w_{sd} = h^*_{sd}$ and

$w_{r_m d} = h^*_{r_m d} \sqrt{\alpha_m}$, the MRC processing result at the destination node D is expressed as

$$x_d^{MRC} = \arg \min_{x \in A_x} \left| w_{sd} y_{sd} + w_{r_m d} y_{r_m d} \right.$$

$$\left. - \left(w_{sd} h_{sd} + w_{r_m d} h_{r_m d} \sqrt{\alpha_m} \right) x \right|^2. \tag{9}$$

To solve (9), h_{sd} and $h_{r_m d} \sqrt{\alpha_m}$ are supposed to be available in the destination node D and both of them can be gotten by the channel training of a trained sequence for good channel estimation.

3. Relay Selection Scheme and Performance Analysis

3.1. Relay Selection Scheme. Based on the ILAR protocol described above, the single relay selection scheme is discussed in this relay selection scheme.

In the selection cooperation (SC) of the relay selection model, all relay nodes listen to the source node S in the first time slot and only those relay nodes of $\gamma_{sr_m} > A$ demodulate the received signal, where A is the threshold of SNR of link S-R$_m$. Therefore, the candidate relay node set is shown as

$$\Omega_m = \left\{ R_m : \gamma_{sr_m} \geq A \right\}. \tag{10}$$

In the second time slot, there are two situations to be discussed.

Situation I. When the set is empty, it means that the relay node does not transmit at all. It corresponds to the situation where the power scaling coefficient α is zero.

Situation II. When the relay nodes have demodulated the information correctly, then only the relay nodes of the highest $\gamma'_{r_m d}$ transmit signals to the destination node D. Assume that the ith relay node is the best choice; there is

$$\gamma'_{r_i d} = \max \left\{ \gamma'_{r_1 d}, \gamma'_{r_2 d}, \dots, \gamma'_{r_M d} \right\}, \tag{11}$$

where $\gamma'_{r_i d} = \alpha_i \gamma_{r_i d}$ is the instantaneous SNR of the received signal from the ith relay node. AS α_i is the power scaling coefficient of the ith relay node for normalization and it has been derived in (8). Under the above circumstances, it is expressed as

$$\alpha_i = \frac{\min \left(\gamma_{sr_i}, \overline{\gamma}_{r_i d} \right)}{\overline{\gamma}_{r_i d}}. \tag{12}$$

3.2. Performance Analysis

3.2.1. BER Analysis. Take a special case for instance, e.g., binary phase shift keying (BPSK) modulation; the signal x is transmitted from the source node S and it can only carry a message either $x = \sqrt{P_x}$ or $x = -\sqrt{P_x}$. Correspondingly, the detected $\hat{x}$ at the relay node R$_i$ can only be $\hat{x} = x$, or $\hat{x} = -x$.

In each case, the received information y_D at the destination node D is represented as

$$y_D = w_{rd} y_{rd} + w_{sd} y_{sd}$$

$$= \begin{cases} w_{sd} h_{sd} + w_{rd} h_{rd} \sqrt{\alpha} x + w_{rd} z_{rd} + w_{sd} z_{sd}, & \hat{x} = x \\ w_{sd} h_{sd} - w_{rd} h_{rd} \sqrt{\alpha} x + w_{rd} z_{rd} + w_{sd} z_{sd}, & \hat{x} = -x. \end{cases} \tag{13}$$

Since BPSK symbols are real value, the real part $y = \text{Re}\{y_D\}$ can be employed in demodulation. It is a real Gaussian random variable with zero mean and variance $\sigma^2 = (|w_{sd}|^2 + |w_{r_i d}|^2) N_0 / 2$. With the weight coefficients $w_{sd} = h^*_{sd}$ and $w_{r_i d} = h^*_{r_i d} \sqrt{\alpha_i}$, the instantaneous BER at the destination node D can be expressed as

$$P^b = \left\{ 1 - Q \left[\sqrt{2\gamma_{sr_i}} \right] \right\} Q \left[\sqrt{2 \left(\gamma_{sd} + \gamma'_{r_i d} \right)} \right]$$

$$+ Q \left[\sqrt{2\gamma_{sr_i}} \right] Q \left[\frac{\sqrt{2} \left(\gamma_{sd} - \gamma'_{r_i d} \right)}{\sqrt{\gamma_{sd} + \gamma'_{r_i d}}} \right], \tag{14}$$

where

$$\gamma'_{r_i d} = \frac{\alpha \left| h_{r_i d} \right|^2 P_x}{N_0} = \alpha_i \gamma_{r_i d}, \tag{15}$$

$$Q(x) = \frac{1}{\sqrt{2\pi}} \int_x^\infty \exp \left(-t^2 / 2 \right) dt. \tag{16}$$

Taking expectation over the instantaneous SNR, the average BEP $E[P^b]$ is obtained, and it follows from the inequality $0 \leq Q(x) \leq 1$ as

$$P^b \leq Q \left[\sqrt{2 \left(\gamma_{sd} + \gamma'_{r_i d} \right)} \right]$$

$$+ Q \left[\sqrt{2\gamma_{sr_i}} \right] Q \left[\frac{\sqrt{2} \left(\gamma_{sd} - \gamma'_{r_i d} \right)}{\sqrt{\gamma_{sd} + \gamma'_{r_i d}}} \right]. \tag{17}$$

3.2.2. Outage Probability. When the transmission rate is larger than the mutual information of the relay system, the system will be interrupted. According to the above relay selection scheme, the relay system will be interrupted in the following two situations. The first one is retransmission situation, i.e., the above Situation I. The second one is the cooperation situation, i.e., the above Situation II. Therefore, the outage probability of the relay system can be written as

$$p^{out} = P_1^{out} + P_2^{out}, \tag{18}$$

where P_1^{out} is the outage probability of retransmission Situation I and P_2^{out} is the outage probability of the cooperation Situation II.

For Situation I, when the candidate relay node set is empty, i.e., all these relay nodes cannot demodulate correctly ($\gamma_{sr_m} > A$), and the source node S retransmits all signals.

When the mutual information I_1 is less than transmission rate R, i.e., $I_1 < R$, the relay system is interrupted and there is the result as

$$I_1 = \frac{1}{2} \cdot lb\left(1 + 2\gamma_{sd}\right) < R. \tag{19}$$

According to (19), there is the conclusion of $\gamma_{sd} < A/2$. Combining the retransmission in Situation I, the outage probability P_1^{out} of Situation I is deduced as

$$
\begin{aligned}
P_1^{out} &= \Pr\left(\Omega_m = \varnothing\right\} \cdot \Pr\left(\gamma_{sd} < \frac{A}{2}\right) \\
&= \prod_{m=1}^{M} \Pr\left(\gamma_{sr_m} < A\right)\Pr\left(\gamma_{sd} < \frac{A}{2}\right) \\
&= \left[1 - \exp\left(-\frac{A}{\overline{\gamma}_{sr}}\right)\right]^{M}\left\{1 - \exp\left[-\frac{A}{(2\gamma_{sd})}\right]\right\},
\end{aligned}
\tag{20}
$$

where $\overline{\gamma}_{sr} = E(\gamma_{sr_m}) = 1/(SNR \cdot \delta_{sr_m}^2)$. At the high SNR, i.e., $\overline{\gamma}_{sr}$ and γ_{sd} approach infinity, (19) can be approximately rewritten as

$$P_1^{out} = \left(\frac{A}{\overline{\gamma}_{sr}}\right)^{M}\left[\frac{A}{(2\gamma_{sd})}\right]. \tag{21}$$

For Situation II, the best relay is selected. When the mutual information I_2 is less than transmission rate R, i.e., $I_2 < R$, the relay system is interrupted, and there is

$$I_2 = \frac{1}{2} \cdot lb\left(1 + \gamma_{sd} + \sqrt{\alpha_i}\gamma_{r_id}\right) < R, \tag{22}$$

where the ith relay is the best relay. According to (22), there is a conclusion of $\gamma_{sd} + \sqrt{\alpha_i}\gamma_{r_id} < A$. Combining the cooperation Situation II, the outage probability P_2^{out} of Situation II is derived as

$$
\begin{aligned}
P_2^{out} &= \Pr\left(\Omega_k\right\} \cdot \Pr\left(\gamma_{sd} + \sqrt{\alpha_i}\gamma_{r_id} < A\right) \\
&= C_M^k \exp\left(-\frac{k\gamma_{sd}}{\overline{\gamma}_{sr}}\right)\left[1 - \exp\left(-\frac{A}{\overline{\gamma}_{sr}}\right)\right]^{M-k} \\
&\quad \times \left\{1 - \exp\left[-\frac{\left(A - \sqrt{\alpha_i}\gamma_{r_id}\right)}{\gamma_{sd}}\right]\right\},
\end{aligned}
\tag{23}
$$

where k is the number of relay nodes with correct demodulation and $C_M^k = M!/[k!(M-k)!]$ is the combination expression in mathematics. At high SNRs, (23) can be rewritten as

$$P_2^{out} = C_M^k\left(\frac{A}{\overline{\gamma}_{sr}}\right)^{M-k}\left[\frac{\left(A - \sqrt{\alpha_i}\gamma_{r_id}\right)}{\gamma_{sd}}\right]. \tag{24}$$

Finally, according to (18), (21), and (24), at high SNR, the outage probability of the relay system can be deduced as

$$
\begin{aligned}
P^{out} &= \left(\frac{A}{\overline{\gamma}_{sr}}\right)^{M}\left(\frac{A}{2\gamma_{sd}}\right) \\
&\quad + C_M^k\left(\frac{A}{\overline{\gamma}_{sr}}\right)^{M-k}\left(\frac{A - \sqrt{\alpha_i}\gamma_{r_id}}{\gamma_{sd}}\right)
\end{aligned}
\tag{25}
$$

3.3. Computational Complexity Analysis. To evaluate the complexity of the proposed ILAR scheme, we compare it with the contrast scheme of the AF, DF, and SDF protocol. The entire relay transmission system is supposed to be made up of a source node, a destination node, and N relay nodes. And the comparison of computational complexity can be analyzed and summarized as follows,

In the AF scheme, N relay nodes all forward the messages from the source node to the destination node by multiplying a specific factor to amplify both the signals and noises. And then there is joint detection of the messages from all source-relay-destination links and the direct source-destination link. So there will be a calculation of the optimally joint maximum ratio combination (MRC) detection with $N + 1$ received user messages. So the additional computations other than the MRC detection are the N times of the multiplication in the amplifying procedure in the relays.

In the DF scheme, N relay nodes all decode with a channel code and then forward the decoded messages from all relay nodes to the destination node. Here the channel code is decided by the channel status and the practical hardware resources available in the source node and relay nodes. And then there is also the joint detection of the messages from all joint source-relay-destination links and the direct source-destination link. So there will be a calculation of the optimally joint MRC detection with $N + 1$ received user messages form the relay nodes. So the additional computations other than the MRC detection are one time of encoding of channel code at the source, N times of decoding of channel code at the relay nodes, respectively.

In the SDF scheme, the relay nodes decode and then forward the decoded source message only if the decoded messages are free of errors. But with half-duplex transmissions, the SDF scheme suffers from a multiplexing loss. And the unsuccessful source messages are supposed to be retransmitted at the next time slot. Given a successful decoding rate of a, the additional computations other than the MRC detection are one time of encoding at the source and an average of N/a time of channel decoding at the relay nodes, respectively.

In the proposed ILAR scheme, there are some negligible calculations about the exchangeable average SNR threshold by (8). Other than these computations, the symbols flows from the source node S are detected at the relay nodes R_m with maximum likelihood (ML) demodulation as in (6). So the additional computations other than the MRC detection are N times of ML demodulation at the relay nodes.

Finally, the computational complexity of the proposed ILAR scheme is concluded at Table 1, as well as those of the contrast AF, DF, and SDF schemes. Generally, the complexity

TABLE 1: Complexity of the proposed ILAR scheme and the contrast schemes.

Forwarding scheme	Pre-process at source node	Decoding/demodulation or transmission at Relay node	Combination at the destination
AF	No	N Multiplications	MRC detection
DF	1 Channel encoding	N Channel decoding	MRC detection
SDF	1 Channel encoding	N Channel decoding	MRC detection
Proposed ILAR	No	N ML-demodulation	MRC detection

of channel decoding is larger than that of the ML demodulation. So from Table 1, we can get the natural order of the computational complexity of the schemes as SDF > DF > ILAR > AF. In addition, the SDF and the ILAR schemes need another time slot to finish the forwarding, which requires much longer time. So the processing delay of the proposed ILAR scheme is still large and it needs to be considered in practice.

4. Numerical Simulations and Result Analyses

Based on the parameter α in (8), the performance of the ILAR scheme is firstly analyzed. Diversity gain G_d is defined as the negative exponent in the average BER, when $\overline{\gamma} \longrightarrow \infty$ [17], i.e., SNR set are sufficiently high. So the relationship among $E[P^b]$, G_c, and $\overline{\gamma}$ is founded as

$$E\left[P^b\right] \approx (G_c\overline{\gamma})^{-G_d}, \qquad (26)$$

where G_c denotes the coding gain. For symbol-by-symbol demodulation of uncoded transmissions, G_c mainly depends on the modulation order and the transmit power. According to the expression of BER in (14) and (17), the BER performance of ILAR scheme is simulated and demonstrated. Suppose that the BPSK and Rayleigh fading channel are applied. The horizontal axis SNR(dB) $= \overline{\gamma}(1 + E[\alpha])$, where $E[\alpha]$ is the average value of α. Simultaneously, the BER performance of other popular protocols, such as the AF, the DF, and the SDF, are also figured out too, which gives a clear comparison among them.

Figure 2 indicates that the relay cooperation of the ILAR protocol has better BER performance than those of other protocols. At BER of 10^{-3}, the proposed scheme of the ILAR protocol has about 7 dB, 3.5 dB, and 3.5 dB gains than those of the AF, the DF, and the SDF protocols, respectively. The scheme of the AF protocol can get full diversity gains, but it does not outperform ILAR, because it cannot utilize the power efficiently due to the noise amplification effect at the relay node. When the channel status of the links connected to the relay nodes is poor, the AF scheme may be even worse than that of the direct link. The scheme of the DF protocol can obtain good performance at high SNRs since it can regenerate message by decoding and at the relay node and then forward to the destination node. But, the DF scheme exhibits worse BER performance at low SNRs due to the error propagation problem at low SNRs. The SDF protocol can be the combination of the AF and DF protocols. By an optimal design, the SDF scheme can even achieve the good performance of the AF scheme at low SNRs and the DF

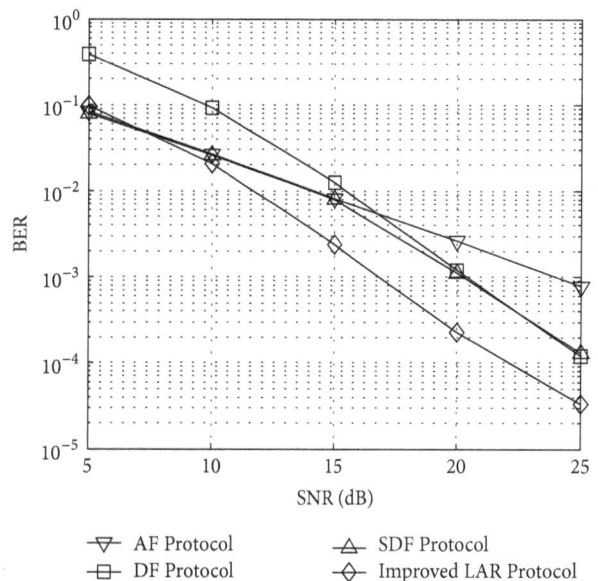

FIGURE 2: The BER performance of the scheme with an ILAR.

scheme at high SNRs. When it comes to adaptive regenerative protocols, the proposed ILAR scheme also outperforms the SDF scheme.

At low SNRs, the SNRs in all relay links are all poor for the proposed ILAR scheme and the contrast schemes, because large noises are in charge of the poor performance rather than the intersymbol interference (ISI) by Rayleigh fading at high SNRs. And the performance of the DF scheme is surely much worse due to the aforementioned error propagation effect by unsuccessful decoding and then encoding. However, compared with the contrast schemes other than the DF scheme, the performance of the proposed ILAR scheme is not improved much and the reasons are discussed as follows: although optimal relay link can be properly chosen by the proposed scheme with (11), the detection performance is still poor due to the similar large noises of the optimal relay link at low SNRs. So the proposed scheme cannot obtain more performance gain than the compared schemes other than the DF one at low SNRs. In other words, the performances are all predominated by the noises rather than the ISI by fading channels. And this phenomenon is also well presented reasonably by the simulations in Figure 2. Therefore, our ILAR scheme is mainly practical at rather high SNRs, e.g., larger than 10 dB under the above simulation parameters from our simulations.

Taking into account of the tradeoff between the redundancy and the detection error probability of the CRC code, although full diversity is achieved, the SDF scheme does not lead to as good BER performance as that of the ILAR scheme over the range of the practical SNRs. This can be explained as follows. Compared with the DF or SDF protocol, the main difference from the ILAR protocol is that the power of the transmitted signals at the relay nodes is exactly matched for the CSI of the link S-R_m and R_m-D. For the DF/SDF protocol, the power scaling coefficient α_i in (12) is just set as a constant number of 1, and the transmitted power at any relay node has nothing to do with the CSI of the channels. Then the channel cannot be properly matched for high power efficiency and good transmission performance. But in the ILAR scheme, just as in (12), the power scaling coefficient α_i is in the range of [0, 1]. When the link S-R_m is much more reliable than the link R_m-D, the relay node can transmit the signals with full power, i.e., $\alpha_i = 1$, because the channel status of the first part of the link, i.e., the link S-R_m, is good enough for the message forwarding at the relay nodes. Otherwise, the transmitted power at the relay node is adjusted by the power scaling coefficient α_i, which falls into the range of (0,1). Also, for the left extreme condition, i.e., $\alpha_i = 0$, the relay node is stood at the idle mode at the second time slot. It means that the relay nodes are not participated in the cooperative transmission. Finally, by the cooperative MRC detection at the destination node D in (9), the proposed ILAR scheme can match the CSI in all of the channels and thus obtain better performance over the traditional schemes of the AF, the DF, and the SDF protocols under the premise of Rayleigh fading channel with the independent and identically distributed (i.i.d.) AWGN parameters.

For the relay selection scheme of the ILAR protocol, the channel parameters are set as follows: the transmission rate R = 1 bit/s and $\delta_{sd} = \delta_{sr_i} = \delta_{r_id} = 1$. The transmit power at the relay node and the source node are both 0.5 W. The channels are i.i.d. Rayleigh channels, with equal AWGN parameters in each branch, are based on the uncoded channels. And the simulations are performed with binary phase shift keying (BPSK) modulation. In Figure 3, the outage probability of the relay selection scheme with the ILAR protocol at different relay number condition is presented.

Figure 3 indicates that the outage probability of the relay system decreases with the relay number M. The reason is that larger number of the relay nodes gives more chance to be chosen, and thus the probability of correct demodulation at each relay node is much higher. Because the channels connected to the relay nodes are i.i.d. Rayleigh channels with equal AWGN parameters in each branch, the performances of the cooperative systems with relay node growth increase gradually. And they almost approach the same performance bound shown in Figure 3, when the number of the relay nodes is larger than or equal to 4. Simultaneously, with the growth of the SNR, the outage probability decreases correspondingly. When the SNR increases, the received SNR at the destination node is much bigger. And just as the theoretical analysis around (22), the relay system with higher SNRs has much lower outage probability.

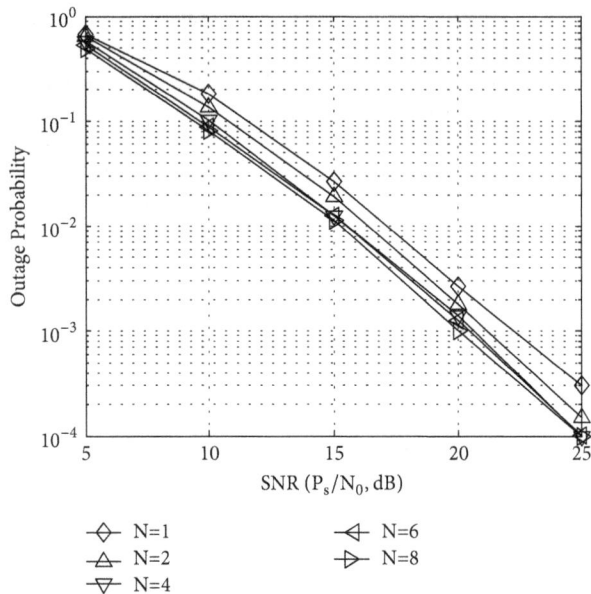

FIGURE 3: The outage probability of the ILAR scheme.

5. Conclusions

In this paper, we present and analyze a power optimized single relay selection scheme with an ILAR protocol. Firstly, in a cooperative communication system, the ILAR protocol is introduced. Then the best relay selection scheme with the proposed ILAR protocol is proposed and analyzed. By the protocol, the BER and outage probability expression are both derived for quantitative analysis. The simulation results preferably show that the scheme of an ILAR protocol has better BER performance than those of other current protocols, such as the AF, the DF, and the SDF protocols. The overall outage probability performance increases with the growth of relay number, but it also indefinitely approaches a performance bound, when the number of the relay nodes is large or equal to 4. Therefore, the proposed scheme can be applied in the optimization of relay selection and forwarding for efficient transmission power.

Conflicts of Interest

The authors declare that there are no conflicts of interest regarding the publication of this paper.

Acknowledgments

This work was supported by the Zhejiang Provincial National Natural Science Foundation (nos. LY17F020024, LY17F010019), the National Natural Science Foundation of China (no. 61471152), the Zhejiang Provincial Research Program for Public Welfare Technology Application (no. LGG18F010011), and the Scientific Research Project of Zhejiang Provincial Department of Education (no. Y201636586).

References

[1] S. T. Rappaport, *Wireless Communications: Principles and Practice*, Prentice Hall PTR, New Jersey, NJ, USA, 1996.

[2] P. W. Wolniansky, G. J. Foschini, G. D. Golden, and R. A. Valenzuela, "V-BLAST: an architecture for realizing very high data rates over the rich-scattering wireless channel," in *Proceedings of the URSI International Symposium on Signals, Systems, and Electronics (ISSSE '98)*, pp. 295–300, IEEE, Pisa, Italy, September-October 1998.

[3] T. M. Cover and A. A. El Gamal, "Capacity theorems for the relay channel," *Institute of Electrical and Electronics Engineers Transactions on Information Theory*, vol. 25, no. 5, pp. 572–584, 1979.

[4] A. Sendonaris, E. Erkip, and B. Aazhang, "User cooperation diversity-part I: system description," *IEEE Transactions on Communications*, vol. 51, no. 11, pp. 1927–1938, 2003.

[5] A. Sendonaris, E. Erkip, and B. Aazhang, "User cooperation diversity. Part II. Implementation aspects and performance analysis," *IEEE Transactions on Communications*, vol. 51, no. 11, pp. 1939–1948, 2003.

[6] T. W. Ban, B. C. Jung, D. K. Sung, and W. Choi, "Performance analysis of two relay selection schemes for cooperative diversity," in *Proceedings of the 18th International Symposium on Personal, Indoor and Mobile Radio Communications*, pp. 1–5, Athens, Greece, September 2007.

[7] E. Beres and R. Adve, "On selection cooperation in distributed networks," in *Proceedings of the 40th Annual Conference on Information Sciences and Systems*, pp. 1056–1061, Princeton, NJ, USA, March 2006.

[8] D. S. Michalopoulos and G. K. Karagiannidis, "Performance analysis of single relay selection in rayleigh fading," *IEEE Transactions on Wireless Communications*, vol. 7, no. 10, pp. 3718–3724, 2008.

[9] S. Guo, J. Tang, Y. Wei, W. Li, and C. Zhang, "Research on relay selection based on statistical channel state information," in *Proceedings of the International Symposium on Instrumentation and Measurement, Sensor Network and Automation (IMSNA '12)*, pp. 176–178, August 2012.

[10] D. B. Da Costa and S. Aissa, "Amplify-and-forward relaying in channel-noise-assisted cooperative networks with relay selection," *IEEE Communications Letters*, vol. 14, no. 7, pp. 608–610, 2010.

[11] J. N. Laneman, D. N. Tse, and G. Wornell, "Cooperative diversity in wireless networks: efficient protocols and outage behavior," *Institute of Electrical and Electronics Engineers Transactions on Information Theory*, vol. 50, no. 12, pp. 3062–3080, 2004.

[12] J. Chambers, O. Alnatouh, and G. Chen, "Outage probability analysis for a cognitive amplify-and-forward relay network with single and multi-relay selection," *IET Communications*, vol. 7, no. 17, pp. 1974–1981, 2013.

[13] E. Zimmermann, P. Herhold, and G. Fettweis, "On the performance of cooperative diversity protocols in practical wireless systems," in *Proceedings of the 58th Vehicular Technology Conference (VTC '03)*, pp. 2212–2216, October 2003.

[14] T. Wang, A. Cano, G. B. Giannakis, and J. N. Laneman, "High-performance cooperative demodulation with decode-and-forward relays," *IEEE Transactions on Communications*, vol. 55, no. 7, pp. 1427–1438, 2007.

[15] M. Janani, A. Hedayat, T. E. Hunter, and A. Nosratinia, "Coded cooperation in wireless communications: space-time transmission and iterative decoding," *IEEE Transactions on Signal Processing*, vol. 52, no. 2, pp. 362–371, 2004.

[16] D. Chen and J. N. Laneman, "Modulation and demodulation for cooperative diversity in wireless systems," *IEEE Transactions on Wireless Communications*, vol. 5, no. 7, pp. 1785–1794, 2006.

[17] T. Wang, G. B. Giannakis, and R. Wang, "Smart regenerative relays for link-adaptive cooperative communications," *IEEE Transactions on Communications*, vol. 56, no. 11, pp. 1950–1960, 2008.

[18] A. Bletsas, A. Khisti, D. P. Reed, and A. Lippman, "A simple cooperative diversity method based on network path selection," *IEEE Journal on Selected Areas in Communications*, vol. 24, no. 3, pp. 659–672, 2006.

18

Design and Verification of Secure Mutual Authentication Protocols for Mobile Multihop Relay WiMAX Networks against Rogue Base/Relay Stations

Jie Huang and Chin-Tser Huang

Department of Computer Science and Engineering, University of South Carolina, Columbia, SC 29201, USA

Correspondence should be addressed to Chin-Tser Huang; huangct@cse.sc.edu

Academic Editor: Jit S. Mandeep

Mobile multihop relay (MMR) WiMAX networks have attracted lots of interest in the wireless communication industry recently because of its scalable coverage, improved data rates, and relatively low cost. However, security of MMR WiMAX networks is the main challenge to be addressed. In this paper, we first identify several possible attacks on MMR WiMAX networks in which a rogue base station (BS) or relay station (RS) can get authenticated and gain control over the connections and show that the current standard does not address this problem well. We then propose a set of new authentication protocols for protecting MMR WiMAX networks from rogue BS attack, rogue RS attack, and suppress-replay attack. Our protocols can provide centralized authentication by using a trusted authentication server to support mutual authentication between RS and BS, between RS and RS, and between mobile station (MS) and RS. Moreover, our protocols can also provide distributed authentication with a license issued by the trusted server. We use a formal tool called Scyther to analyze and verify the security properties of our protocols. The results show that our protocols can counter rogue BS and RS attacks and suppress-replay attack and are not susceptible to any known attacks.

1. Introduction

In the era of the Internet of Things (IoT) when every device is connected to the Internet, one of the most important technical enablers is the wireless technologies and infrastructure that make connecting different things (devices) possible. Among them, WiMAX (Worldwide Interoperability for Microwave Access) plays a very important role as it delivers lower-cost, longer-range, and high-bandwidth mobile broadband access for mobile clients and devices to connect to the Internet from anywhere at any time.

WiMAX is a broadband wireless access technology designed for the advancement of IEEE 802.16 standard [1]. It is considered to be a replacement for WiFi-based mobile broadband connection, because of its better coverage and faster speed. WiMAX is ideal for high data-rate IP applications such as video conferencing, VoIP, online gaming, and HD video streaming. Another wireless broadband technology, LTE (Long Term Evolution) [1], can also provide high-speed data transmission for mobile phones and data terminals and has been adopted by many cellular service providers. In recent years, some people think that LTE has won over WiMAX in the standard war. However, according to the report from Intel Capital [2], WiMAX does not fade away and is still considered a very good option by many service providers, including Egyptian telecom startup, Orascom Telecom, and Netherlands startup Enertel Holding [2].

In the early days when WiMAX was designed, WiMAX faced the paradox that increasing data rate will reduce reliability, and increasing minimum reliability service will reduce the coverage area [3]. One possible solution is to deploy more base stations (BSs) closely, but the high cost of deploying BSs will give away the original economic competitiveness of WiMAX. Therefore researchers switched to a more viable approach, which is to insert relatively cheaper fixed relay stations into the cell. This kind of networks is called multihop relay networks. In June 2009, the IEEE 802.16 Relay Task

Group (TG) proposed the IEEE 802.16j-2009 amendment, whose main purpose is to expand the previous single-hop 802.16 standard to include multihop capabilities [4], enable the operations of multihop communications based on relay stations (RSs), specify the mobile multihop relay (MMR) deployment, and define two new types of elements, the multihop relay base station (MR-BS) and the relay station (RS). In 2012, 802.16 Working Group announced the latest version of the standard, IEEE 802.16-2012 [5], which incorporated 802.16j along with two other amendments, 802.16h and 802.16m.

Security has been an open challenge in WiMAX since its commencement because of the open-air nature of wireless communications. To overcome the potential attacks to WiMAX, IEEE 802.16 standard specifies a security sublayer in the MAC (Media Access Control) layer. IEEE 802.16-2009 offers an improved authentication and authorization mechanism compared to its previous versions such as IEEE 802.16-2004. IEEE 802.16-2009 provides better encryption methods, more secure key management protocol, and an EAP-based authentication strategy. However, in an MMR WiMAX network, more security issues are exposed since messages have to be transmitted through one or more relay stations, which makes it more difficult to ensure the authenticity of messages and devices involved in the transmission. Therefore the latest standard IEEE 802.16-2012 defines an air interface between an MR-BS and a RS with the following additional security functionalities [5]:

(i) *Trust within a certain cell*, that is, MR-BS and a group of RSs in the MR cell maintain a set of trusted relationships, called Security Zone, in order to satisfy requirements of multihop relay system operations.

(ii) *Centralized security control in MR-BS*, that is, MR-BS is in charge of the generation of the security association materials between RS and MR-BS.

(iii) *Transparency for mobile station (MS) connected to the network through one or more RSs*, that is, any intermediate RS does not try to decrypt the user data or authenticate the MAC management message it receives from the MS but simply relays it to the next node.

(iv) One RS does not have Authorization Key (AK) security context of any other RS.

(v) The intermediate RS authenticates management messages it receives from other RSs using relay-specific shard keys.

(vi) *Protection of nonauthenticated Pairwise Master Key (PKM) messages by MR-BS and the access RS*, that is, any nonauthenticated PKM messages which are transmitted between MS and MR-BS through the access RS will be protected by the HMAC/CMAC based on the shared security associations established between MR-BS and the access RS.

However, even with these additional functionalities, researchers in [6] found that the security mechanism in IEEE 802.16j (part of current IEEE 802.16-2012 standard) is still not adequate in that it has vulnerabilities in its weak protection of some PKM messages and security zone key update and is susceptible to DoS attacks on BS and rogue BS and rogue RS attacks.

In our previous conference paper [7], we focused on addressing the rogue BS and rogue RS attacks in MMR WiMAX networks with centralized authentication schemes and designed a set of protocols to address the aforementioned attacks. However, a formal verification we conducted later using the Scyther tool [8] shows that our previous protocols allow for a new attack called suppress-replay attack, which exploits the asynchronization of time stamps in BS and RS. Moreover, our previous work did not provide a distributed authentication scheme, which is considered to be a more favorable way to conduct authentication nowadays because of the multitude of distributed mobile clients and devices.

In order to address these new issues found in our previous paper, we propose in this paper a complete authentication solution to address the rogue BS and rogue RS attacks in MMR WiMAX networks, the suppress-replay attack found in our previous scheme, and a distributed authentication scheme between RSs. We present three different mutual authentication protocols which utilize a trusted authentication server to support three possible scenarios in which the RS or MS connects to an MMR WiMAX network. Our protocols are conformant to the security requirements of IEEE 802.16-2012 standard. We also verify the correctness of our protocols by utilizing the former verification tool for security protocols called Scyther. According to the verification result, our protocols are able to counter against all the attacks we discussed in our paper.

The remainder of this paper is organized as follows. In Section 2, we give an overview of related works. Section 3 describes the possible rogue BS and rogue RS attacks on access service in MMR networks. In Section 4, we present our new schemes for securing the original authentication protocols for MMR networks against rogue BS and rogue RS. In Section 5, we give a formal analysis and verification of our protocols using the Scyther tool. Finally, we conclude our paper and discuss the future work in Section 6.

2. Related Work

A number of papers have been published regarding the security issues of WiMAX networks since IEEE 802.16 standard was developed. Xu et al. give a detailed analysis on privacy and key management protocols of the standard in [9, 10]. Several other papers addressed the security issues of one-way authentication and rogue base station attack such as [9, 11, 12]. However, these publications considered only the single-hop WiMAX when the MMR WiMAX network had not come to existence.

The authentication issue has been studied in several other types of multihop networks, such as wireless mesh networks [13], cellular networks [14], and sensor networks [15]. However, the MMR WiMAX network is still very recent and has its own unique characteristics that need to be investigated separately.

Distributed and centralized authentication approaches are two major options when it comes to authentication protocol design. In [16], Yang et al. discussed security issues in WiMAX MMR networks and the pros and cons of the two major types of authentication protocol design. In [17], Jin et al. propose an improved mutual authentication scheme in multihop WiMAX networks, in which they improve the X.509 certificate by using ECC algorithm instead of RSA, and modify the flow of mutual authentication to improve the security in multihop WiMAX networks. In [18, 19], Khan et al. proposed a modified PKM protocol using distributed authentication and localized key management scheme. In [3], Tie and Yi proposed a multihop ticket based handover authentication which adopted the idea from Kerberos and used a ticket to allow MS, RS, and BS to mutually authenticate each other. However, the authors in the aforementioned papers did not take rogue access node attack into consideration.

In order to solve the problems like security zone key update, DDoS attack, and rogue RS attack, in [20], the authors propose a design of hybrid authentication and key distribution scheme to support the IEEE 802.16j (part of current IEEE 802.16-2012 standard) MMR requirements. Although the authors claim that this hybrid design is robust enough to prevent rogue node attack, they only consider the case when a rogue RS tries to join the network at initial phase, and they do not take rogue BS attack into account. The latter case will cause more severe damage to the network since a rogue BS can take control of the whole area within its communication range if it successfully joins the network as a legitimate BS. In another paper [21], the authors present a distributed scheme using decode and forward relays with localized authentication, which helps to authenticate MS and RS at initial network entry. However, this scheme still cannot solve the problem of rogue BS attack. In [22], the authors proposed a self-testing approach to defend against rogue BS attack of intelligent terminal. However, their work did not focus on MMR networks and thus cannot address the other security issues in MMR networks we discussed in Section 1. In [23], the authors discussed detection of rogue BS attack in WiMAX networks; however, their discussion did not address the security issues existing specifically in MMR networks.

3. Rogue BS and Rogue RS Attacks in MMR WiMAX Networks

Rogue stations have been one of the most common threats in wireless networks [23–25]. In order to design a secure wireless authentication protocol in MMR networks, rogue stations threat must be considered and addressed.

Denial of Service (DoS) jamming is a type of the attacks that can involve a rogue station. In MMR, a rogue station can be a rogue BS or a rogue RS. In a DoS jamming attack, by jamming a legitimate BS, connectivity between client and a legitimate BS can be interrupted, which makes it possible for a rogue BS to stand in and impersonate the legitimate BS with fake credentials, trying to convince a joining client or RS to connect with it, so as to cause DoS or even redirect the traffic and hijack the communications.

Another form of attack that might involve a rogue station, especially a rogue RS, is man-in-the-middle (MITM) attack [25]. When a client MS or RS initiates a connection, the rogue RS will intercept the connection, and then complete the connection to the intended legitimate RS and proxy all communications to the intended legitimate RS. The rogue RS is now in a position to inject data, modify messages and communications, or eavesdrop on a session that would normally be difficult to decode, such as encrypted sessions.

One form of MITM attack involves asynchronization. If the clocks of the client and the legitimate station are not synchronized, it is possible for the rogue station to launch a suppress-replay attack [26]. In a suppress-replay attack, the rogue station can intercept messages that carry a timestamp corresponding to a future time due to an unsynchronized clock and extract from them the component containing the future timestamp. Then, the rogue station can combine the extracted timestamp component with valid components from other messages to create a fake message and replay it later when the timestamp in the fake message becomes current with respect to the clock of the legitimate recipient station.

An example to demonstrate the suppress-replay attack is given as follows. Suppose we have a client MS whose clock is 2 minutes ahead of the one in the legitimate station BS. At client side time 1:10 pm and time 1:11 pm (which corresponds to BS time 1:08 pm and time 1:09 pm), client MS sends two messages, message 110 and message 111, to the legitimate BS trying to initiate a connection; each message contains its current MS timestamp and component with necessary authentication credentials to prove MS's identity; we call this component *Authentication Component (AC)* here. When an attacker intercepts these two messages, the attacker can extract the AC from message 110 and the 1:11 pm timestamp from message 111 and then combine these two parts together to create a new message $111'$. The attacker will then send out this newly created fake message $111'$ when the time on the legitimate station BS becomes 1:11 pm. IEEE 802.16-2012 uses PKMv2 to counter possible rogue BS attack by using mutual authentication. However, there is an implicit assumption in PKMv2 that BS is always trustworthy; thus PKMv2 does not provide any protection measure to detect and counter the attack from a compromised BS.

Moreover, the distributed security mode in MMR WiMAX networks also makes rogue RS attack more possible. This is because the authentication procedure between RS nodes is not performed by a centralized server but is based on the trust between nodes. If one node is compromised, its trust with other nodes is also compromised.

Another issue with PKMv2 is that the preassumed trust of BS cannot prevent suppress-replay attack mentioned above. To address all of the aforementioned attacks in MMR, careful protocol design is required in order to achieve secure authentication in MMR networks.

4. Proposed Secure Authentication Protocol

We have introduced the related work on security issues in MMR WiMAX networks and have shown that current standards are not sufficient for addressing the rogue BS and rogue RS attacks. In this section, we present a set of new secure protocols to provide robust authentication in MMR WiMAX networks. Specifically, our protocols can defend against rogue BS and RS attack by using a trusted authentication server to provide dual authentication and security zone key. Our protocols support three scenarios of network access: RS connects to an MMR network through BS; RS connects to an MMR network through other RS; and MS connects to an MMR network through RS.

4.1. Assumptions. The proposed protocols are based on the following assumptions:

(a) Before the initial access authentication process, each RS, BS, and mobile user (MS) is preregistered with the Authentication Server (AS) by providing their MAC addresses and other necessary credentials. Each RS, BS, and MS shares its own public key (K_{RS}, K_{BS}, K_{MS}) with AS, and each RS, BS, and MS also gets AS's public key from AS. These keys were obtained during the preregistration phase.

(b) AS is trusted by all the nodes in the MMR WiMAX network. It is believed by the nodes in this network that AS maintains a correct database of all legitimate registered nodes' MAC addresses, each node's corresponding public key, and other credentials. It is easier to ensure the physical security of AS because AS can always be indoor.

4.2. Notations. Before we describe the details of our protocols, we specify the simplified notation for each element used in the protocol:

MAC$_X$: X'S MAC address (X can be either RS, BS, or MS),

Seq$_X$: a sequence number generated by station X,

Nnc$_X$: a nonce generated by station X,

K$_X$: X's secret key shared with AS,

K$_{X\_PUB}$: X's public key stored in X's certificate,

K$_{X\_PRV}$: X's private key corresponding to the public key stored in X's certificate,

PMK: Pairwise Master Key,

SZK: a group key used in the security zone among BS and many RSs,

AK: Authorization Key,

MD[M]: message digest of message M,

CERT$_X$: X's digital certificate with X's public key included,

E$_{K_X}$[M]: M encrypted with X's shared secret key,

E$_{K_{SZK}}$[M]: M encrypted with security zone key,

E$_{K_{X\_PUB}}$[M]: M encrypted with X's public key stored in X's certificate,

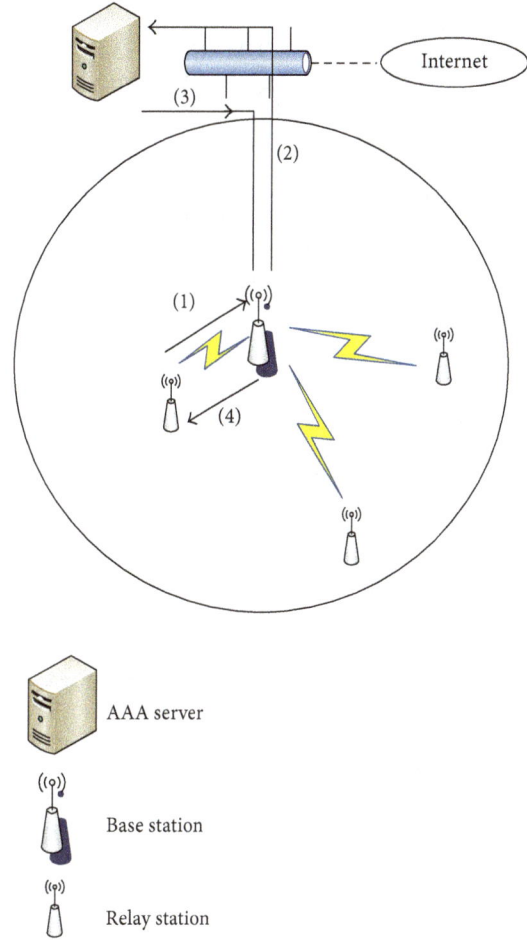

FIGURE 1: RS connects to MMR networks via BS.

E$_{K_{X\_PRV}}$[M]: M encrypted with X's private key corresponding to the public key stored in X's certificate,

License$_X$: a signature issued to legitimate BS/RS/MS; the signature is generated by AS using AS's private key,

EXPR$_X$: expiration time for License$_X$.

Here the format of License$_X$ is $\{E_{K_{AS\_PRV}}[MAC_X \parallel CERT_X \parallel EXPR_X]\}$.

4.3. Scenario 1: RS Connects to the Network through BS. The first scenario in which a RS needs to connect to an MMR network via BS is shown in Figure 1.

A RS broadcasts the AUTH-REQ message when it wants to connect to an MMR WiMAX network. Normally the BS which is the closest to this RS will handle this message and send it to AS. AS will then perform the authentication and send back an AUTH-REPLY message. The detailed message format is specified as follows:

(1) RS → BS:

MAC$_{RS}$ $\parallel$ Seq$_{RS}$ $\parallel$ MD[MAC$_{RS}$ $\parallel$ Seq$_{RS}$ $\parallel$ K$_{RS}$] $\parallel$ CERT$_{RS}$

(2) BS → AS:

$MAC_{RS} \parallel Seq_{RS} \parallel MD[MAC_{RS} \parallel Seq\_RS \parallel K_{RS}] \parallel$ $CERT_{RS} \parallel MAC_{BS} \parallel Seq_{BS} \parallel MD[MAC_{BS} \parallel Seq\_BS \parallel MAC_{RS} \parallel K_{BS}] \parallel CERT_{BS}$

(3) AS → BS:

$E_{K_{BS}}\{MAC_{AS} \parallel MAC_{RS} \parallel Seq\_BS \parallel PMK \parallel License\_BS \parallel SZK \parallel MD[PMK \parallel SZK \parallel K_{BS}]\} \parallel E_{K_{RS}}\{MAC_{AS} \parallel MAC_{BS} \parallel Seq\_RS \parallel PMK \parallel License\_RS \parallel SZK \parallel MD[PMK \parallel SZK \parallel K_{RS}]\}$

(4) BS → RS:

$E_{K_{RS}}\{MAC_{AS} \parallel MAC_{BS} \parallel Seq\_RS \parallel PMK \parallel License\_RS \parallel SZK \parallel MD[PMK \parallel SZK \parallel K_{RS}]\} \parallel MD[PMK \parallel AK \parallel SZK]$

As we mentioned before, each RS is preregistered with AS. Therefore AS can match the MAC address in the message with the corresponding shared secret key in the database. When AS receives message (2), AS can use MAC address, sequence number, and the shared secret key to calculate the corresponding message digest in order to verify the authenticity of the sender (both BS and RS), since only AS and the corresponding legitimate node know the shared secret key. After successful authentication of BS and RS, AS will use the shared keys of BS and RS to encrypt the authentication reply message and send it back to BS and the requesting RS. If both BS and RS are legitimate, in message (3) AS will generate a PMK and a SZK and include them in the reply message. After BS receives and decrypts the AUTH-REPLY message from AS, it can be sure about the authenticity of the requesting RS. Thus BS can decide whether to grant RS the access to the network or not. BS also gets PMK from AS and uses it to generate AK. PMK is encrypted with BS's shared key with AS. AS also sends the message digest of PMK in this message to BS, which is meant to allow BS to verify the integrity of this PMK it receives. In message (3), BS receives its License$_{BS}$ from AS, which works as a unique digital signature to prove its identity to other nodes before the license expires. This signature is generated by encrypting the concatenation of BS's MAC address, BS's certificate (e.g., X.509 certificate), and the expiration time of this license, using AS's private key. As what has been described in our assumptions, every legitimate node in this network has the AS's public key, so when BS gets its license, it can use AS's public key to verify this license's authenticity. Similar license also appears in message (4) in which it provides legitimate RS with a license to prove its identity to other nodes in the future. The usage of signature here in messages (3) and (4) can protect the network from malicious node with a fake license, since it should be computationally infeasible for a party who does not possess the private key to generate a valid signature.

After successful authentication with BS/RS, AS assigns a security zone key, denoted as SZK, for nodes in the security zone to secure future communications between them. At AS, for each AUTH-REQ from RS, AS will check with BS's MAC address in its received message to decide which security zone this requesting RS belongs to. After authentication is successfully completed, a zone key SZK which corresponds to the right BS will be assigned to the RS.

In message (4), along with PMK, BS also sends the message digest of AK which it generated from PMK; the purpose is to let RS verify the integrity of the PMK it receives. If RS can use the PMK it received to generate the same AK which is consistent with the message digest MD[AK] included in message (4), it can believe that the message is not compromised and the AK is good to use.

We analyze why our protocol can protect the MMR WiMAX network against rogue BS or rogue RS attacks. In the case of rogue BS attack, since BS does not have RS's secret key shared with AS, it cannot decrypt or modify the message destined to RS. Hence if RS gets the message from AS which notifies RS of this illegitimate BS, RS will not try to get access through this rogue BS again. Therefore the rogue BS will not have the opportunity to take control of RS in the network.

In the case of rogue RS attack, when an access RS becomes a rogue one, it would not receive the AUTH-REPLY message with a license included. In the case of message hijacking or man-in-the-middle attack, even if an attacker can get the message which was supposed to be transmitted to the legitimate BS or RS, the attacker still cannot decrypt the message to get the license since he does not have the secret key K_{BS} or K_{RS}. Therefore the attacker cannot obtain a valid license without getting authenticated by AS, which it cannot pass.

4.4. Scenario 2: RS Connects to the Network through Other RS. The second scenario in which a RS requests to connect to an MMR network through an edge RS is shown in Figure 2. In this scenario, the requesting RS wants to set up connection with the BS via one or more intermediate RSs. Our protocol for this scenario contains three authentication phases: (i) initial verification of edge RS; (ii) dual authentication of BS and the requesting RS by AS; and (iii) distributed authentication when requesting RS holds a valid license.

In authentication phase (i), when the requesting RS_A tries to get access to the network through another RS_B (edge RS), RS_A needs to verify the authenticity of this intermediate RS_B first. The message format of phase (i) is specified as follows:

(1) Requesting RS_A → Edge RS_B:

$MAC_{RS(A)} \parallel CERT_{RS(A)} \parallel Seq\_RS(A) \parallel MD[MAC_{RS(A)} \parallel Seq\_RS(A)]$

(2) Edge RS_B → Requesting RS_A:

$MAC_{RS(B)} \parallel CERT_{RS(B)} \parallel License\_RS(B) \parallel MD[MAC_{RS(B)} \parallel Seq\_RS(A)] \parallel E_{K_{RS(B)\_PRV}}[E_{K_{RS(A)\_PUB}}[Seq\_RS(A) \parallel Nnc\_RS(B)]]$

(3) Requesting RS_A → Edge RS_B:

$E_{K_{RS(A)\_PRV}}[E_{K_{RS(B)\_PUB}}[Nnc\_RS(B) \parallel Nnc\_RS(A)]]$

(4) Edge RS_B → Requesting RS_A:

$E_{K_{RS(B)\_PRV}}[E_{K_{RS(A)\_PUB}}[Nnc\_RS(A)]]$

In messages (1) and (2), the requesting relay station RS_A and the edge relay station RS_B exchange their certificate, such that they know each other's public key. In message (2), based on our assumption that each registered RS has the AS's public key, the requesting RS_A can use it to verify whether the edge

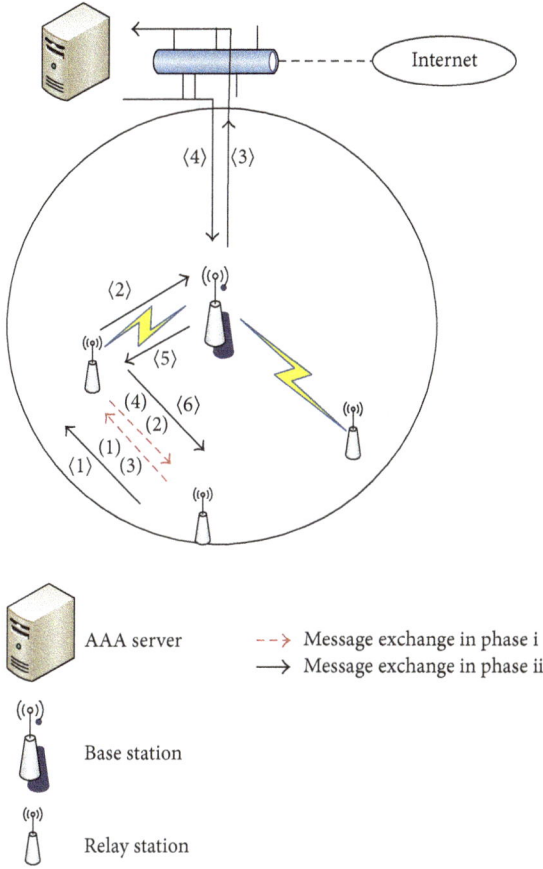

FIGURE 2: RS connects to MMR networks via an edge RS.

Legend:
AAA server

$--\rightarrow$ Message exchange in phase i
$\longrightarrow$ Message exchange in phase ii

Base station

Relay station

RS_B is a legitimate relay station by checking that the license is currently valid. However, if the authentication phase (i) stops at message (2), then it will be susceptible to a *replay attack* described as follows: a malicious relay station RS_E listening in the middle makes a copy of message (1) sent by the requesting RS_A, makes a separate run of this protocol by forwarding the copy of message (1) to a legitimate edge relay station RS_B (i.e., pretending RS_E itself is the requesting RS_A), and replays the message (2) received from the RS_B to RS_A. When RS_A receives the message, it will be convinced that the malicious RS_E is a legitimate edge relay station. To address this problem of replay attack, we add messages (3) and (4), which is in challenge-response style. In message (3), RS_A concatenates $Nnc_{RS(B)}$ (derived from decrypting message (2)) with a nonce of its choice $Nnc_{RS(A)}$, encrypts it using the public key of RS_B and the private key of RS_A, and sends it to RS_B. RS_B decrypts the received message (3) and extracts $Nnc_{RS(A)}$, encrypts it using the public key of RS_A and the private key of RS_B, and sends it to RS_A in message (4). When RS_A verifies that the $Nnc_{RS(A)}$ received in message (4) is the same as the nonce it chooses in message (3), RS_A can believe that RS_B is really the authentic edge relay station, rather than a malicious relay station of a replay attack.

After edge RS_B's authenticity has been verified in phase (i), the requesting RS_A starts phase (ii), in which both BS and RS_A need to be authenticated by AS. The message format of phase (ii) is specified as follows:

⟨1⟩ Requesting $RS_A \rightarrow$ Edge RS_B:

$MAC_{RS(A)} \parallel Seq_{RS(A)} \parallel CERT_{RS(A)} \parallel MD[MAC_{RS(A)} \parallel Seq_{RS(A)} \parallel K_{RS(A)}]$

⟨2⟩ Edge $RS_B \rightarrow$ (possibly other RS in between) $\rightarrow$ BS:

$E_{K_{SZK}}\{MAC_{RS(A)} \parallel Seq_{RS(A)} \parallel CERT_{RS(A)} \parallel MD[MAC_{RS(A)} \parallel Seq_{RS(A)} \parallel K_{RS(A)}]\}$

⟨3⟩ BS $\rightarrow$ AS:

$MAC_{RS(A)} \parallel Seq_{RS(A)} \parallel CERT_{RS(A)} \parallel MD[MAC_{RS(A)} \parallel Seq_{RS(A)} \parallel K_{RS(A)}] \parallel MAC_{BS} \parallel Seq_{BS} \parallel MD[MAC_{BS} \parallel MAC_{RS(A)} \parallel Seq_{BS} \parallel K_{BS}] \parallel CERT_{BS}$

⟨4⟩ AS $\rightarrow$ BS:

$E_{K_{BS}}\{MAC_{AS} \parallel MAC_{RS(A)} \parallel Seq_{BS} \parallel PMK \parallel License_{BS} \parallel SZK \parallel MD[PMK \parallel SZK \parallel K_{BS}]\} \parallel E_{K_{RS(A)}}\{MAC_{AS} \parallel MAC_{BS} \parallel Seq_{RS(A)} \parallel PMK \parallel License_{RS(A)} \parallel SZK \parallel MD[PMK \parallel SZK \parallel K_{RS(A)}]\}$

⟨5⟩ BS $\rightarrow$ Edge RS_B:

$E_{K_{SZK}}\{E_{K_{RS(A)}}\{MAC_{AS} \parallel MAC_{BS} \parallel Seq_{RS(A)} \parallel PMK \parallel License_{RS(A)} \parallel SZK \parallel MD[PMK \parallel SZK \parallel K_{RS(A)}]\} \parallel MD[PMK \parallel AK \parallel SZK]\}$

⟨6⟩ Edge $RS_B \rightarrow$ Requesting RS_A:

$E_{K_{RS(A)}}\{MAC_{AS} \parallel MAC_{BS} \parallel Seq_{RS(A)} \parallel PMK \parallel License_{RS(A)} \parallel SZK \parallel MD[PMK \parallel SZK \parallel K_{RS(A)}]\} \parallel MD[PMK \parallel AK \parallel SZK]$

If the authentication between RS_A and AS is successful and RS_A is regarded to be legitimate, then RS_A will be assigned with the security zone key SZK which RS_A and RS_B can use to secure future communications between them.

Comparing the message format in phase (ii) with the message format in the first scenario in which RS connects to the network directly through BS, we can see that the major contents in the messages are very similar except that the security zone key SZK is being used here to encrypt all the messages that are transmitted in this security zone (messages ⟨2⟩ and ⟨5⟩). This design is in accordance with the aforementioned security requirement of IEEE 802.16-2012 standard, which requires that trust is maintained within the security zone in order to support multihop relay system operations. At edge RS_B, RS_B decrypts the message it received with SZK and then forwards the decrypted messages to the requesting RS_A. In this case the message is still secured because it is still encrypted by AS with RS_A's public key.

If the requesting RS_A has been authenticated by AS within a valid period of time before it tries to connect to the current network, RS_A can skip phases (i) and (ii) and directly enter an optional phase (iii) in which a more efficient distributed authentication will be performed. The message format is described as follows:

(1) Requesting $RS_A \rightarrow$ All neighborhood nodes:

$MAC_{RS(A)} \parallel Seq_{RS(A)} \parallel CERT_{RS(A)} \parallel License_{RS(A)} \parallel MD[MAC_{RS(A)} \parallel Seq_{RS(A)} \parallel CERT_{RS(A)} \parallel License_{RS(A)}]$

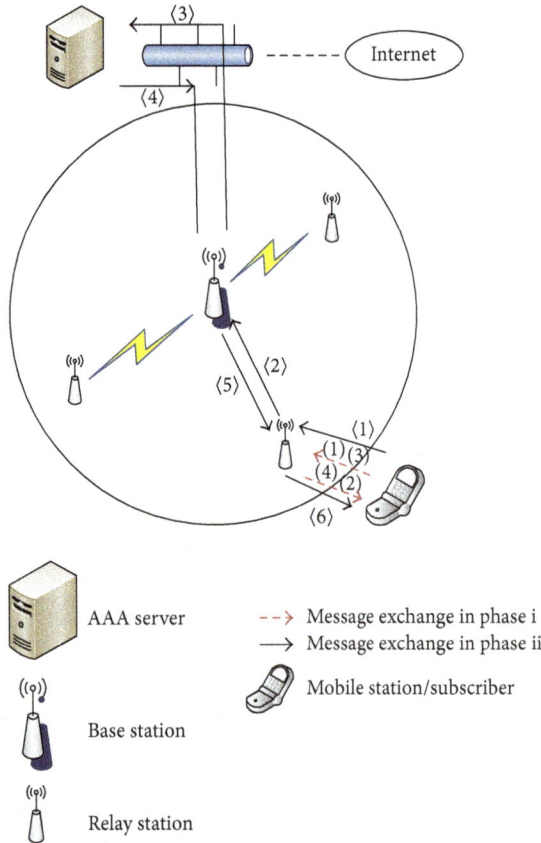

FIGURE 3: MS connects to MMR networks via an edge RS.

(2) Edge RS_B → Requesting RS_A:

$MAC_{RS(B)}$ ‖ $CERT_{RS(B)}$ ‖ $License_{RS(B)}$ ‖ MAC_{BS} ‖ $E_{K_{RS_A\text{-}PUB}}[SZK]$ ‖ $MD[MAC_{RS(B)}$ ‖ $Seq_{RS(A)}$ ‖ $CERT_{RS(B)}$ ‖ $License_{RS(B)}$ ‖ MAC_{BS} ‖ $SZK]$

In phase (iii), RS_A first broadcasts to all its neighborhood nodes message (1), which contains RS_A's MAC address, sequence number, digital certificate, its license, and the message digest. Once a legitimate edge RS_B receives this message, it can use AS's public key to verify the authenticity of RS_A. If RS_A is verified to be authentic, then edge RS_B can send RS_A message (2) which includes RS_B's license, certificate, security zone key, and the MAC address of BS which is in charge of this security zone. When RS_A gets message (2) and has the SZK and BS's MAC address, it can establish secure communications with BS to generate and exchange AK and TEK.

4.5. Scenario 3: MS Connects to the Network through RS.
The third scenario in which a mobile user (MS) needs to connect to an MMR network through an edge RS is shown in Figure 3. In this scenario we have two cases to consider: the first one is when MS connects to the network for the first time through one or more RS; the second one is when MS has been authenticated by AS within a valid period of time before it tries to connect to the current network. For both cases, the message formats are similar to the message formats from the

second scenario when a RS tries to get access to the network via other RS.

For MS which joins the network for the first time through an edge RS, there are two phases whose message format is specified as follows:

Phase (i) (initial verification of the edge RS)

(1) MS → Edge RS:

MAC_{MS} ‖ $CERT_{MS}$ ‖ Seq_{MS} ‖ $MD[MAC_{MS}$ ‖ $Seq_{MS}]$

(2) Edge RS → MS:

MAC_{RS} ‖ $CERT_{RS}$ ‖ $License_{RS}$ ‖ $MD[MAC_{RS}$ ‖ $Seq_{MS}]$ ‖ $E_{K_{RS\text{-}PRV}}[E_{K_{MS\text{-}PUB}}[Seq_{MS}$ ‖ $Nnc_{RS}]]$

(3) MS → Edge RS:

$E_{K_{MS\text{-}PRV}}[E_{K_{RS\text{-}PUB}}[Nnc_{RS}$ ‖ $Nnc_{MS}]]$

(4) Edge RS → MS:

$E_{K_{RS\text{-}PRV}}[E_{K_{MS\text{-}PUB}}[Nnc_{MS}]]$

Note that the four messages in phase (i) are similar to the four messages in phase (i) of Scenario 2 presented in the previous Section 4.4, because we need to prevent the possible replay attack launched by a malicious relay station. We will not repeat the explanation of the replay attack here because it has been explained in detail in the previous subsection.

Phase (ii) (dual authentication of MS and BS)

⟨1⟩ MS → Edge RS:

MAC_{MS} ‖ Seq_{MS} ‖ $CERT_{MS}$ ‖ $MD[MAC_{MS}$ ‖ Seq_{MS} ‖ $K_{MS}]$

According to IEEE 802.16-2012 standard, here relay stations do not try to decrypt the user date or authenticate the MAC management message they receive from mobile stations but simply relay it. And RS does not have any key information associate with the MS:

⟨2⟩ Edge RS → (possibly other RS in between) → BS:

$E_{K_{SZK}}\{MAC_{MS}$ ‖ Seq_{MS} ‖ $CERT_{MS}$ ‖ $MD[MAC_{MS}$ ‖ Seq_{MS} ‖ $K_{MS}]\}$

⟨3⟩ BS → AS:

MAC_{MS} ‖ Seq_{MS} ‖ $CERT_{MS}$ ‖ $MD[MAC_{MS}$ ‖ Seq_{MS} ‖ $K_{MS}]$ ‖ MAC_{BS} ‖ Seq_{BS} ‖ $MD[MAC_{MS}$ ‖ Seq_{BS} ‖ $K_{BS}]$ ‖ $CERT_{BS}$

⟨4⟩ AS → BS:

$E_{K_{BS}}\{MAC_{AS}$ ‖ MAC_{MS} ‖ Seq_{BS} ‖ PMK ‖ $License_{BS}$ ‖ SZK ‖ $MD[PMK$ ‖ SZK ‖ $K_{BS}]\}$ ‖ $E_{K_{MS}}\{MAC_{AS}$ ‖ MAC_{BS} ‖ Seq_{MS} ‖ PMK ‖ $License_{MS}$ ‖ SZK ‖ $MD[PMK$ ‖ SZK ‖ $K_{MS}]\}$

⟨5⟩ BS → (possibly other RS in between) → Edge RS:

$E_{K_{SZK}}\{E_{K_{MS}}\{MAC_{AS}$ ‖ MAC_{BS} ‖ Seq_{MS} ‖ PMK ‖ $License_{MS}$ ‖ SZK ‖ $MD[PMK$ ‖ SZK ‖ $K_{MS}]\}$ ‖ $MD[PMK$ ‖ AK ‖ $SZK]\}$

⟨6⟩ Edge RS → MS:

$E_{K_{MS}}\{MAC_{AS} \parallel MAC_{BS} \parallel Seq_{MS} \parallel PMK \parallel License_{\_MS} \parallel SZK \parallel MD[PMK \parallel SZK \parallel K_{MS}]\} \parallel MD[PMK \parallel AK \parallel SZK]$

In message ⟨6⟩, after MS gets PMK and generates related AK and TEK, MS can verify its AK with the AK in the received message to check their consistency. TEK will be used to secure future communications between BS and MS.

For MS which has been authenticated by AS within a valid period of time before it tries to connect to the current network, a distributed authentication will be performed. The message format is specified as follows:

(1) MS → All neighborhood nodes:

$MAC_{MS} \parallel Seq_{MS} \parallel CERT_{MS} \parallel License_{\_MS} \parallel MD[MAC_{MS} \parallel Seq_{MS} \parallel CERT_{MS} \parallel License_{\_MS}]$

(2) Edge RS → MS:

$MAC_{RS} \parallel CERT_{RS} \parallel License_{\_RS} \parallel MAC_{BS} \parallel E_{K_{MS\_PUB}}[SZK] \parallel MD[MAC_{RS} \parallel Seq_{\_MS} \parallel CERT_{RS} \parallel License_{\_RS} \parallel MAC_{BS} \parallel SZK]$

MS broadcasts message (1) to all its neighborhood nodes. Message (1) contains MS's MAC address, sequence number, digital certificate, its license, and the message digest. Once a legitimate edge RS receives this message, it can use the AS's public key to verify the authenticity of MS. If MS is verified to be authentic, then edge RS can send MS message (2) which includes RS's license, certificate, security zone key between RS and MS, and the MAC address of BS which is in charge of this security zone. When MS gets message (2) and has the SZK and BS's MAC address, with SZK it can establish a secure communication with BS before it has TEK to encrypt its messages.

5. Formal Analysis of Proposed Protocols

Next, we present the formal analysis and verification of the proposed protocols using a tool called Scyther [8]. Scyther is an automated tool for verification, falsification, and analysis of security protocols [8]. Its effectiveness and correctness have been proved and its operational semantics can be found in [27] for interested readers. Scyther can verify protocols with unbounded number of sessions and roles, if computational resources allow [8]. It is also currently the only existing tool capable of verifying synchronization [27]. Synchronization is an important property in mutual authentication protocols. It indicates that all the messages are transmitted exactly in the order as described by the protocol, which can be used to detect suppress-replay attacks [8]. Hence we include it as one of the properties we need to verify in our proposed protocols, and we choose Scyther since it is the only tool which can do such verification [8]. To have the most accurate verification results, we choose the latest stable version Scyther v1.1.3 released in 2014.

5.1. Model Description. In our model, we describe the behavior of the protocol entities in terms of their roles, that is, an initiator or a responder (receiver) or both. For example, for protocol 1, we have three agents, RS, BS, and AS. All of them play two roles: message initiator and message responder. In Scyther, a *run* is a unique execution of a role that is performed by an agent; that is, each agent executes its runs to implement the protocol and preserve the secrecy of the credentials (e.g., keys, licenses, and sequence number) it claims to achieve. Each agent has several variable definitions for the credentials used in the messages this agent will send or receive when playing different roles. Each agent also has a sequence of events that entails what messages this agent will send or receive, along with its claims. In Scyther, a claim is defined as *claim (A, c, P)*, which means that agent A expects goal c to hold with parameter P. If an attack exists, then the claim the agent wants to achieve will not hold. Such a claim is called a falsified claim [8].

A claim has the locality property, so once agents get the messages, they will be able to view the state of the system from a local perspective. Therefore the protocol needs to make sure that the agent is able to have the knowledge of some properties of global state of the system from the local perspective; for example, the agent is able to know that something is beyond the intruder's knowledge, or that a specific agent is active. However, for the same protocol, the claim on the same secret credential based on different agents might not always hold; that is, it is possible that for agent A as the message initiator and agent B as the message responder *claim (A, c, P)* holds while *claim (B, c, P)* does not hold [8].

In order to verify a security protocol using Scyther, we need to specify an attacker and a set of agents executing several runs. Scyther will trace all possible attacks that the attacker might launch and determine whether a security claim the agent holds is true or attacks exist. In the verification of our proposed protocols, we focus on two security properties: *secrecy of keys* and *authentication*. To verify these two properties, we use two types of claims in our model: secrecy claim and authentication claim [8].

(i) Secrecy Claim. A secrecy claim is defined as *claim (A, Secret, P)*. It is a statement that the credentials P included in this claim by A will not be obtained or spoofed by the attacker. Secrecy means that the information in question is not to the knowledge of an attacker, even if it is transmitted over an untrusted network.

When agents are communicating data (public or secret credentials) with untrusted agents, the transmitted data is also open to the attacker. Although transmitted data is public in the air now, this does not imply that our protocol is broken. Rather, what we need is a secrecy claim saying that if an agent only speaks with trusted agents, then the data being transmitted or shared is kept in secret [8].

(ii) Authentication Claim. In Scyther, authentication focuses on the verification that when a role in a protocol is executed, we can guarantee that in the current network there exists at least one entity that is communicating with this role. However, only knowing that some entity is communicating with the role is not enough to guarantee the correctness and robustness of an authentication protocol; we want to use

Scyther to verify and show a stronger guarantee that when the protocol is being executed, the intended entity is aware of the communication, and the messages exchanged between the entity and the role follow the protocol description.

An authentication claim is defined as *claim (A, Nisynch)*. It means agent *A* sends and receives messages in the order which is exactly the same as what has been described in the protocol [8].

5.2. Security Properties to Be Verified. We aim to verify that our proposed protocols have the following three security properties: *information confidentiality*, *no theft of service possible* [28], and *message sequence synchronization*. Successful verification of these three properties can prove the robustness of our proposed protocols against the possible rogue RS and rogue BS attack described in Section 3. In our verification, we include an additional restriction that only claims concerning sessions between trusted agents are evaluated.

To illustrate our claims in the verification, we define a term *KeyFields*. Each data message exchanged between agents is composed of a set of key elements, for instance, pmk, ak, and szk. *KeyFields* is denoted as the set of elements. The three properties are explained in detail as follows.

5.2.1. Information Confidentiality. This property is satisfied if the access nodes, that is, RS and MS, can make sure that all exchanged keys are kept in secret. This property requires that each individual element α in *KeyFields* should be kept in secret.

The general formalization of the information confidentiality property is given below:

$$\forall \alpha \in KeyFields: claim\ (MS/RS, Secret, \alpha)\ holds.$$

This formalization can be expanded into the following six claims:

Claim 1: *claim* (MS, *Secret*, pmk) holds.

Claim 2: *claim* (MS, *Secret*, ak) holds.

Claim 3: *claim* (MS, *Secret*, szk) holds.

Claim 4: *claim* (RS, *Secret*, pmk) holds.

Claim 5: *claim* (RS, *Secret*, ak) holds.

Claim 6: *claim* (RS, *Secret*, szk) holds.

5.2.2. No Theft of Service Possible. This property is satisfied if (i) AS can be ensured that neither an unauthorized BS nor an unauthorized RS can be able to impersonate a legitimate one and get access to the network, and (ii) BS has the guarantee that an unauthenticated RS cannot gain access to the services provided, nor could it impersonate a legitimate user. A service should always be bound to an authenticated user. This property is similar to the information confidentiality property but involves different agents of the protocol. Its formal definition is given as follows:

$$\forall \alpha \in KeyFields: claim\ (AS/BS, Secret, \alpha)\ holds.$$

This formalization is expanded into claims 7–13:

TABLE 1

Key element	Description
PMK	Pairwise Master Key
AK	Authorization Key
SZK	A group key used in the security zone among BS and many RSs
k(R, A)	K_{RS}, RS's secret key shared with AS
k(B, A)	K_{BS}, BS's secret key shared with AS

Claim 7: *claim* (AS, *Secret*, pmk) holds.

Claim 8: *claim* (AS, *Secret*, szk) holds.

Claim 9: *claim* (AS, *Secret*, k(R, A)) holds.

Claim 10: *claim* (AS, *Secret*, k(B, A)) holds.

Claim 11: *claim* (BS, *Secret*, pmk) holds.

Claim 12: *claim* (BS, *Secret*, ak) holds.

Claim 13: *claim* (BS, *Secret*, szk) holds.

5.2.3. Message Sequence Synchronization. This property is satisfied if all of RS, BS, and AS can be ensured that the corresponding send and receive messages are executed in the order exactly the same as what has been specified in the protocol. Its formal definition is given as follows:

Claim 14: *claim* (RS, *Nisynch*) holds.

Claim 15: *claim* (BS, *Nisynch*) holds.

Claim 16: *claim* (AS, *Nisynch*) holds.

Claim 17: *claim* (MS, *Nisynch*) holds.

5.3. Formal Verifications. Table 1 shows the message elements that are used in our formal model.

Recall that, according to our protocol description in Section 4, several keys will be transmitted. Hence we need to verify our protocols to see whether they can satisfy our secrecy claim and authentication claim.

Below are the specific verification results from Scyther. There are three parts in each of the following figures of verification results. The first part is "Claim" which contains several columns indicating which element α in KeyFields should be kept in secret in a specific node. For example, "R1, Secret pmk" represents key pmk in node R1 should be kept secret. The second part is "Status" which indicates the status of verification with a variety of possible attacks. "Ok" means the specific claim holds (i.e., passes the verification). "Failed" means possible attack(s) exists in this claim, indicating that the design of the authorization protocol might have some potential flaws. The third part is "Comments"; this part provides more specific information regarding the status. The expected results of our protocols are that all claims should hold; that is, we expect to see all status as "Ok" and no attacks are found.

5.3.1. Protocol 1: RS Connects to the Network through BS. From the verification results in Figure 4, we can see that all of our

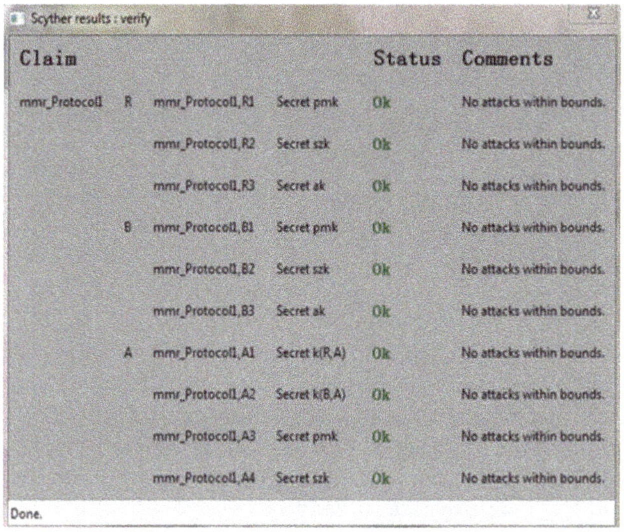

FIGURE 4: Verification result for Protocol 1.

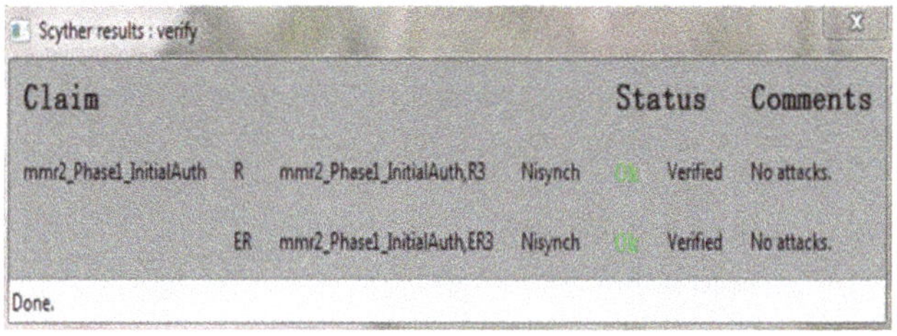

FIGURE 5: Verification result for Protocol 2-Phase 1, initial authentication.

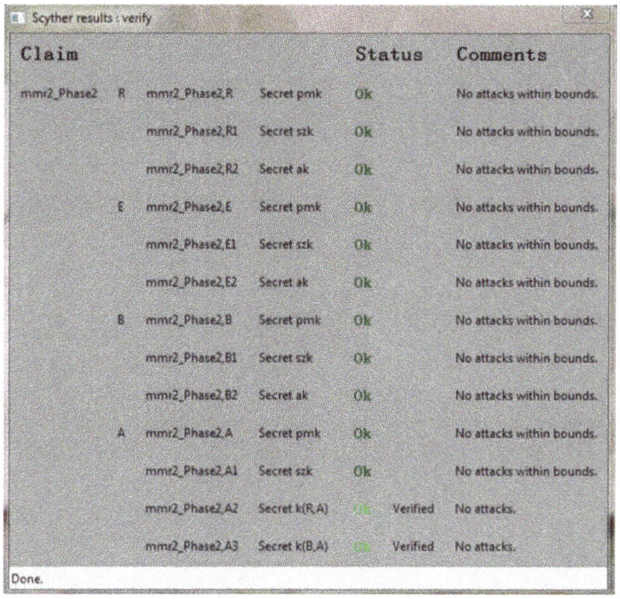

FIGURE 6: Verification result for Protocol 2-Phase 2, mutual authentication.

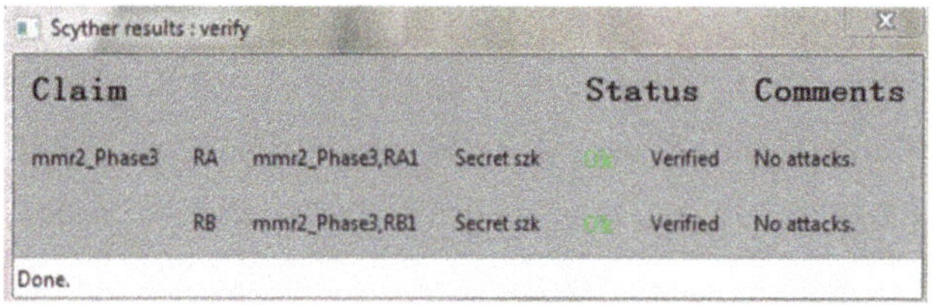

FIGURE 7: Verification result for Protocol 2-Phase 3, distributed authentication.

properties from Claim 4 to Claim 13 in Section 5.2 regarding key elements hold and no possible attacks are detected. This proves that our goals of *information confidentiality* and *no theft of service possible* are satisfied. Therefore, all the key information can be transmitted and exchanged safely.

5.3.2. Protocol 2: RS Connects to the Network through Other RS(s)

(i) Verification Results for Phase 1. From the verification results in Figure 5, we can see that Claim 14 holds and no possible attacks are detected. This proves that our goal of message sequence synchronization is satisfied, and the requesting RS can successfully authenticate a legitimate edge RS.

(ii) Verification Results for Phase 2. The verification results for phase 2 are shown in Figure 6, from which we can see that all of our claims from Claim 4 to Claim 13 regarding key element are held and no possible attacks are detected. This proves that our goals of *information confidentiality* and *no theft of service possible* are satisfied. Therefore all the key information can be transmitted and exchanged safely.

(iii) Verification Results for Phase 3. The verification results for phase 3 are shown in Figure 7, from which we can see that Claim 6 regarding the confidentiality of SZK holds and

no possible attacks are detected. This proves that our goals of information confidentiality and no theft of service possible are satisfied.

5.3.3. Protocol 3: MS Connects to the Network through RS(s)

(i) Verification Results for Phase 1. The verification results for phase 1 are shown in Figure 8, from which we can see that Claims 17 and 14 hold and no possible attacks are detected. This proves that our goal of message sequence synchronization is satisfied, and the MS can successfully authenticate a legitimate edge RS.

(ii) Verification Results for Phase 2. From Figure 9, we can see that all of our claims from Claim 1 to Claim 13 regarding key element are held and no possible attacks are detected. This proves that our goals of *information confidentiality* and *no theft of service possible* are satisfied. Therefore all the key information can be transmitted and exchanged safely.

(iii) Verification Results for Phase 3. The verification results for phase 3 are shown in Figure 10, from which we can see that Claims 6 and 3 regarding the confidentiality of SZK hold and no possible attacks are detected. This proves that our goals of

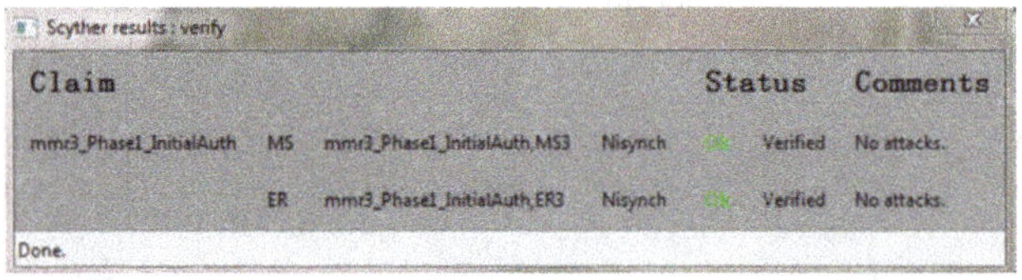

FIGURE 8: Verification result for Protocol 3-Phase 1, initial authentication.

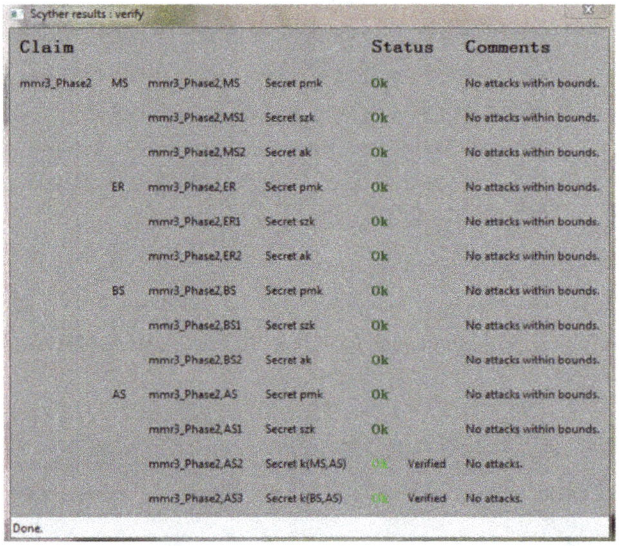

FIGURE 9: Verification result for Protocol 3-Phase 2, initial authentication.

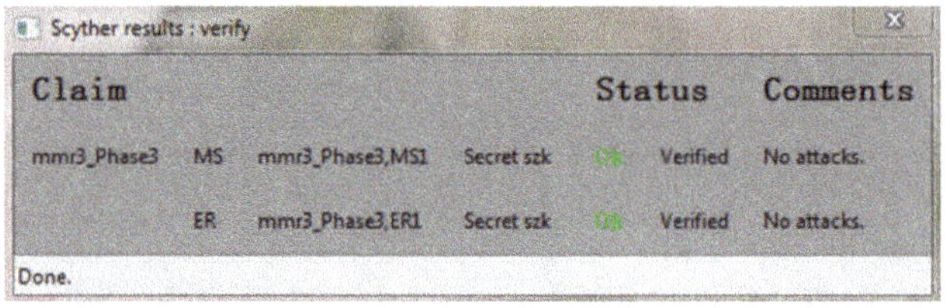

FIGURE 10: Verification result for Protocol 3-Phase 3, distributed authentication.

the complete mutual authentication in AS; and a node with a valid license is trustworthy to give away SZK or provide authentication message forwarding. This method is useful for reducing the overhead especially when fast handoff happens frequently. To verify the security properties of our protocols, we apply the Scyther tool to conduct a formal verification. The verification does not find any attack with our protocols, which proves that our protocols satisfy three desirable security goals, namely, *information confidentiality*, *no theft of service possible*, and *message sequence synchronization*.

In the future work, we will perform a numerical analysis on the authentication performance of our protocols in terms of key processing time, response time, and total overhead of provisioning the dual authentication, the security zone key, and the license.

Competing Interests

The authors declare that they have no competing interests.

References

[1] 3GPP.org, "LTE (Long Term Evolution)," http://www.3gpp.org/technologies/keywords-acronyms/98-lte.

[2] Kyle, "Intel Capital: WiMAX Is Not Dead," http://www.nibletz.com/international/intel-capital-wimax-is-not-dead/.

[3] L. Tie and Y. Yi, "Extended security analysis of multi-hop ticket based handover authentication protocol in the 802.16j network," in *Proceedings of the 8th International Conference on Wireless Communications, Networking and Mobile Computing (WiCOM '12)*, pp. 1–10, Shanghai, China, September 2012.

[4] S. W. Peters and R. W. Heath, "The future of WiMAX: multihop relaying with IEEE 802.16j," *IEEE Communications Magazine*, vol. 47, no. 1, pp. 104–111, 2009.

[5] IEEE Standard for Air Interface for Broadband Wireless Access Systems, "IEEE Std 802.16-2012 (Revision of IEEE Std 802.16-2009)," pp. 1–2542, August 17, 2012.

[6] X. Dai and X. Xie, "Analysis and research of security mechanism in IEEE 802.16j," in *Proceedings of the International Conference on Anti-Counterfeiting, Security and Identification (ASID '10)*, pp. 33–36, IEEE, Chengdu, China, July 2010.

[7] J. Huang and C.-T. Huang, "Secure mutual authentication protocols for mobile multi-hop relay WiMAX networks against Rogue base/relay stations," in *Proceedings of the IEEE International Conference on Communications (ICC '11)*, pp. 1–5, IEEE, Kyoto, Japan, June 2011.

[8] C. Cremers, "The scyther tool: verification, falsification, and analysis of security protocols," in *Computer Aided Verification:*

information confidentiality and *no theft of service possible* are satisfied.

As seen in the above formal analysis, our claims for the secrecy and uniqueness of the exchanged key elements, the no theft of service possible, and message sequence synchronization are satisfied in all of the three protocols.

6. Concluding Remarks

In this paper, we present a set of new secure authentication protocols to address the attack of rogue BS/RS attack in the MMR WiMAX networks. First, relay station and mobile user authentication provide access control. Second, in order to protect the MMR network from rogue BS attack, an authentication server is used to conduct mutual authentication for both BS and RS/MS in the network. In this way, a rogue station can be detected at early stage and its harm to this network can be minimized. Third, a security zone key is generated and securely delivered to an authenticated BS by AS. BS can then distribute this SZK to legitimate RS within the network; in this way the communications within the network can be secured using SZK. Fourth, in order to reduce the overhead introduced by mutual authentications between nodes, a license and a distributed authentication are used as a pass ticket for nodes which have been authenticated within a valid period of time. Such a node with a valid license can get access to the network without having to go through

20th International Conference, CAV 2008 Princeton, NJ, USA, July 7–14, 2008 Proceedings, vol. 5123 of *Lecture Notes in Computer Science*, pp. 414–418, Springer, Berlin, Germany, 2008.

[9] S. Xu, M. Matthews, and C.-T. Huang, "Security issues in privacy and key management protocols of IEEE 802.16," in *Proceedings of the 44th ACM Southeast Regional*, pp. 113–118, Melbourne, Fla, USA, March 2006.

[10] S. Xu and C.-T. Huang, "Attacks on PKM protocols of IEEE 802.16 and its later versions," in *Proceedings of the 3rd International Symposium on Wireless Communication Systems (ISWCS '06)*, pp. 185–189, IEEE, Valencia, Spain, September 2006.

[11] D. Johnston and J. Walker, "Overview of IEEE 802.16 security," *IEEE Security and Privacy*, vol. 2, no. 3, pp. 40–48, 2004.

[12] F. Yang, H. Zhou, L. Zhang, and J. Feng, "An improved security scheme in WMAN based on IEEE standard 802.16," in *Proceedings of the International Conference on Wireless Communications, Networking and Mobile Computing*, pp. 1191–1194, September 2005.

[13] K. Khan and M. Akbar, "Authentication in multi-hop wireless mesh networks," *Transactions on Science, Engineering and Technology*, vol. 16, pp. 178–183, 2006.

[14] M. E. Mahmoud and X. Shen, "Anonymous and authenticated routing in multi-hop cellular networks," in *Proceedings of the IEEE International Conference on Communications (ICC '09)*, pp. 1–6, IEEE, Dresden, Germany, June 2009.

[15] S. Zhu, S. Setia, S. Jajodia, and P. Ning, "An interleaved hop-by-hop authentication scheme for filtering of injected false data in sensor networks," in *Proceedings of the IEEE Symposium on Security and Privacy*, pp. 259–271, May 2004.

[16] Y. Fan and Q. Yali, "Two different schemes of authentication in IEEE 802.16j multi-hop relay network," in *Proceedings of the 8th International Conference on Wireless Communications, Networking and Mobile Computing (WiCOM '12)*, pp. 1–4, Shanghai, China, September 2012.

[17] H. X. Jin, L. Tu, G. Yang, and Y. Yang, "An improved mutual authentication scheme in multi-hop WiMax network," in *Proceedings of the International Conference on Computer and Electrical Engineering (ICCEE '08)*, pp. 296–299, Phuket, Thailand, December 2008.

[18] A. S. Khan, N. Fisal, Z. A. Bakar et al., "Secure authentication and key management protocols for mobile multihop WiMAX networks," *Indian Journal of Science and Technology*, vol. 7, no. 3, pp. 282–295, 2014.

[19] A. S. Khan, H. Lenando, and J. Abdullah, "Lightweight message authentication protocol for mobile multihop relay networks," *International Review on Computers and Software*, vol. 9, no. 10, pp. 1720–1730, 2014.

[20] Y. Lee, H. K. Lee, G. Y. Lee, H. J. Kim, and C. K. Jeong, "Design of hybrid authentication scheme and key distribution for mobile multi-hop relay in IEEE 802.16j," in *Proceedings of the Euro American Conference on Telematics And Information Systems: New Opportunities to Increase Digital Citizenship (EATIS '09)*, Prague, Czech Republic, June 2009.

[21] A. S. Khan, N. Fisal, S. Kamilah, and M. Abbas, "Efficient distributed authentication key scheme for multi-hop relay in IEEE 802.16j network," *International Journal of Engineering Science and Technology*, vol. 2, pp. 2192–2199, 2010.

[22] D. Zhu, N. Pang, and Z. Fan, "A self-testing approach defending against rogue base station hijacking of intelligent terminal," in *Proceedings of the International Conference on Applied Science and Engineering Innovation*, Zhengzhou, China, May 2015.

[23] M. Barbeau and J.-M. Robert, "Rogue-base station detection in WiMax/802.16 wireless access networks," *Annals of Telecommunications*, vol. 61, no. 11-12, pp. 1300–1313, 2006.

[24] T. Shon and W. Choi, "An analysis of mobile WiMAX security: vulnerabilities and solutions," in *Proceedings of the 1st International Conference on Network-Based Information Systems*, Regensburg, Germany, September 2007.

[25] M. Maxim and D. Pollino, *WiMAX Security*, RSA Press, McGraw-Hill/Osborne, Berkeley, Calif, USA, 2002.

[26] M. Khosrowpour, "Managing information technology resources in organizations in the next millennium," in *Proceedings of the Information Resources Management Association International Conferances*, May 1999.

[27] C. J. Cremers, S. Mauw, and E. P. de Vink, "Injective synchronisation: an extension of the authentication hierarchy," *Theoretical Computer Science*, vol. 367, no. 1-2, pp. 139–161, 2006.

[28] E. Kaasenbrood, *WiMAX Security—a formal and informal analysis [M.S. thesis]*, Eindhoven University of Technology, Department of Mathematics and Computer Science, Groningen, The Netherlands, 2006.

19

First-Order Statistical Characteristics of Macrodiversity System with Three Microdiversity MRC Receivers in the Presence of k-μ Short-Term Fading and Gamma Long-Term Fading

Branimir Jaksic,[1] **Mihajlo Stefanovic,**[2] **Danijela Aleksic,**[2] **Dragan Radenkovic,**[2] **and Sinisa Minic**[3]

[1]*Faculty of Technical Sciences, University of Pristina, Kneza Milosa 7, 38220 Kosovska Mitrovica, Serbia*
[2]*Faculty of Electrical Engineering, University of Nis, Aleksandra Medvedeva 14, 18000 Nis, Serbia*
[3]*Teachers College, University of Pristina, Nemanjina, 38218 Leposavic, Serbia*

Correspondence should be addressed to Branimir Jaksic; branimir.jaksic@pr.ac.rs

Academic Editor: George S. Tombras

Macrodiversity system with macrodiversity SC receiver and three microdiversity MRC (maximum ratio combining) receivers is considered. Independent k-μ short-term fading and correlated Gamma long-term fading are present at the inputs of microdiversity MRC receivers. For this model, the probability density function and the cumulative density function of microdiversity MRC receivers and macrodiversity SC receiver output signal envelopes are calculated. Influences of Gamma shadowing severity, k-μ multipath fading severity, Rician factor and correlation coefficient at probability density function, and cumulative density function of macrodiversity SC receiver output signal envelopes are graphically presented.

1. Introduction

Fading is a basic type of nuisance in wireless mobile telecommunication systems. Depending on propagation environment and different communications cases various types of fading can arise. Short-term fading is a result of signal propagation on multipath. The interaction of waves between transmitter and receiver (reflection, diffraction, and scattering) induces large numbers of sent copies signals on the input of receivers. Propagation environment can be linear and nonlinear. Nonlinear environment is defined as correlated surfaces in which dissipation field is not equal [1–3].

Long term fading arises because of shadow effect. Various objects create shadow effect in areas between transmitter and receiver. In most cases, long term fading is correlated. Changing of signal power due to the influence of shadow effect is slow in comparison to the signal envelope changing into short term fading. The signal envelope is variable due to short term fading, and the signal envelope power is variable due to the long term fading [1, 4].

The signal from the transmitter to the receiver can be propagated over one, two, or more clusters. Cluster is defined as waves which arrive at the inputs of receivers with approximately same delay. When the number of clusters increases, the fading severity decreases. Each cluster is formed by a pair of Gaussian components at the receiver [2, 5, 6].

The statistical behavior of signal in such systems can be described by different distributions as Rayleigh, Rice, Nakagami-m, Weibull, or k-μ. k-μ distribution can be used to describe the variation of the signal envelope in linear environments, with dominant component, several clusters in propagation environment, and equal components in quadrature of signal. k-μ distribution has two parameters. The parameter k is Rician factor. Rician factor is defined as ratio of dominant components power and scattering components power. System performance is better for higher values of Rician factor. Rician factor increases as dominant components power increases or scattering components power decreases. The parameter μ is related to the number of clusters in propagation channels. The k-μ distribution is general distribution and several another distributions (Rician, Nakagami-*m*, and Rayleigh) can be obtained from *k-μ* distribution as special cases [7, 8].

Various diversity techniques to reduce the impact of short-term fading and long-term fading on the system performance can be used. The most commonly used are spatial diversion techniques. Spatial diversity techniques are implemented with multiple antennas mounted on the receiver. By using spatial diversity technique the reliability of the system and the channel capacity increase without increasing the transmitter power and frequency band expansion. There are several spatial diversity combining techniques that can be used to reduce the influence of fading and cochannel interference on system performances. The most commonly used diversity techniques are MRC (maximum ratio combining), EGC (equal gain combining), and SC (selection combining) [2, 7].

MRC diversity technique gives the best results. This technique effectively reduces the influence of *k-μ* short term fading on the system performance and provides the greatest diversity gain. The ratio of signal power and noise at the output of the MRC receiver is equal to the sum of the ratio signal power and noise at its inputs. If noise power is the same in all diversity branches, then squared output signal is equal to the sum of the squares of the signal at its input. This method requires that the signals at the inputs are brought to the same phase. Because of that this method of combining is very complex and expensive for real implementation [8, 9].

There are more works in open technical literature considering second order statistics of diversity systems. In [10], average crossing rate and average fade duration of macrodiversity SC receiver with two microdiversity maximum ratio combining (MRC) receivers operating over Gamma shadowed Nakagami-*m* multipath fading are calculated. Macrodiversity SC receiver selects microdiversity receiver with higher input power to serve to user. In [11], average fade duration and level crossing rate of macrodiversity SC receiver with two microdiversity MRC receivers operating over Rician fading multipath environment are derived.

2. System Model

In this paper, macrodiversity system with macrodiversity SC (selection combining) receiver and three microdiversity MRC (maximum ratio combining) receivers is analyzed. At the inputs of microdiversity MRC receivers are independent *k-μ* short term fading and Gamma long term fading. Gamma long term fading is correlated. The correlation coefficient decreases with increasing distance between the antennas.

Microdiversity MRC receiver reduces *k-μ* short term fading effects and macrodiversity SC receivers reduce Gamma long term fading effects on system performances. Obtained macrosystem is predicted for one cell in the cellular mobile radio systems. Microdiversity receivers are placed on the base stations for mobile users of this cell. Macrodiversity systems used signals from several base stations placed in one cell (two or more) [2].

The system that is being considered is shown in Figure 1. The signals at the inputs and the outputs of the MRC receivers

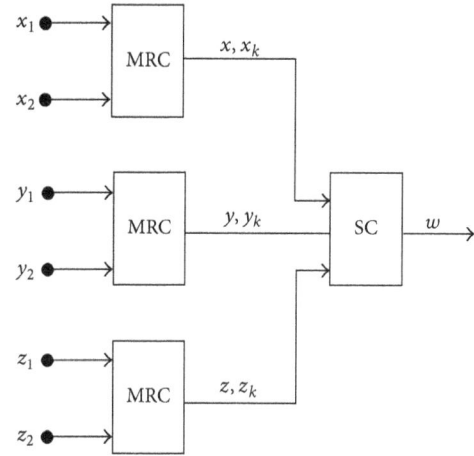

FIGURE 1: Macrodiversity system with three microdiversity MRC receivers and one macrodiversity SC receiver.

are denoted as in Figure 1. The signal at the output of macrodiversity system is denoted with *w*.

The square of the macrodiversity system output signal envelope is equal to sum of signal squares from its inputs. The macrodiversity SC receiver output signal envelope is equal to the microdiversity MRC receiver output signal envelope whose signal power at the input microdivesity MRC receiver is greater than the signal power at the input of the other two microdivesity MRC receivers.

Squared *k-μ* distribution is equal to sum of *2μ* squares of Gaussian random independent with same variance variables. In this way we can determine the probability density function of *k-μ* distribution. With these probability density functions we can determine cumulative density function of *k-μ* random variables, characteristic function of the random variable *k-μ*, and moments of the *k-μ* random variable.

Probability density function of macrodiversity SC receiver output signal envelope is equal to the probability density function of microdiversity MRC receiver output signal envelope with highest signal power at this input.

The cumulative density function is equal to integrating of probability density function. Thus can be calculated the cumulative density function of output signal envelope for the first, second, and third microdiversity MRC receiver. The cumulative density function of macrodiversity SC receiver output signal envelope can be determined by integrating the probability density function of macrodiversity SC receiver output signal envelope. By using the cumulative density function can be determined outage probability [3].

3. Probability Density Function of Macrodiversity SC Receiver Output Signal Envelope

Squared *k-μ* x_1 signal is equal to

$$x_1^2 = x_{11}^2 + x_{12}^2 + \cdots + x_{12\mu}^2. \tag{1}$$

Squared k-μ x_2 signal is equal to

$$x_2^2 = x_{21}^2 + x_{22}^2 + \cdots + x_{22\mu}^2. \tag{2}$$

The squared signal x is equal to

$$
\begin{aligned}
x^2 &= x_1^2 + x_2^2 \\
&= x_{11}^2 + x_{12}^2 + \cdots + x_{12\mu}^2 + x_{21}^2 + x_{22}^2 + \cdots + x_{22\mu}^2 \tag{3} \\
&= x_k.
\end{aligned}
$$

Random variable x_k has a χ^2 distribution:

$$
\begin{aligned}
& p_{x_k}\left(x_k\right) \\
&= \frac{1}{2\sigma^2}\left(\frac{x_k}{\lambda}\right)^{(2\mu-1)/2} e^{-(\lambda+x_k)/2\sigma^2} I_{2\mu+1}\left(\frac{\lambda^{1/2} x_k^{1/2}}{\sigma^2}\right), \tag{4}
\end{aligned}
$$

where $I_n(\cdot)$ is modified Bessel function of the first kind, order n and argument x, σ signal variance, μ the number of clusters for the signal, and λ the mean signal power.

Relations between x_k and x are

$$
\begin{aligned}
x^2 &= x_k, \\
x &= x_k^{1/2}. \tag{5}
\end{aligned}
$$

Probability density function of microdiversity MRC receiver output signal envelope x is

$$p_x(x) = \frac{dx_k}{dx} p_{x_k}\left(x^2\right), \tag{6}$$

where

$$\frac{dx_k}{dx} = 2x. \tag{7}$$

By substituting (7) and (4) in (6) the development of Bessel functions is obtained:

$$
\begin{aligned}
p_x(x) &= \frac{1}{\sigma_1^2 \lambda^{2\mu-1}} \sum_{i_1=0}^{\infty} \left(\frac{\lambda^{1/2}}{2\sigma_1^2}\right)^{2i_1+2\mu-1} \frac{1}{i_1! \Gamma\left(i_1+2\mu\right)} \\
&\quad \cdot x^{2i_1+4\mu-1} e^{-(\lambda+x^2)/2\sigma_1^2}. \tag{8}
\end{aligned}
$$

One takes

$$2\sigma_1^2 = \Omega_1. \tag{9}$$

After substituting (9) in (8), probability density function of microdiversity MRC receiver output signal envelope becomes

$$
\begin{aligned}
p_x(x) &= \frac{2}{\Omega_1 \lambda^{2\mu-1}} \sum_{i_1=0}^{\infty} \lambda^{i_1+\mu-1/2} \Omega_1^{-i_1-\mu+1/2} \frac{1}{i_1! \Gamma\left(i_1+2\mu\right)} \\
&\quad \cdot x^{2i_1+4\mu-1} e^{-(\lambda+x^2)/\Omega_1}, \tag{10}
\end{aligned}
$$

where Ω_1 means signal power.

In a similar way we can obtain the probability density function of microdiversity MRC receiver output signal envelope for second and third microdiversity MRC receiver:

$$
\begin{aligned}
p_y(y) &= \frac{2}{\Omega_2 \lambda^{2\mu-1}} \sum_{i_2=0}^{\infty} \lambda^{i_2+\mu-1/2} \Omega_2^{-i_2-\mu+1/2} \frac{1}{i_2! \Gamma\left(i_2+2\mu\right)} \\
&\quad \cdot y^{2i_2+4\mu-1} e^{-(\lambda+y^2)/\Omega_2}, \tag{11}
\end{aligned}
$$

$$
\begin{aligned}
p_z(z) &= \frac{2}{\Omega_3 \lambda^{2\mu-1}} \sum_{i_3=0}^{\infty} \lambda^{i_3+\mu-1/2} \Omega_3^{-i_3-\mu+1/2} \frac{1}{i_3! \Gamma\left(i_3+2\mu\right)} \\
&\quad \cdot z^{2i_3+4\mu-1} e^{-(\lambda+z^2)/\Omega_3}. \tag{12}
\end{aligned}
$$

Signal envelopes powers at inputs in microdiversity receivers are correlated. Signal envelope powers Ω_1, Ω_2, and Ω_3 follow Gamma distribution:

$$
\begin{aligned}
& p_{\Omega_1 \Omega_2 \Omega_3}\left(\Omega_1 \Omega_2 \Omega_3\right) \\
&= \frac{1}{\Gamma(c)\left(1-\rho^2\right) \rho^{c-1} \Omega_0^{c+2}}\left(\Omega_1 \Omega_3\right)^{(c-1)/2} \\
&\quad \cdot e^{-(\Omega_1+\Omega_3+\Omega_2(1+\rho))/\Omega_0(1-\rho)} I_{c-1}\left(\frac{2\sqrt{\rho}}{\Omega_0(1-\rho)} \sqrt{\Omega_1 \Omega_2}\right) \\
&\quad \cdot I_{c-1}\left(\frac{2\sqrt{\rho}}{\Omega_0(1-\rho)} \sqrt{\Omega_2 \Omega_3}\right), \tag{13}
\end{aligned}
$$

where $I_n(\cdot)$ is modified Bessel function of the first kind, order n, and argument x; Ω_0 is mean square of signal power variation, ρ correlation coefficient, and c Gamma shadowing severity.

Probability density function of macrodiversity SC receiver output signal envelope w is

$$
\begin{aligned}
p_w(w) &= \int_0^{\infty} d\Omega_1 \int_0^{\Omega_1} d\Omega_2 \\
&\quad \cdot \int_0^{\Omega_1} d\Omega_3 p_x\left(\frac{w}{\Omega_1}\right) p_{\Omega_1 \Omega_2 \Omega_3}\left(\Omega_1 \Omega_2 \Omega_3\right) + \int_0^{\infty} d\Omega_2 \\
&\quad \cdot \int_0^{\Omega_2} d\Omega_1 \int_0^{\Omega_2} d\Omega_3 p_y\left(\frac{w}{\Omega_2}\right) p_{\Omega_1 \Omega_2 \Omega_3}\left(\Omega_1 \Omega_2 \Omega_3\right) \\
&\quad + \int_0^{\infty} d\Omega_3 \int_0^{\Omega_3} d\Omega_1 \\
&\quad \cdot \int_0^{\Omega_3} d\Omega_2 p_z\left(\frac{w}{\Omega_3}\right) p_{\Omega_1 \Omega_2 \Omega_3}\left(\Omega_1 \Omega_2 \Omega_3\right) = I_1 + I_2 \\
&\quad + I_3 = 2I_1 + I_2. \tag{14}
\end{aligned}
$$

Functions $p_x(w/\Omega_1)$, $p_y(w/\Omega_2)$, and $p_z(w/\Omega_3)$ are given in (10), (11), and (12), respectively.

Integral I_1 is equal to

$$I_1 = \frac{2}{\lambda^{2\mu-1}} \frac{1}{\Gamma(c)(1-\rho^2)\rho^{c-1}\Omega_0^{c+2}}$$

$$\cdot \sum_{i_1=0}^{\infty} \frac{1}{i_1!\Gamma(i_1+2\mu)} \lambda^{i_1+\mu-1/2} x^{2i_1+4\mu-1}$$

$$\cdot \sum_{j_1=0}^{\infty} \left(\frac{\sqrt{\rho}}{\Omega_0(1-\rho)}\right)^{2j_1+c-1} \frac{1}{j_1!\Gamma(j_1+c)}$$

$$\cdot \sum_{j_2=0}^{\infty} \left(\frac{\sqrt{\rho}}{\Omega_0(1-\rho)}\right)^{2j_2+c-1} \frac{1}{j_2!\Gamma(j_2+c)} \tag{15}$$

$$\cdot \int_0^{\infty} d\Omega_1 \Omega_1^{-1-i_1-\mu+1/2+j_1+c-1} e^{-(\lambda+x^2)/\Omega_1-\Omega_1/\Omega_0(1-\rho)}$$

$$\cdot \int_0^{\Omega_1} d\Omega_2 \Omega_2^{j_1+j_2+c-1} e^{-\Omega_2(1+\rho)/\Omega_0(1-\rho)}$$

$$\cdot \int_0^{\Omega_1} d\Omega_3 \Omega_3^{j_2+c-1} e^{-\Omega_3/\Omega_0(1-\rho)}.$$

After using [12] for resolving the second and third integrals in (15), I_1 becomes

$$I_1 = \frac{2}{\lambda^{2\mu-1}} \frac{1}{\Gamma(c)(1-\rho^2)\rho^{c-1}\Omega_0^{c+2}}$$

$$\cdot \sum_{i_1=0}^{\infty} \frac{1}{i_1!\Gamma(i_1+2\mu)} \lambda^{i_1+\mu-1/2} x^{2i_1+4\mu-1}$$

$$\cdot \sum_{j_1=0}^{\infty} \left(\frac{\sqrt{\rho}}{\Omega_0(1-\rho)}\right)^{2j_1+c-1} \frac{1}{j_1!\Gamma(j_1+c)}$$

$$\cdot \sum_{j_2=0}^{\infty} \left(\frac{\sqrt{\rho}}{\Omega_0(1-\rho)}\right)^{2j_2+c-1} \frac{1}{j_2!\Gamma(j_2+c)} \tag{16}$$

$$\cdot \left(\frac{\Omega_0(1-\rho)}{1+\rho}\right)^{j_1+j_2+c} (\Omega_0(1-\rho))^{j_2+c}$$

$$\cdot \int_0^{\infty} d\Omega_1 \Omega_1^{-1-i_1-\mu+1/2+j_1+c-1} e^{-(\lambda+x^2)/\Omega_1-\Omega_1/\Omega_0(1-\rho)}$$

$$\cdot \gamma\left(j_1+j_2+c, \frac{1+\rho}{\Omega_0(1-\rho)}\Omega_1\right)$$

$$\cdot \gamma\left(j_2+c, \frac{1}{\Omega_0(1-\rho)}\Omega_1\right),$$

where $\gamma(\cdot)$ represents the lower incomplete Gamma function. After the developing Gamma function,

$$\gamma(n,x) = \frac{1}{n} x^n e^{-x} {}_1F_1(1, n+1, x)$$

$$= \frac{1}{n} x^n e^{-x} \sum_{i=0}^{\infty} \frac{n!}{(n+i)!} x^i, \tag{17}$$

I_1 becomes

$$I_1 = \frac{2}{\lambda^{2\mu-1}} \frac{1}{\Gamma(c)(1-\rho^2)\rho^{c-1}\Omega_0^{c+2}}$$

$$\cdot \sum_{i_1=0}^{\infty} \frac{1}{i_1!\Gamma(i_1+2\mu)} \lambda^{i_1+\mu-1/2} x^{2i_1+4\mu-1}$$

$$\cdot \sum_{j_1=0}^{\infty} \left(\frac{\sqrt{\rho}}{\Omega_0(1-\rho)}\right)^{2j_1+c-1} \frac{1}{j_1!\Gamma(j_1+c)}$$

$$\cdot \sum_{j_2=0}^{\infty} \left(\frac{\sqrt{\rho}}{\Omega_0(1-\rho)}\right)^{2j_2+c-1} \frac{1}{j_2!\Gamma(j_2+c)}$$

$$\cdot \left(\frac{\Omega_0(1-\rho)}{1+\rho}\right)^{j_1+j_2+c} (\Omega_0(1-\rho))^{j_2+c} \tag{18}$$

$$\cdot \int_0^{\infty} d\Omega_1 \Omega_1^{-1-i_1-\mu+1/2+j_1+c-1} e^{-(\lambda+x^2)/\Omega_1-\Omega_1/\Omega_0(1-\rho)}$$

$$\cdot \frac{1}{j_1+j_2+c} \left(\frac{(1+\rho)\Omega_1}{\Omega_0(1-\rho)}\right)^{j_1+j_2+c}$$

$$\cdot e^{-(1+\rho)\Omega_1/\Omega_0(1-\rho)} \sum_{j_3=0}^{\infty} \frac{(j_1+j_2+c)!}{(j_1+j_2+c+j_3)!}$$

$$\cdot \left(\frac{(1+\rho)\Omega_1}{\Omega_0(1-\rho)}\right)^{j_3} \frac{1}{j_2+c} \left(\frac{\Omega_1}{\Omega_0(1-\rho)}\right)^{j_2+c}$$

$$\cdot e^{-\Omega_1/\Omega_0(1-\rho)} \sum_{j_4=0}^{\infty} \frac{(j_2+c)!}{(j_2+c+j_4)!} \left(\frac{\Omega_1}{\Omega_0(1-\rho)}\right)^{j_4}.$$

By using [12] the following is obtained:

$$I_1 = \frac{2}{\lambda^{2\mu-1}} \frac{1}{\Gamma(c)(1-\rho^2)\rho^{c-1}\Omega_0^{c+2}} \sum_{i_1=0}^{\infty} \frac{1}{i_1!\Gamma(i_1+2\mu)}$$

$$\cdot \lambda^{i_1+\mu-1/2} x^{2i_1+4\mu-1} \sum_{j_1=0}^{\infty} \left(\frac{\sqrt{\rho}}{\Omega_0(1-\rho)}\right)^{2j_1+c-1} \frac{1}{j_1!\Gamma(j_1+c)}$$

$$\cdot \sum_{j_2=0}^{\infty} \left(\frac{\sqrt{\rho}}{\Omega_0(1-\rho)}\right)^{2j_2+c-1}$$

$$\cdot \frac{1}{j_2!\Gamma(j_2+c)} \left(\frac{\Omega_0(1-\rho)}{1+\rho}\right)^{j_1+j_2+c} (\Omega_0(1-\rho))^{j_2+c}$$

$$\cdot \frac{1}{j_1+j_2+c} \left(\frac{(1+\rho)}{\Omega_0(1-\rho)}\right)^{j_1+j_2+c} \tag{19}$$

$$\cdot \sum_{j_3=0}^{\infty} \frac{(j_1+j_2+c)!}{(j_1+j_2+c+j_3)!} \left(\frac{(1+\rho)}{\Omega_0(1-\rho)}\right)^{j_3} \frac{1}{j_2+c}$$

$$\cdot \frac{1}{(\Omega_0(1-\rho))^{j_2+c}} \sum_{j_4=0}^{\infty} \frac{(j_2+c)!}{(j_2+c+j_4)!} \frac{1}{(\Omega_0(1-\rho))^{j_4}}$$

$$\cdot \left(\frac{(\lambda+x^2)\Omega_0(1-\rho)}{(3+\rho)}\right)^{-1/4-i_1-\mu/2+j_1+j_2+j_3/2+j_3/2+3c/2}$$

$$\cdot K_{-1/2-i_1-\mu+2j_1+2j_2+j_3+j_4+3c}\left(2\sqrt{\frac{(\lambda+x^2)(3+\rho)}{\Omega_0(1-\rho)}}\right),$$

where $K_n(x)$ is modified Bessel function of the second kind, order n, and argument x.

Integral I_2 is equal to

$$I_2 = \frac{2}{\lambda^{2\mu-1}} \frac{1}{\Gamma(c)(1-\rho^2)\rho^{c-1}\Omega_0^{c+2}}$$

$$\cdot \sum_{i_2=0}^{\infty} \frac{1}{i_2!\Gamma(i_2+2\mu)} \lambda^{i_2+\mu-1/2} y^{2i_2+4\mu-1}$$

$$\cdot \sum_{j_1=0}^{\infty} \left(\frac{\sqrt{\rho}}{\Omega_0(1-\rho)}\right)^{2j_1+c-1} \frac{1}{j_1!\Gamma(j_1+c)}$$

$$\cdot \sum_{j_2=0}^{\infty} \left(\frac{\sqrt{\rho}}{\Omega_0(1-\rho)}\right)^{2j_2+c-1} \frac{1}{j_2!\Gamma(j_2+c)} \qquad (20)$$

$$\cdot \int_0^{\infty} d\Omega_2 \Omega_2^{-1/2-i_2-\mu+j_1+c-1} e^{-(\lambda+y^2)/\Omega_2 - \Omega_2(1+\rho)/\Omega_0(1-\rho)}$$

$$\cdot \int_0^{\Omega_2} d\Omega_1 \Omega_1^{j_1+c-1} e^{-\Omega_1/\Omega_0(1-\rho)}$$

$$\cdot \int_0^{\Omega_2} d\Omega_3 \Omega_3^{j_2+c-1} e^{-\Omega_3/\Omega_0(1-\rho)}.$$

After the using procedure for solving I_1, I_2 becomes

$$I_2 = \frac{2}{\lambda^{2\mu-1}} \frac{1}{\Gamma(c)(1-\rho^2)\rho^{c-1}\Omega_0^{c+2}}$$

$$\cdot \sum_{i_2=0}^{\infty} \frac{1}{i_2!\Gamma(i_2+2\mu)} \lambda^{i_2+\mu-1/2} y^{2i_2+4\mu-1}$$

$$\cdot \sum_{j_1=0}^{\infty} \left(\frac{\sqrt{\rho}}{\Omega_0(1-\rho)}\right)^{2j_1+c-1} \frac{1}{j_1!\Gamma(j_1+c)}$$

$$\cdot \sum_{j_2=0}^{\infty} \left(\frac{\sqrt{\rho}}{\Omega_0(1-\rho)}\right)^{2j_2+c-1} \frac{(\Omega_0(1-\rho))^{j_1+j_2+2c}}{j_2!\Gamma(j_2+c)} \frac{1}{j_1+c}$$

$$\cdot \frac{1}{(\Omega_0(1-\rho))^{j_1+c}} \sum_{j_3=0}^{\infty} \frac{(j_1+c)!}{(j_1+c+j_3)!} \frac{1}{(\Omega_0(1-\rho))^{j_3}} \frac{1}{j_2+c} \qquad (21)$$

$$\cdot \frac{1}{(\Omega_0(1-\rho))^{j_2+c}} \sum_{j_4=0}^{\infty} \frac{(j_2+c)!}{(j_2+c+j_4)!} \frac{1}{(\Omega_0(1-\rho))^{j_4}}$$

$$\cdot \left(\frac{(\lambda+y^2)\Omega_0(1-\rho)}{(3+\rho)}\right)^{-1/4-i_2/2-\mu/2+j_1+j_2/2+j_3/2+j_4/2+3c/2}$$

$$\cdot K_{-1/2-i_2-\mu+2j_1+j_2+j_3+j_4+3c}\left(2\sqrt{\frac{(\lambda+y^2)(3+\rho)}{\Omega_0(1-\rho)}}\right).$$

After replacing the integrals I_1 and I_2 in (14), we obtain an expression for PDF of macrodiversity SC receiver output signal envelope:

$$p = \frac{8}{\lambda^{2\mu-1}} \frac{1}{\Gamma(c)(1-\rho^2)\rho^{c-1}\Omega_0^{c+2}} \sum_{i_1=0}^{\infty} \frac{\lambda^{i_1+\mu-1/2}}{i_1!\Gamma(i_1+2\mu)}$$

$$\cdot \sum_{j_1=0}^{\infty} \left(\frac{\sqrt{\rho}}{\Omega_0(1-\rho)}\right)^{2j_1+c-1} \frac{1}{j_1!\Gamma(j_1+c)}$$

$$\cdot \sum_{j_2=0}^{\infty} \left(\frac{\sqrt{\rho}}{\Omega_0(1-\rho)}\right)^{2j_2+c-1} \frac{1}{j_2!\Gamma(j_2+c)}$$

$$\cdot \left[x^{2i_1+4\mu-1} \left(\frac{\Omega_0(1-\rho)}{1+\rho}\right)^{j_2+c} \frac{1}{j_1+j_2+c} \right.$$

$$\cdot \sum_{j_3=0}^{\infty} \frac{(j_1+j_2+c)!}{(j_1+j_2+c+j_3)!} \left(\frac{(1+\rho)}{\Omega_0(1-\rho)}\right)^{j_3} \frac{1}{j_2+c}$$

$$\cdot \frac{1}{(\Omega_0(1-\rho))^{j_2+c}} \sum_{j_4=0}^{\infty} \frac{(j_2+c)!}{(j_2+c+j_4)!} \frac{1}{(\Omega_0(1-\rho))^{j_4}}$$

$$\cdot \left(\frac{(\lambda+x^2)\Omega_0(1-\rho)}{(3+\rho)}\right)^{-1/4-i_1/2-\mu/2+j_1+j_2+j_3/2+j_3/2+3c/2}$$

$$\cdot K_{-1/2-i_1-\mu+2j_1+2j_2+j_3+j_4+3c}\left(2\sqrt{\frac{(\lambda+x^2)(3+\rho)}{\Omega_0(1-\rho)}}\right)$$

$$+ y^{2i_1+4\mu-1} \frac{(\Omega_0(1-\rho))^{j_2+c}}{j_1+c}$$

$$\cdot \sum_{j_3=0}^{\infty} \frac{(j_1+c)!}{(j_1+c+j_3)!} \frac{1}{(\Omega_0(1-\rho))^{j_2+j_3+c}} \frac{1}{j_2+c}$$

$$\cdot \sum_{j_4=0}^{\infty} \frac{(j_2+c)!}{(j_2+c+j_4)!} \frac{1}{(\Omega_0(1-\rho))^{j_4}}$$

$$\cdot \left(\frac{(\lambda+y^2)\Omega_0(1-\rho)}{(3+\rho)}\right)^{-1/4-i_1/2-\mu/2+j_1+j_2/2+j_3/2+j_4/2+3c/2}$$

$$\cdot K_{-1/2-i_1-\mu+2j_1+j_2+j_3+j_4+3c}\left(2\sqrt{\frac{(\lambda+y^2)(3+\rho)}{\Omega_0(1-\rho)}}\right) \Bigg].$$

$$(22)$$

4. Cumulative Density Function of Macrodiversity SC Receiver Output Signal Envelope

Probability density function of microdiversity MRC receiver output signal envelope x is

$$F_x(x) = \int_0^x dt\, p_x(t) = \frac{1}{\Omega_1 \lambda^{2\mu-1}} e^{-\lambda/\Omega_1} \sum_{i_1=0}^{\infty} \lambda^{i_1+\mu-1/2}$$

$$\cdot \Omega_1^{-i_1-\mu+1/2} \frac{1}{i_1!\Gamma(i_1+2\mu)} \Omega_1^{i_1+2\mu} \gamma\left(i_1+2\mu, \frac{x^2}{\Omega_1}\right),$$

$$(23)$$

where $p_x(t)$ is probability density function given by (10).

In a similar way we can obtain the cumulative density function of microdiversity MRC receiver output signal envelope for second and third microdiversity MRC receiver:

$$F_y(y) = \frac{1}{\Omega_2 \lambda^{2\mu-1}} e^{-\lambda/\Omega_2} \sum_{i_2=0}^{\infty} \lambda^{i_2+\mu-1/2} \Omega_2^{-i_2-\mu+1/2}$$
$$\cdot \frac{1}{i_2! \Gamma(i_2+2\mu)} \Omega_2^{i_2+2\mu} \gamma\left(i_2+2\mu, \frac{y^2}{\Omega_2}\right), \tag{24}$$

$$F_z(z) = \frac{1}{\Omega_3 \lambda^{2\mu-1}} e^{-\lambda/\Omega_3} \sum_{i_3=0}^{\infty} \lambda^{i_3+\mu-1/2} \Omega_1^{-i_3-\mu+1/2}$$
$$\cdot \frac{1}{i_3! \Gamma(i_3+2\mu)} \Omega_1^{i_3+2\mu} \gamma\left(i_3+2\mu, \frac{z^2}{\Omega_3}\right). \tag{25}$$

Cumulative density function of macrodiversity SC receiver output signal envelope w is

$$F_w(w) = \int_0^{\infty} d\Omega_1 \int_0^{\Omega_1} d\Omega_2$$
$$\cdot \int_0^{\Omega_1} d\Omega_3 F_x\left(\frac{w}{\Omega_1}\right) p_{\Omega_1 \Omega_2 \Omega_3}(\Omega_1 \Omega_2 \Omega_3)$$
$$+ \int_0^{\infty} d\Omega_2 \int_0^{\Omega_2} d\Omega_1$$
$$\cdot \int_0^{\Omega_2} d\Omega_3 F_y\left(\frac{w}{\Omega_2}\right) p_{\Omega_1 \Omega_2 \Omega_3}(\Omega_1 \Omega_2 \Omega_3) \tag{26}$$
$$+ \int_0^{\infty} d\Omega_3 \int_0^{\Omega_3} d\Omega_1$$
$$\cdot \int_0^{\Omega_3} d\Omega_2 F_z\left(\frac{w}{\Omega_3}\right) p_{\Omega_1 \Omega_2 \Omega_3}(\Omega_1 \Omega_2 \Omega_3) = I_1 + I_2$$
$$+ I_3 = 2I_1 + I_2.$$

Functions $F_x(w/\Omega_1)$, $F_y(w/\Omega_2)$, and $F_z(w/\Omega_3)$ are given by (23), (24), and (25), respectively, and $p_{\Omega_1 \Omega_2 \Omega_3}(\Omega_1 \Omega_2 \Omega_3)$ with (13).

Integral I_1 is equal to

$$I_1 = \frac{1}{\lambda^{2\mu-1}} \frac{1}{\Gamma(c)(1-\rho^2)\rho^{c-1}\Omega_0^{c+2}}$$
$$\cdot \sum_{i_1=0}^{\infty} \frac{1}{i_1! \Gamma(i_1+2\mu)} \lambda^{i_1+\mu-1/2} \sum_{j_1=0}^{\infty} \left(\frac{\sqrt{\rho}}{\Omega_0(1-\rho)}\right)^{2j_1+c-1}$$
$$\cdot \frac{1}{j_1! \Gamma(j_1+c)} \sum_{j_2=0}^{\infty} \left(\frac{\sqrt{\rho}}{\Omega_0(1-\rho)}\right)^{2j_2+c-1} \frac{1}{j_2! \Gamma(j_2+c)}$$
$$\cdot \int_0^{\infty} d\Omega_1 \Omega_1^{-1-i_1-\mu+1/2+i_1+2\mu+j_1+c-1} e^{-\lambda/\Omega_1 - \Omega_1/\Omega_0(1-\rho)} \gamma\Big(i_1$$
$$+ 2\mu, \frac{w^2}{\Omega_1}\Big) \int_0^{\Omega_1} d\Omega_2 \Omega_2^{j_1+j_2+c-1} e^{-\Omega_2(1+\rho)/\Omega_0(1-\rho)}$$
$$\cdot \int_0^{\Omega_1} d\Omega_3 \Omega_3^{j_2+c-1} e^{-\Omega_3/\Omega_0(1-\rho)}. \tag{27}$$

After using [12] for resolving the second and third integrals in (27), I_1 becomes

$$I_1 = \frac{1}{\lambda^{2\mu-1}} \frac{1}{\Gamma(c)(1-\rho^2)\rho^{c-1}\Omega_0^{c+2}}$$
$$\cdot \sum_{i_1=0}^{\infty} \frac{1}{i_1! \Gamma(i_1+2\mu)} \lambda^{i_1+\mu-1/2}$$
$$\cdot \sum_{j_1=0}^{\infty} \left(\frac{\sqrt{\rho}}{\Omega_0(1-\rho)}\right)^{2j_1+c-1} \frac{1}{j_1! \Gamma(j_1+c)}$$
$$\cdot \sum_{j_2=0}^{\infty} \left(\frac{\sqrt{\rho}}{\Omega_0(1-\rho)}\right)^{2j_2+c-1} \frac{1}{j_2! \Gamma(j_2+c)}$$
$$\cdot \left(\frac{\Omega_0(1-\rho)}{1+\rho}\right)^{j_1+j_2+c} (\Omega_0(1-\rho))^{j_2+c}$$
$$\cdot \int_0^{\infty} \Omega_1 \Omega_1^{-1/2+\mu+j_1+c-1} e^{-\lambda/\Omega_1 - \Omega_1/\Omega_0(1-\rho)}$$
$$\cdot \gamma\left(i_1+2\mu, \frac{w^2}{\Omega_1}\right) \gamma\left(j_1+j_2+c, \frac{(1+\rho)\Omega_1}{\Omega_0(1-\rho)}\right)$$
$$\cdot \gamma\left(j_2+c, \frac{\Omega_1}{\Omega_0(1-\rho)}\right). \tag{28}$$

After developing Gamma function defined by (17) and using [12] the following is obtained:

$$I_1 = \frac{1}{\lambda^{2\mu-1}} \frac{1}{\Gamma(c)(1-\rho^2)\rho^{c-1}\Omega_0^{c+2}}$$
$$\cdot \sum_{i_1=0}^{\infty} \frac{1}{i_1! \Gamma(i_1+2\mu)} \lambda^{i_1+\mu-1/2} \sum_{j_1=0}^{\infty} \left(\frac{\sqrt{\rho}}{\Omega_0(1-\rho)}\right)^{2j_1+c-1}$$
$$\cdot \frac{1}{j_1! \Gamma(j_1+c)} \sum_{j_2=0}^{\infty} \left(\frac{\sqrt{\rho}}{\Omega_0(1-\rho)}\right)^{2j_2+c-1} \frac{1}{j_2! \Gamma(j_2+c)}$$
$$\cdot \left(\frac{\Omega_0(1-\rho)}{1+\rho}\right)^{j_1+j_2+c} (\Omega_0(1-\rho))^{j_2+c} \frac{1}{i_1+2\mu}$$
$$\cdot w^{2i_1+4\mu} \sum_{j_3=0}^{\infty} \frac{(i_1+2\mu)!}{(i_1+2\mu+j_3)!} w^{2j_3}$$
$$\cdot \frac{1}{j_1+j_2+c} \left(\frac{(1+\rho)}{\Omega_0(1-\rho)}\right)^{j_1+j_2+c} \tag{29}$$
$$\cdot \sum_{j_4=0}^{\infty} \frac{(j_1+j_2+c)!}{(j_1+j_2+c+j_4)!} \left(\frac{(1+\rho)}{\Omega_0(1-\rho)}\right)^{j_4} \frac{1}{j_2+c}$$
$$\cdot \frac{1}{(\Omega_0(1-\rho))^{j_2+c}} \sum_{j_5=0}^{\infty} \frac{(j_2+c)!}{(j_2+c+j_5)!} \frac{1}{(\Omega_0(1-\rho))^{j_5}}$$
$$\cdot \left(\frac{(\lambda+w^2)\Omega_0(1-\rho)}{(3+\rho)}\right)^{-1/4-\mu/2+j_1+j_2-j_3/2+j_4/2+j_5/2+3c/2}$$
$$\cdot K_{-1/2-\mu-i_1+2j_1+2j_2-j_3+j_4+j_5+3c}\left(2\sqrt{\frac{(\lambda+w^2)(3+\rho)}{\Omega_0(1-\rho)}}\right).$$

Integral I_2 is equal to

$$I_2 = \frac{1}{\lambda^{2\mu-1}} \frac{1}{\Gamma(c)(1-\rho^2)\rho^{c-1}\Omega_0^{c+2}} \sum_{i_2=0}^{\infty} \frac{1}{i_2!\Gamma(i_2+2\mu)} \lambda^{i_2+\mu-1/2}$$

$$\cdot \sum_{j_1=0}^{\infty} \left(\frac{\sqrt{\rho}}{\Omega_0(1-\rho)}\right)^{2j_1+c-1} \frac{1}{j_1!\Gamma(j_1+c)}$$

$$\cdot \sum_{j_2=0}^{\infty} \left(\frac{\sqrt{\rho}}{\Omega_0(1-\rho)}\right)^{2j_2+c-1} \frac{1}{j_2!\Gamma(j_2+c)} \tag{30}$$

$$\cdot \int_0^{\infty} d\Omega_2 \Omega_2^{-1-i_2-\mu+1/2+i_2+2\mu+j_1+j_2+c-1} e^{-\lambda/\Omega_2-\Omega_2(1+\rho)/\Omega_0(1-\rho)} \gamma\left(i_2\right.$$

$$\left.+2\mu, \frac{w^2}{\Omega_2}\right) \int_0^{\Omega_2} d\Omega_1 \Omega_1^{j_1+c-1} e^{-\Omega_1/\Omega_0(1-\rho)}$$

$$\cdot \int_0^{\Omega_2} d\Omega_3 \Omega_3^{j_2+c-1} e^{-\Omega_3/\Omega_0(1-\rho)}.$$

After using the procedure for solving I_1, I_2 becomes

$$I_2 = \frac{1}{\lambda^{2\mu-1}} \frac{1}{\Gamma(c)(1-\rho^2)\rho^{c-1}\Omega_0^{c+2}} \sum_{i_2=0}^{\infty} \frac{1}{i_2!\Gamma(i_2+2\mu)} \lambda^{i_2+\mu-1/2}$$

$$\cdot \sum_{j_1=0}^{\infty} \left(\frac{\sqrt{\rho}}{\Omega_0(1-\rho)}\right)^{2j_1+c-1} \frac{1}{j_1!\Gamma(j_1+c)}$$

$$\cdot \sum_{j_2=0}^{\infty} \left(\frac{\sqrt{\rho}}{\Omega_0(1-\rho)}\right)^{2j_2+c-1} \frac{1}{j_2!\Gamma(j_2+c)} (\Omega_0(1-\rho))^{j_1+j_2+2c}$$

$$\cdot \frac{1}{i_2+2\mu} w^{2i_2+4\mu} \sum_{j_3=0}^{\infty} \frac{(i_2+2\mu)!}{(i_2+2\mu+j_3)!} w^{2j_3} \frac{1}{j_1+c} \tag{31}$$

$$\cdot \frac{1}{(\Omega_0(1-\rho))^{j_1+c}} \sum_{j_4=0}^{\infty} \frac{(j_1+c)!}{(j_1+c+j_4)!} \frac{1}{(\Omega_0(1-\rho))^{j_4}} \frac{1}{j_2+c}$$

$$\cdot \frac{1}{(\Omega_0(1-\rho))^{j_2+c}} \sum_{j_5=0}^{\infty} \frac{(j_2+c)!}{(j_2+c+j_5)!} \frac{1}{(\Omega_0(1-\rho))^{j_5}}$$

$$\cdot \left(\frac{(\lambda+w^2)\Omega_0(1-\rho)}{(3+\rho)}\right)^{-1/4-\mu/2-i_2/2+j_1+j_2-j_3/2+j_4/2+j_5/2+3c/2}$$

$$\cdot K_{-1/2-\mu-i_2+2j_1+2j_2-j_3+j_4+j_5+3c} \left(2\sqrt{\frac{(\lambda+w^2)(3+\rho)}{\Omega_0(1-\rho)}}\right).$$

After replacing the integrals I_1 and I_2 in (26), we obtain an expression for CDF of macrodiversity SC receiver output signal envelope:

$$F_w(w) = 2 \frac{1}{\lambda^{2\mu-1}} \frac{1}{\Gamma(c)(1-\rho^2)\rho^{c-1}\Omega_0^{c+2}} \sum_{i_1=0}^{\infty} \frac{\lambda^{i_1+\mu-1/2}}{i_1!\Gamma(i_1+2\mu)}$$

$$\cdot \sum_{j_1=0}^{\infty} \left(\frac{\sqrt{\rho}}{\Omega_0(1-\rho)}\right)^{2j_1+c-1} \frac{1}{j_1!\Gamma(j_1+c)}$$

$$\cdot \sum_{j_2=0}^{\infty} \left(\frac{\sqrt{\rho}}{\Omega_0(1-\rho)}\right)^{2j_2+c-1} \frac{1}{j_2!\Gamma(j_2+c)} \frac{w^{2i_1+4\mu}}{i_1+2\mu}$$

$$\cdot \sum_{j_3=0}^{\infty} \frac{(i_1+2\mu)!}{(i_1+2\mu+j_3)!} w^{2j_3} \left[\left(\frac{\Omega_0(1-\rho)}{1+\rho}\right)^{j_1+2j_2+2c}\right.$$

$$\cdot \frac{1}{j_1+j_2+c} \left(\frac{(1+\rho)}{\Omega_0(1-\rho)}\right)^{j_1+j_2+c}$$

$$\cdot \sum_{j_4=0}^{\infty} \frac{(j_1+j_2+c)!}{(j_1+j_2+c+j_4)!} \left(\frac{(1+\rho)}{\Omega_0(1-\rho)}\right)^{j_4} \frac{1}{j_2+c}$$

$$\cdot \frac{1}{(\Omega_0(1-\rho))^{j_2+c}} \sum_{j_5=0}^{\infty} \frac{(j_2+c)!}{(j_2+c+j_5)!} \frac{1}{(\Omega_0(1-\rho))^{j_5}}$$

$$\cdot \left(\frac{(\lambda+w^2)\Omega_0(1-\rho)}{(3+\rho)}\right)^{-1/4-\mu/2+j_1+j_2-j_3/2+j_4/2+j_5/2+3c/2}$$

$$\cdot K_{-1/2-\mu-i_1+2j_1+2j_2-j_3+j_4+j_5+3c} \left(2\sqrt{\frac{(\lambda+w^2)(3+\rho)}{\Omega_0(1-\rho)}}\right)$$

$$+ (\Omega_0(1-\rho))^{j_1+j_2+2c} \frac{1}{j_1+c} \frac{1}{(\Omega_0(1-\rho))^{j_1+c}}$$

$$\cdot \sum_{j_4=0}^{\infty} \frac{(j_1+c)!}{(j_1+c+j_4)!} \frac{1}{(\Omega_0(1-\rho))^{j_4}} \frac{1}{j_2+c} \frac{1}{(\Omega_0(1-\rho))^{j_2+c}}$$

$$\cdot \sum_{j_5=0}^{\infty} \frac{(j_2+c)!}{(j_2+c+j_5)!} \frac{1}{(\Omega_0(1-\rho))^{j_5}}$$

$$\cdot \left(\frac{(\lambda+w^2)\Omega_0(1-\rho)}{(3+\rho)}\right)^{-1/4-\mu/2-i_1/2+j_1+j_2-j_3/2+j_4/2+j_5/2+3c/2}$$

$$\cdot K_{-1/2-\mu-i_1+2j_1+2j_2-j_3+j_4+j_5+3c} \left(2\sqrt{\frac{(\lambda+w^2)(3+\rho)}{\Omega_0(1-\rho)}}\right) \Bigg]. \tag{32}$$

5. Numerical Results

Probability density function of macrodiversity SC receiver output signal envelope versus SC receiver output signal envelope is plotted in Figure 2 for several values of k-μ fading severity parameter μ.

In Figure 3, cumulative distribution function of macrodiversity SC receiver output signal envelope versus SC receiver output signal envelope is plotted for different k-μ fading severity parameter μ, Rician factor k, Gamma fading severity c, and correlation coefficient ρ.

In Figure 4, cumulative distribution function of macrodiversity SC receiver output signal envelope versus Rician factor k is plotted for different k-μ fading severity parameter μ and correlation coefficient ρ.

6. Conclusion

In this paper analysis of diversity system with three microdiversity MRC receivers and one macrodiversity SC receiver was done. At the inputs of microdiversity MRC receivers exist independent k-μ short term fading and correlated Gamma long term fading. Macrodiversity SC receiver reduces k-μ

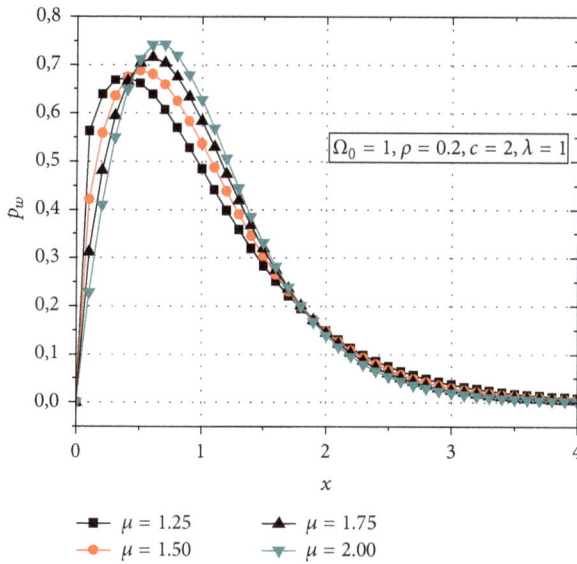

FIGURE 2: Probability density function of microdiversity MRC receiver.

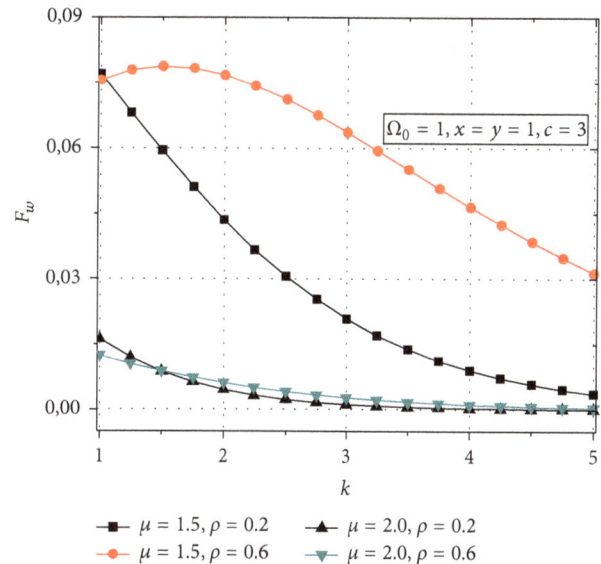

FIGURE 4: Cumulative distribution function of macrodiversity SC receiver output signal envelope versus Rician factor k.

derived from k-μ distribution by setting for $\mu = 1$. The k-μ distribution approximates Rayleigh distribution by setting $k = 0$ and $\mu = 1$. System performances are on low level if we increase fading severity. With less k factor fading becomes of high severity.

Increase of Rician k factor contributes to the decrease of cumulative density function. Cumulative density function of macrodiversity SC receiver output signal envelope decreases significantly faster for lower values of μ parameters, in comparison to higher value of parameter μ. With the increasing of correlation coefficient, the value of cumulative density function of macrodiversity SC receiver output signal envelope decreases. Signal envelope increases the value of cumulative density function growth and tends to 1. With increase of Rician k factor and the number of clusters μ, cumulative probability of signal grows faster.

Competing Interests

The authors declare that they have no competing interests.

References

[1] G. L. Stuber, *Mobile Communication*, Kluwer Academic, Dordrecht, The Netherlands, 2nd edition, 2003.

[2] S. Panic, M. Stefanovic, J. Anastasov, and P. Spalevic, *Fading and Interference Mitigation in Wireless Communications*, CRC Press, Boca Raton, Fla, USA, 2013.

[3] M. K. Simon and M. S. Alouini, *Digital Communication over Fading Channels*, John Wiley & Sons, New York, NY, USA, 2000.

[4] M. Milišić, M. Hamza, and M. Hadžialić, "BEP/SEP and outage performance analysis of L-branch maximal-ratio combiner for κ-μ fading," *International Journal of Digital Multimedia Broadcasting*, vol. 2009, Article ID 573404, 8 pages, 2009.

[5] G. Malmgren, "On the performance of single frequency networks in correlated shadow fading," *IEEE Transactions on Broadcasting*, vol. 43, no. 2, pp. 155–160, 1997.

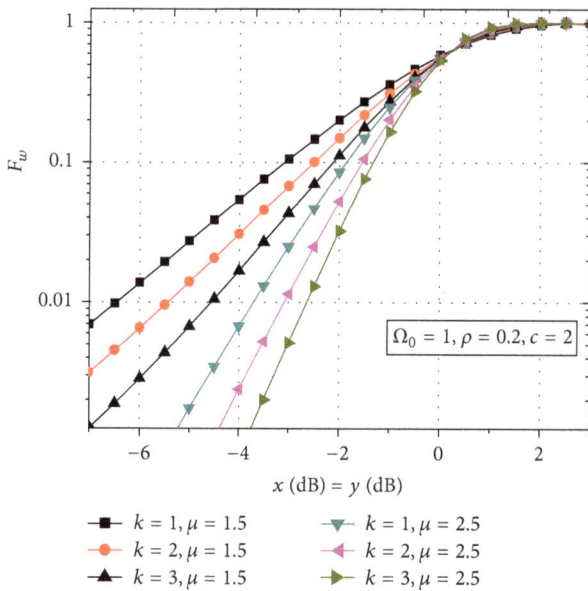

FIGURE 3: Cumulative distribution function of macrodiversity SC receiver.

short term fading effects and microdiversity SC receivers reduce Gamma long term fading effects on system performances. For this system, the probability density function of macrodiversity SC receiver output signal envelope and the cumulative density function of macrodiversity SC receiver output signal envelope was calculated. Probability density function and cumulative density function are very important statistical characteristics. By using them we can calculate outage probability and average bit error probability.

By setting $k = 0$, the k-μ distribution reduces to Nakagami-m distribution and Rician distribution can be

[6] M. D. Yacoub, "The η-μ distribution and the κ – μ distribution," *IEEE Antennas and Propagation Magazine*, vol. 49, no. 1, pp. 68–81, 2007.

[7] J. Proakis, *Digital Communications*, McGraw-Hill, New York, NY, USA, 4th edition, 2001.

[8] P. M. Shankar, "Analysis of microdiversity and dual channel macrodiversity in shadowed fading channels using a compound fading model," *International Journal of Electronics and Communications*, vol. 62, no. 6, pp. 445–449, 2008.

[9] J. Zhang and V. Aalo, "Effect of macrodiversity on average-error probabilities in a Rician fading channel with correlated lognormal shadowing," *IEEE Transactions on Communications*, vol. 49, no. 1, pp. 14–18, 2001.

[10] D. M. Stefanović, S. R. Panić, and P. Ć. Spalević, "Second-order statistics of SC macrodiversity system operating over Gamma shadowed Nakagami-m fading channels," *AEU—International Journal of Electronics and Communications*, vol. 65, no. 5, pp. 413–418, 2011.

[11] N. M. Sekulović and M. Č. Stefanović, "Performance analysis of system with micro- and macrodiversity reception in correlated gamma shadowed rician fading channels," *Wireless Personal Communications*, vol. 65, no. 1, pp. 143–156, 2012.

[12] I. S. Gradshteyn and I. M. Ryzhik, *Table of Integrals, Series, and Products*, Academic Press, San Diego, Calif, USA, 2000.

Permissions

List of Contributors

Changlun Zhang, Chao Li and Jian Zhang
Science School, Beijing University of Civil Engineering and Architecture, Beijing 100044, China

Md. Jaminul Haque Biddut, Nazrul Islam, Md. Maksudul Karim and Mohammad Badrul Alam Miah
Department of Information and Communication Technology (ICT), Mawlana Bhashani Science and Technology University, Santosh, Tangail 1902, Bangladesh

Lingwei Xu
College of Information Science and Engineering, Ocean University of China, Qingdao 266100, China

Hao Zhang
College of Information Science and Engineering, Ocean University of China, Qingdao 266100, China
Department of Electrical and Computer Engineering, University of Victoria, Victoria, BC, Canada V8W 2Y2

T. Aaron Gulliver
Department of Electrical and Computer Engineering, University of Victoria, Victoria, BC, Canada V8W 2Y2

Xihai Zhang, Junlong Fang and FanfengMeng
School of Electrical and Information Engineering, Northeast Agricultural University, Harbin 150030, China

Lu Cao and Mingshan Xie
College of Information Science & Technology, Hainan University, Haikou 570228, China

Zhuhua Hu, Yong Bai and Mengxing Huang
College of Information Science & Technology, Hainan University, Haikou 570228, China
State Key Laboratory of Marine Resource Utilization in South China Sea, Hainan University, Haikou 570228, China

Chengtie Li, Jinkuan Wang and Mingwei Li
School of Information Science and Engineering, Northeastern University, Shenyang, Liaoning 110819, China

Amar Sharma, Puneet Khanna and Arun Kumar
Electronics and Communication Engineering, IFTM University, Moradabad, India

Kshitij Shinghal
Electronics and Communication Engineering, MIT, Moradabad, India

Lahcen Amhaimar, Ali Elyaakoubi, Abdelmoumen Kaabal, Kamal Attari and Adel Asselman
Optics & Photonics Team, Faculty of Sciences, Abdelmalek Essaadi University, Tétouan, Morocco

Saida Ahyoud
Information Technology and Systems Modeling Team, Faculty of Sciences, Abdelmalek Essaadi University, Tétouan, Morocco

Yasmin M. Amin and Amr T. Abdel-Hamid
Department of Networks Engineering, German University in Cairo, Cairo 11835, Egypt

Wei Cheng, Yong Li, Yi Jiang and Xipeng Yin
School of Electronics and Information, Northwestern Polytechnical University, Xi'an, Shaanxi 710072, China

Nam Nguyen, Mohammad Arifuzzaman and Takuro Sato
Graduate School of Global Information and Telecommunication Studies, Waseda University, Building No. 29-7, 1-3-10 Nishi-Waseda, Shinjuku-ku, Tokyo 169-0051, Japan

Peng Du
College of Automation, Nanjing University of Posts and Telecommunications, Nanjing 210023, China

Yuan Zhang
National Mobile Communications Research Laboratory, Southeast University, Nanjing, China

Kefei Mao, Jianwei Liu and Jie Chen
School of Electronic and Information Engineering, Beihang University, Beijing 100191, China

Jie Jiang, Yun Liu, Fuxing Song, Ronghao Du and Mengsen Huang
Department of Electronic and Information Engineering, Beijing Jiaotong University, Beijing 100044, China

Xingwang Wang
School of Computer Engineering and Science, Shanghai University, Shanghai 200444, China
Shanghai Vocational Technical College of Agriculture & Forestry, Shanghai 201699, China

Nan Zhao, Minghu Wu, Wei Xiong and Cong Liu
Hubei Collaborative Innovation Center for High-Efficiency Utilization of Solar Energy, Hubei University of Technology, Wuhan 430068, China

Jie Li, Jianrong Bao, Shenji Luan, Bin Jiang and Chao Liu
Information Engineering School, Hangzhou Dianzi University, Hangzhou, Zhejiang, 310018, China

Jie Huang and Chin-Tser Huang
Department of Computer Science and Engineering, University of South Carolina, Columbia, SC 29201, USA

Branimir Jaksic
Faculty of Technical Sciences, University of Pristina, Kneza Milosa 7, 38220 Kosovska Mitrovica, Serbia

Mihajlo Stefanovic, Danijela Aleksic and Dragan Radenkovic
Faculty of Electrical Engineering, University of Nis, Aleksandra Medvedeva 14, 18000 Nis, Serbia

Sinisa Minic
Teachers College, University of Pristina, Nemanjina, 38218 Leposavic, Serbia

Index